Proceedings in Adaptation, Learning and Optimization

Volume 12

The role of adaptation, learning and optimization are becoming increasingly essential and intertwined. The capability of a system to adapt either through modification of its physiological structure or via some revalidation process of internal mechanisms that directly dictate the response or behavior is crucial in many real world applications. Optimization lies at the heart of most machine learning approaches while learning and optimization are two primary means to effect adaptation in various forms. They usually involve computational processes incorporated within the system that trigger parametric updating and knowledge or model enhancement, giving rise to progressive improvement. This book series serves as a channel to consolidate work related to topics linked to adaptation, learning and optimization in systems and structures. Topics covered under this series include:

- complex adaptive systems including evolutionary computation, memetic computing, swarm intelligence, neural networks, fuzzy systems, tabu search, simulated annealing, etc.
- machine learning, data mining & mathematical programming
- hybridization of techniques that span across artificial intelligence and computational intelligence for synergistic alliance of strategies for problem-solving.
- aspects of adaptation in robotics
- agent-based computing
- autonomic/pervasive computing
- dynamic optimization/learning in noisy and uncertain environment
- systemic alliance of stochastic and conventional search techniques
- all aspects of adaptations in man-machine systems.

This book series bridges the dichotomy of modern and conventional mathematical and heuristic/meta-heuristics approaches to bring about effective adaptation, learning and optimization. It propels the maxim that the old and the new can come together and be combined synergistically to scale new heights in problem-solving. To reach such a level, numerous research issues will emerge and researchers will find the book series a convenient medium to track the progresses made.

** Indexing: The books of this series are submitted to ISI Proceedings, DBLP, Google Scholar and Springerlink **

More information about this series at http://www.springer.com/series/13543

Hiroshi Sato · Saori Iwanaga · Akira Ishii
Editors

Proceedings of the 23rd Asia Pacific Symposium on Intelligent and Evolutionary Systems

Springer

Editors
Hiroshi Sato
Department of Computer Science
National Defense Academy of Japan
Yokosuka-shi, Japan

Saori Iwanaga
Faculty of Maritime Safety Technology
Japan Coast Guard Academy
Wakabacho, Japan

Akira Ishii
Department of Applied
Mathematics and Physics
Tottori University
Tottori, Japan

ISSN 2363-6084 ISSN 2363-6092 (electronic)
Proceedings in Adaptation, Learning and Optimization
ISBN 978-3-030-37441-9 ISBN 978-3-030-37442-6 (eBook)
https://doi.org/10.1007/978-3-030-37442-6

This Springer imprint is published by the registered company Springer Nature Switzerland AG
The registered company address is: Gewerbestrasse 11, 6330 Cham, Switzerland

Preface

This PALO volume constitutes the proceedings of the 23rd Asia Pacific Symposium on Intelligent and Evolutionary Systems (IES2019). The symposium first took place in 1997 in Canberra, and since then, the series has become a prestigious incubator of research ideas as well as a facilitator of research collaborations. The symposium aims to bring together researchers and practitioners from countries of the Asia-Pacific region in the fields of intelligent systems and evolutionary computation to present ongoing work, exchange ideas, and discuss future collaboration.

The 23rd edition of the conference, IES2019, was held in Tottori, Japan, during December 6–8, 2019, and organized by Tottori University.

There were 23 accepted papers selected to be included in the final proceedings. Each submission was reviewed by 2–4 members from our international program committee. The program also includes two invited talks.

We would like to thank the steering, organizing, and program committees for their efforts in supporting the symposium. The support and assistance from Tottori University, Tottori Prefecture, Springer, and EasyChair are gratefully acknowledged.

October 2019

Hiroshi Sato
Saori Iwanaga
Akira Ishii

Organization

Program Committee

Sung-Bae Cho	Yonsei University, Korea
Masanori Goka	Fukuyama University, Japan
Hiroyuki Iizuka	Hokkaido University, Japan
Akira Ishii	Tottori University, Japan
Saori Iwanaga	Japan Coast Guard Academy, Japan
Yasushi Kambayashi	Nippon Institute of Technology, Japan
Yoshiaki Katada	Setsunan University, Japan
Hidenori Kawamura	Hokkaido University, Japan
Tan Kay Chen	City University of Hong Kong
Masao Kubo	National Defense Academy of Japan, Japan
George Leu	University of New South Wales (ADFA), Australia
Yoshiyuki Matsumura	Shinshu University, Japan
Naoki Mori	Osaka Prefecture University, Japan
Keita Nakamura	University of Aizu, Japan
Keitaro Naruse	University of Aizu, Japan
Van-Giang Nguyen	Le Quy Don Technical University, Vietnam
Jun Ogawa	Yamagata University, Japan
Kazuhiro Ohkura	Hiroshima University, Japan
Yukiko Orito	Hiroshima University, Japan
Hiroshi Sato	National Defense Academy of Japan, Japan
Keiji Suzuki	Future University Hakodate, Japan
Keiki Takadama	The University of Electro-Communications, Japan
Yasumasa Tamura	Tokyo Institute of Technology
Kazuaki Yamada	Toyo University, Japan
Masahito Yamamoto	Hokkaido University, Japan

Tomohisa Yamashita	Hokkaido University, Japan
Toshiyuki Yasuda	University of Toyama, Japan
Keisuke Yoneda	Kanazawa University, Japan

Additional Reviewers

Goka, Masanori
Iizuka, Hiroyuki
Ishii, Akira
Kambayashi, Yasushi
Katada, Yoshiaki
Kubo, Masao
Leu, George
Matsumura, Yoshiyuki
Mori, Naoki
Nakamura, Keita
Nguyen, Van-Giang
Ogawa, Jun
Ohkura, Kazuhiro
Orito, Yukiko
Sato, Hiroshi
Suzuki, Keiji
Takadama, Keiki
Tamura, Yasumasa
Yamada, Kazuaki
Yamamoto, Masahito
Yamashita, Tomohisa
Yasuda, Toshiyuki
Yoneda, Keisuke

Contents

Ensemble Learning to Generate Virtual Data for Pattern Recognition and Its Possibility of Application to Strategy Inference in Table Tennis

Isao Hayashi(✉) and Honoka Irie

Graduate School of Informatics, Kansai University,
2-1-1, Ryozenji-cho, Takatsuki, Osaka 569-1095, Japan
{ihaya,honoka}@cbii.kutc.kansai-u.ac.jp

Abstract. Pattern recognition techniques are a class of method used to categorize observations across multiple classes. The structure and learning method of a model are selected so as to maximize the classification accuracy obtained over a certain data set. Methods also exist to improve accuracy by suitably structuring the training data [1]. Ensemble learning [2–5] can be seen as one such method.

In this talk, we discuss a new method that improves the structure of data in order to improve the accuracy of a classifier. In contrast to AdaBoost, a conventional boosting method that increases the weight for misclassified data and adds data over learning iterations, we propose Bagging/Boosting methods that generate virtual data using fuzzy membership functions [6] centered on misclassified data. We call this the pdi-Bagging/Boosting method (Possibilistic Data Interpolation Bagging/Boosting method) [1,7–10]. The newly generated virtual data around the misclassified data is used exclusively for training – not for validation or testing. Model accuracy is improved by adding virtual data to the original training data because the pattern recognition model is identified by the training data. On the other hand, the accuracy of the identified model is evaluated on the original test data which is not augmented with virtual data. In other words, the virtual data is not involved in the evaluation of the model at all.

We apply the pdi-Bagging/Boosting method to Trapezoidal fuzzy inference. Trapezoidal fuzzy inference is a general form of triangular fuzzy reasoning, that has proven to be effective at solving various types of inference problems [11,12]. When used as a clustering method of bagging/boosting algorithm, fuzzy inference allows efficiently for adjusting cluster boundaries with each new data [13,14]. In trapezoidal fuzzy inference, both the membership function for the antecedent part and the singleton real value of the consequent part need to be learned [15–20]. In this talk, we frame the problem of parameter adjustment for fuzzy rules not as a tuning problem but rather, as a design problem [21]. We focus in particular on how to learn coefficients both for the membership functions of the antecedent part and the singletons of the consequent part, how to set initial values, and how to determine the learning sequence of consequent and antecedent parts.

H. Sato et al. (Eds.): IES 2019, PALO 12, pp. 1–3, 2020.
https://doi.org/10.1007/978-3-030-37442-6_1

We are developing a system for automatically extracting rules of table tennis strategy from video using a clustering method [22,23]. We analyzed videos from women's table tennis singles tournament at the 2016 Rio de Janeiro Olympics, including 16 matches from the third round to the final. This video corpus contains a total of 407 plays, and 7,434 separate ball trajectories. In addition, we are collaborating with OMRON Corporation which has been developing a table tennis robot, "FORPHEUS" [24]. In this talk, the possibility of strategy extraction from video or images using ensemble learning algorithm in the near future is discussed, as well as the possibility to characterize players' strategies based on the learned relationship between player characteristics and ball position in image-based coordinates [25]. Through future work, we hope to integrate this new capability into an "AI strategy coach" that can help improve player strategies [26–28].

Keywords: Ensemble learning · Pattern recognition · Fuzzy inference · Virtual data · Table Tennis strategy · AI coach

References

1. Hayashi, I.: Data science of "Shinayakana" decision on advanced computational intelligence. In: Proceedings of the 5th International Workshop on Frontier of Science and Technology (FST 2017) and the 5th International Workshop on Advanced Computational Intelligence and Intelligent Informatics (IWACIII 2017) (2017)
2. Freund, Y.: Boosting a weak learning algorithm by majority. Inf. Comput. **121**(2), 256–285 (1995)
3. Freund, Y., Schapire, R.E.: A decision-theoretic generalization of on-line learning and an application to boosting. J. Comput. Syst. Sci. **55**(1), 119–139 (1997)
4. Dietterich, T.G.: Ensemble methods in machine learning. In: The First International Workshop on Multiple Classifier System (MCS 2000), pp. 1-15 (2000)
5. Schapire, R.E.: The boosting approach to machine learning: an overview. In: Denison, D.D., Hansen, M.H., Holmes, C., Mallick, B., Yu, B. (eds.) Nonlinear Estimation and Classification. Springer, Berlin (2003)
6. Zadeh, L.A.: Fuzzy Sets, Fuzzy Logic, Fuzzy Systems. World Scientific Press (1996)
7. Hayashi, I., Tsuruse, S.: A proposal of boosting algorithm by probabilistic data interpolation for brain-computer interface. In: Proceedings of the 26th Fuzzy System Symposium, pp. 288-291 (2010). (in Japanese)
8. Hayashi, I., Tsuruse, S.: A proposal of boosting algorithm by possibilistic data interpolation and its application to brain-computer interface. J. Jpn. Soc. Fuzzy Theory Intell. Inform. **28**(1), 501–510 (2016). (in Japanese)
9. Hayashi, I., Tsuruse, S., Suzuki, J., Kozma, R.T.: A proposal for applying PDI-boosting to brain-computer interfaces. In: Proceedings of 2012 IEEE International Conference on Fuzzy Systems (FUZZ-IEEE 2012) in 2012 IEEE World Congress on Computational Intelligence (WCCI 2012) (2012)
10. Irie, H., Hayashi, I.: Performance evaluation of PDI-bagging by generation of correct-error virtual data. In: Proceedings of the 29th Symposium on Fuzzy, Artificial Intelligence, Neural Networks and Computational Intelligence, Paper ID:No. A3-3 (2019). (in Japanese)

11. Mamdani, E.H.: Application of fuzzy algorithms for control of simple dynamic plant. Proc. Inst. Electr. Eng. **121**(12), 1585–1588 (1974)
12. Kosko, B.: Neural Networks and Fuzzy Systems: A Dynamical Approach to Machine Intelligence. Prentice Hall International Inc., Singapore (1992)
13. Ishibuchi, H., Nozaki, K., Tanaka, H.: Pattern classification by fuzzy rules. J. Jpn. Soc. Fuzzy Theory Syst. **5**(1), 74–84 (1993). (in Japanese)
14. Nakashima, T., Shoji, Y.: The effect of data partition in constructing fuzzy ensemble classifiers. In: Proceedings of the 25th Fuzzy System Symposium, No. 3E2-01 (2009). (in Japanese)
15. Ichihashi, H., Watanabe, T.: Learning controlby fuzzy model using a simplified fuzzy reasoning. J. Jpn. Soc. Fuzzy Theory Syst. **2**(3), 429–437 (1990). (in Japanese)
16. Nomura, H., Hayashi, I., Wakami, N.: A self-turning method of fuzzy reasoning by delta rule and its application to a moving obstacle avoidance. J. Jpn. Soc. Fuzzy Theory Syst. **4**(2), 379–388 (1992). (in Japanese)
17. Juang, C.F., Wang, C.Y.: A self-Generating fuzzy system with ant and particle swarm cooperative optimization. Expert Syst. Appl. **36**(3), 5362–5370 (2009)
18. Habbi, H., Kidouche, M., Zelmat, M.: Data-driven fuzzy models for nonlinear identification of a complex heat exchanger. Appl. Math. Model. **35**, 1470–1482 (2011)
19. Habbi, H., Boudouaoui, Y., Karaboga, D., Ozturk, C.: Self-generated fuzzy systems design using artificial Bee colony optimization. Inf. Sci. **295**, 145–159 (2015)
20. Gorzalczany, M.B., Rudzinski, F.: Interpretable and accurate medical data classification - a multi-objective genetic-fuzzy optimization approach. Expert Syst. Appl. **71**, 26–39 (2017)
21. Irie, H., Hayashi, I.: Characterization of hyperparameter of learning type trapezoidal fuzzy inference. In: Proceedings of the 34th Fuzzy System Symposium, pp. 259-264 (2018). (in Japanese)
22. Hayashi, I., Maeda, T., Fujii, M., Wang, S., Tasaka, T.: Acquisition of embodied knowledge on sport skill using TAM network. In: 2009 IEEE International Conference on Fuzzy Systems (FUZZ-IEEE 2009), pp. 1038-1043 (2009)
23. Hayashi, I., Fujii, M., Maeda, T., Leveille, J., Tasaka, T.: Extraction of knowledge from the topographic attentive mapping network and its application in skill analysis of Table Tennis. J. Hum. Kinet. **55**, 39–54 (2017). https://doi.org/10.1515/hukin-2017-0005
24. OMRON Corporation: FORPHEUS. https://www.omron.com/innovation/forpheus.html. Accessed 11 Oct 2019
25. Hayashi, I., Ogino, M., Irie, H., Tamaki, S., Yoshida, K., Kondric, M.: Development of image processing system to realize Table Tennis strategy board. In: Proceedings of Japan Table Tennis Association Sports Science and Medicine Committee International Meeting 2018 (JTTA-SSMC 2018), p. 21 (2018)
26. Mochizuki, Y., Himeno, R., Omura, K.: Artificial skill and a new principle in sports. Syst. Control Inf. **46**(8), 498–505 (2002)
27. Perl, J., Baca, A.: Application of neural networks to analyze performance in sports. In: The 8th Annual Congress of the European College of Sport Science (2003)
28. Hayashi, I., Ogino, M., Irie, H., Tamaki, S., Yoshida, K., Kondric, M.: AI coach: learning table tennis strategy rules from video. In: Proceedings of the 16th International Table Tennis Federation (ITTF) Sports Science Congress (ITTF-SSC 2019), p. 76 (2019)

The Rise of Smart Contract and Its Complex Impacts on the Digital Ecosystem

Yuji Aruka(✉)

Faculty of Commerce, Chuo University, Hachioji-City, Tokyo, Japan
aruka@tamacc.chuo-u.ac.jp

Abstract. Nowadays, we are forced to join in a more and more complicated contract and a faster and faster transaction in the market. We are also seriously faced with security problems. As the society evolves, the related systems generally become complicated. The economic system, more specifically, the market system is not exempted from this law. This exactly means the advent of a new formula of it currently corresponding to the extent of complications. Without any incessant improvement of the digital technologies, our performance would not be smoothly achieved.

Given an idea of **smart contract**, blockchain becomes a broader computational technology to realize a set of smarter systems/transactions associated with the Distributed Ledger Technologies (DLT). Such a well-known instance is the smart contracts on the Ethereum platform, that is associated with account information each. The smart contract tied together with DLT actually enables to design an economically well-behaved set of Peer to Peer (P2P) system, which ranges over from the micro market to the international currency transaction system. The introduction of this technology may suggest what a truly decentralized future is. This must cast a new light on the problem how to design either auction or vote of P2P, i.e., without auctioneer or chairperson. Either of them will possibly mean the change of existing social consensus.

The DLT originally was designed by the presumption that someone always is malicious, as Byzantine Generals Problem suggests. The programming of smart contract is written in Turing complete language. A smarter solution against a complicated system actually then requires a stronger security against their vulnerabilities. Thus, the diffusion of blockchain technologies into the contract systems will demand a fully protective infrastructure. As Stephen Wolfram suggests, furthermore, "[a]s the number of DLTs grows, it becomes harder and harder to learn and build ecosystem tools to make monitoring transactions, smart contracts and services possible." In other words, the diffusion of blockchain will bring newly some coordination problems as well as the new statistical examinations.

In this paper, after looking briefly the current features of digital contract, we will argue some evolutionary impacts on the digital ecosystem in the context of either smart contracts on the Ethereum platform or the Byzantine Generals Problem.

H. Sato et al. (Eds.): IES 2019, PALO 12, pp. 4–16, 2020.
https://doi.org/10.1007/978-3-030-37442-6_2

Keywords: Blockchain · Byzantine Generals Problem · Distributed Ledger Technologies (DLT) · Smart contract · P2P based auction

JEL Classification. A10 · D10 · D20 · D50 · D70

1 The Landscape of the Modern Socio-Economics System

The new century has already passed by the first 1/5th time window. Compared with the last century, the socio-economic system essentially is evolving, in particular, geometrically in the sphere of digital technology, which could replace the traditional dominant governance with a new one. It will take much time to examine the detailed process currently running. So we like to depict a rough sketch of the landscape on the current configuration of digital technologies wired to our socio-economic system. Here we focus on the difference of speeds around the digital technologies suited for economic transactions.

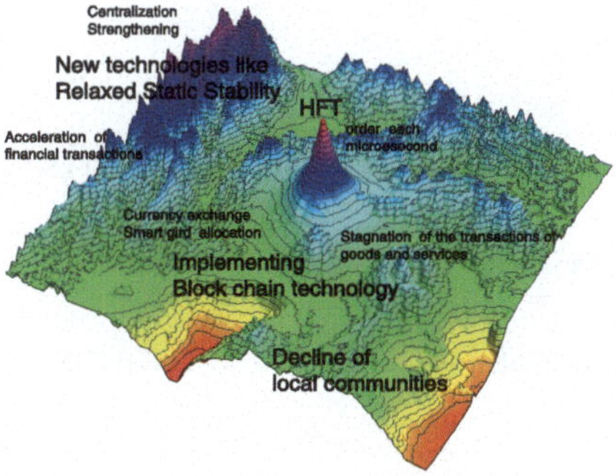

Fig. 1. The landscape of the socio-economic system in view of different speeds

1.1 The Separation of the Financial Transaction from the Real Economy

As shown by Mirrowski (2007), the standard implication of the original market is lost when the market is layer-structured by generating markets on the original market. The financial market since the so-called financial Big-Bang, as characterized by generating derivative commodities, are typical of hierarchical markets. In particular, in the latter market, a self-organizing growth has been implemented. This implementation has promoted more and more to separate relatively the real market from the financial market. Thus the stock prices have been inactive/sluggish against the changes of the real economy but rather reactive excessively to political issues. In this point of view, the world is divided into two distinct domains which mutually are partly disjoint but incessantly influential. It is noted from Table 1 that the contribution from “Finance and Insurance”

Table 1. The input-output table of 13 sector, 2015, Japan at production prices *cited from 2015 Input-Output Tables, e-Stat, Portal Site of Official Statistics of Japan: https://www.e-stat.go.jp/en/stat-search/files?page=1&layout=datalist&toukei=00200603&tstat=000001130583&cycle=0&year=20150&month=0

Industry	Value added (Million yen)
Agriculture, forestry and fishery	11310425
Mining	21116028
Manufacturing	202892254
Construction	3699380
Electricity, gas and heat supply	20506251
Commerce	32587127
Finance and Insurance	17327829
Estate	11884118
Transport and postal services	34909649
Information and communications	28703703
Public administration	1157289
Services	78757323
Activities not elsewhere classified	4728298
Total of intermediate sectors	469579674
Total of gross value added sectors	548238714

industry to GDP merely is 3.2% in Japan and the like[1]. If the prosperity in the financial domain should be directly wired with the real domain, GDP could be much more increased.

[1] Even in the US, the figure is not so much different from this tendency.

1.2 The Hyper-speed Domain

We firstly exposit the first upper part of Fig. 1. This diagram is depicted by taking notice of the different speeds of the transaction roughly divided into two domains: the domain dominated by the hyper speed and the other one by the slow speed.

During the last decade, the high frequency transaction, i.e., HFT, has been dominant. After Lehmann shock, there happened a flash crash firstly in 2010[2], which implies an instant crash of stock prices, accompanying a quick counterbalancing restoration of the prices. Hyper-acceleration of the order speed is making the occurrence of flash crashes invisible. However, this could not remove flash crash but moreover spread to the currency market.

This observation may call us "the relaxed static stability". Depending on the idea of "fail safe", structural stability has been important in designing planes and ships up to now. On the contrary, a modern stealth fighter like Lockheed F35 is designed by the idea of relaxing "static stability". It is well known that bicycle behaves much more quickly than tricycle, and tries to restore its instability by resort to its counter motion. The counter motion will be controlled by computational powers. Thus we expect that some nonlinear effect is implemented behind the flash crashes, always accompanying a countervailing power. So we demonstrate the nonlinear force working in the hyper-speed domain.

Within such a hyper-structured hierarchy, the working of each market can only be examined by computer simulation. However, there is little sign that economics tries to solve this problem. The advent of HFT has led to the settlement of microsecond payments, which means that the order speed and the physical distance between the order terminal and the transaction server can dominate the settlement. The continuous double auction depends on the double priority principles of time and price. Usually, the time principle is preferably applied. Thus the physical speed of order, instead of economic motivation, dominates the settlement. Under the HFT, the economic meaning of the auction rule has been then considerably changed. Even the traditional stock market rule of membership has virtually been altered by the lending of high-performance servers in the stock exchange.

1.3 The Slow-Speed Domain

Secondly, we examine the second domain of Fig. 1. This is relevant to the main interest of this paper. In contrast with the hyper speed domain, it matters "the slow speed" to employ blockchain technology[3]. Here the word "slow" may be interpreted as "non-hyper-speed" or "non-microsecond". In view of a certain scaling, blockchain may be enough speedy excepting the mining speed of Bitcoin.

[2] The May 6, 2010, Flash Crash, also known as the Crash of 2:45, the 2010 Flash Crash or simply the Flash Crash, was a United States trillion-dollar stock market crash. At this moment, about 9% of the stock prices was lost.

[3] The latter speed is not necessarily slow, if we compare it with the speed of our ordinary life. Sometimes it may be enough speedy.

In this sense, it is easy to identify the crypt-currency using blockchain technology with a slow speed technology. More interestingly, there currently is existing a world wide nodes network at the basis of **P2P transaction**.

It is noted that a large-scaled network system of P2P mode actually is working, as the live node map in Fig. 2a shows. It is also noted that a P2P network, due to its own attribute of P2P mode, will take much time for propagation of any value transfer, as Fig. 2 shows.

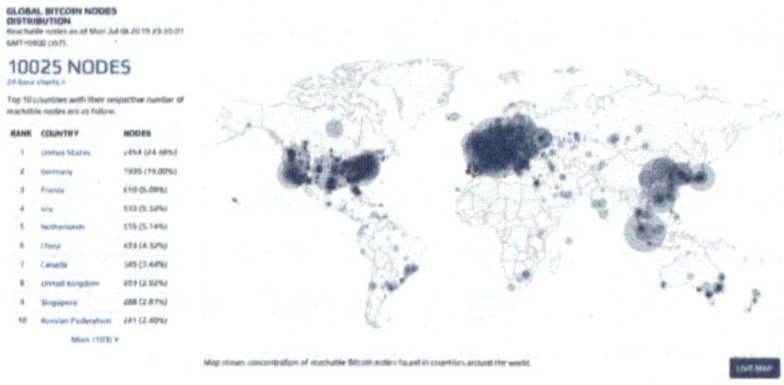

(a) The node distribution of Bitcoin *https://bitnodes.earn.com/nodes/live-map/

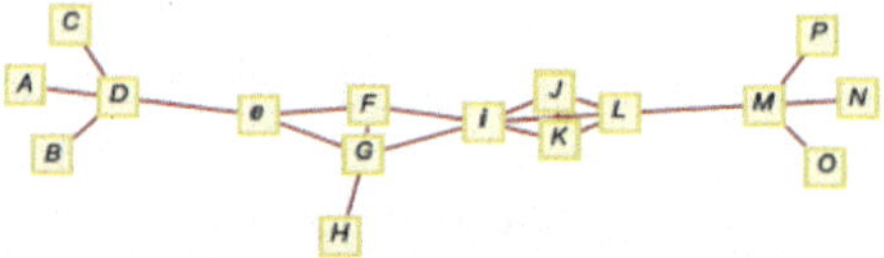

(b) A conceptual propagation process

Fig. 2. Node distributions

2 A Simple Design of Auction on the Ethereum Platform

2.1 The Distributive Ledger System of the P2P Type

In our argument, the distributive ledger system matters. This system is as usual compared with a centralized ledger system, shown in Fig. 3. Here the process is intermediated by mining process using the Hash function, as the table below of the system above. In the former system, key and ledger are managed by the central authority. However, in the latter system, they are distributed and managed by each entity.

It is noted that blockchain is not a technology only simply for crypt-currencies. Blockchain is also used a censor technology, e.g., in order to retrieve a defective lot in a factory system of the manufacturing industry. Even if a block chain technology is to be applied to the financial field, its usage is not limited to crypt-currency like Bitcoin or Ethereum. If you turn your eyes to finance, you will see two perspectives on the use of blockchain. When talking about blockchain, naturally, we suppose *permission-less* mining process. However, we can also establish a *permission-ed* mining process. There are actually somewhat a hybrid form to use blockchain within the permission-ed framework, like **Libra** by

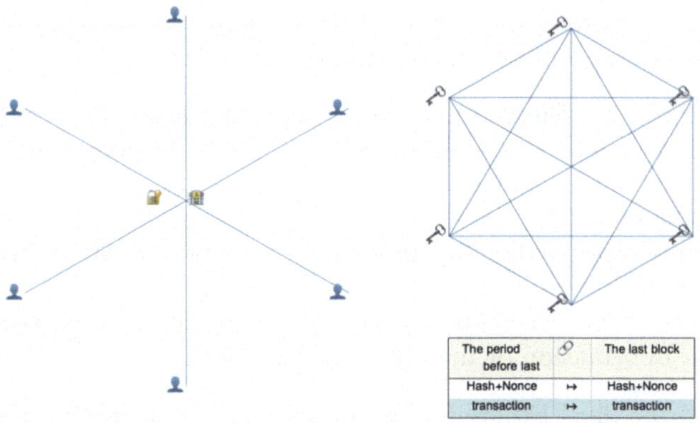

Fig. 3. Centralized network and distributive ledger system of the P2P type

FACEBOOK. Among them, we note the IBM blockchain applied to the currency exchange, which provides the financial institutions with a better opportunity by way of a world wire network (API): either € or $ are transformed into "a stable coin" as the digital asset, which can achieve a simultaneous exchange between the two. Here messaging, clearing, and settlement for the desired exchange will be integrated by blockchain technology without falsifying bookkeeping. The latter usage may be classified into a hybrid application of blockchain in a sense that blockchain is incorporated as a sub-system of the currency exchange. This kind of usage of blockchain will also improve the transaction efficiency.

In this section, we will argue other examples of blockchain like auction[4]. Furthermore, we may also refer to the decline of local communities shown in Fig. 1. The decline of local communities are still in progress during the modern process of expanding inequalities of income and wealth. Needless to say, at this moment, the field to establish a local currency peculiar to its local community may be also one to be applied by blockchain technology to promote the transactions of its local community more efficiently.

2.2 Smart Contracts on the Ethereum

In 1993, Nick Szabo described how users could input data or value/then regain from the transfer system as if it were a digital vending machine. **Ethereum** shares the same function as Bitcoin in a sense that **the network** can transfer value from one to another. Different from Bitcoin, Ethereum, provides each node with additional information account, which could function as the distributive ledger for each, while Bitcoin only transfer the value of currency.

Ethereum is called a system written in Turing complete language[5]. As a Turing Complete system means a system in which a program can be written

[4] We may argue voting mechanism in a similar context. See Sects. 3.4 and 4.

[5] See, for example: https://stackoverflow.com/questions/7284/what-is-turing-complete.

that will find an answer, Ethereum, if it is, allows to program **autonomous agents**. The contract will be called **smart contract**[6].

- Smart contracts are fulfilled with multi-signature accounts, which imply that "funds are spent only when a required percentage of people agree".
- Smart contracts can manage agreements between users, say, if one buys insurance from the other.
- Smart contracts provide utility to other contracts (similar to how a software library works).
- Smart contracts can store information about an application, such as domain registration information or membership records.

In other words, thus, Ethereum becomes a platform suited for creating smart contracts. Hence, using Ethereum platform, if they use a smart contract, users can send 1 ether to their friend. A rough sketch of the smart contracted transaction system based on the Ethereum platform will be depicted as shown in Fig. 4.

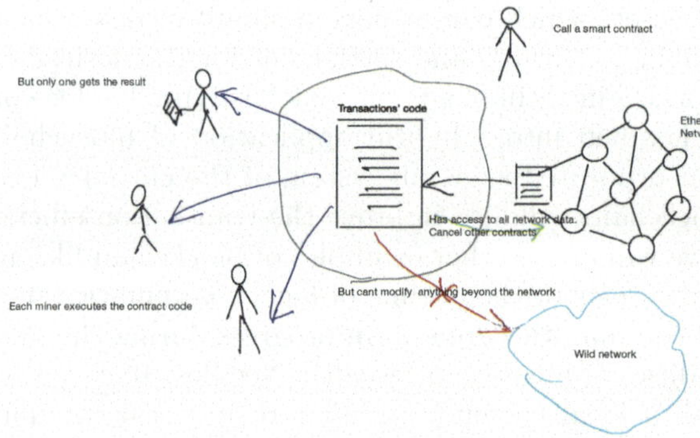

*Cited from `https://vas3k.com/blog/ethereum/` but the chart was originally produced.

Fig. 4. Smart contracts on the Ethereum platform

The idea of smart contracts will be applied to a simple auction of P2P type.

2.3 A Simple Design of Auction on the Ethereum Platform

In order to spare the time for exposition, firstly, we depict a flowchart of a simple auction on the Ethereum platform. We suppose the auction to be 1 owner and 3 bidders in our simple auction. Correspondingly, 3 accounts/ledgers each bidder. An elementary process of P2P auction will be designed as follows:

[6] The items described below is cited from https://www.coindesk.com/information/ethereum-smart-contracts-work.

⟨1⟩ Generation: **Owner** asks their bidders
⟨2⟩ Bidder 1 bids 1 ether;
⟨3-1⟩ Bidder 2 bids 2 ether
⟨3-2⟩ Bidder 1 refunds 1 ether
⟨4⟩ **Malicious Generation** ⟨Bidder 3⟩ generates 3 ether
⟨5-1⟩ Malicious Bidder bids 3 ether
⟨5-2⟩ Bidder 2 refunds 2 ether
⟨6-1⟩ Bidder 2 bids 4 ether
⟨6-2⟩ Bidder 3 refunds 3 ether
⟨7⟩ An infinite loop will be formed ....

The flow described above will be depicted in Fig. 5. Due to an intermediary of **Malicious Generation**, Bidders may be induced to give a higher bidder. In order to interrupt a chain reaction, the process ⟨6-1, 2⟩, by *mining*, is indispensable.

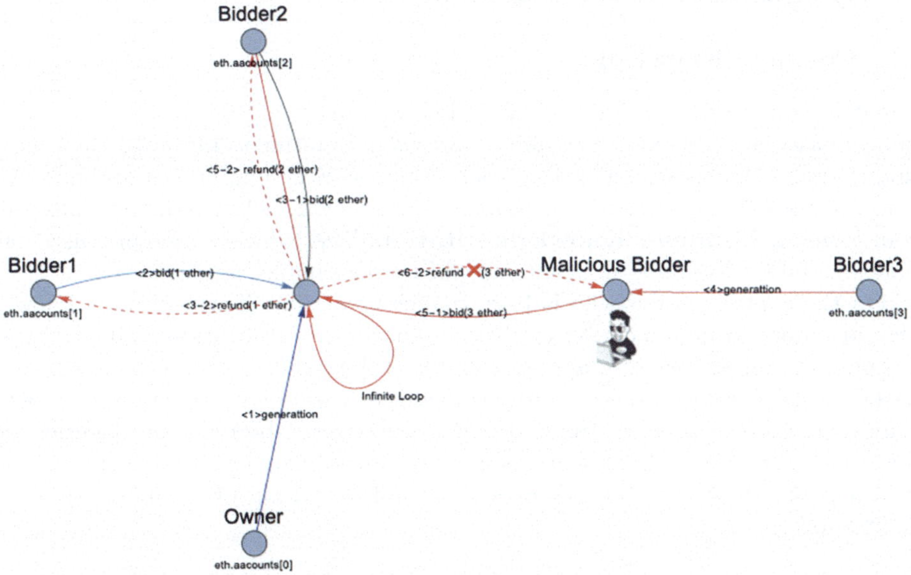

Fig. 5. The design of a P2P auction taking notice of malicious bidder *The chart was originally produced by referring to Fig. 10.1 of Takago (2017).

It is noted that the auction above is of P2P type. It is interesting for us to mention two important remarks around the auction on Ethereum platform. Due to the P2P attribute, there no longer exists an auctioneer at Exchange[7].

Malicious trader P2P can work well autonomously without any coordination of a centralized management. Instead of an auctioneer, the verification process of

[7] In the modern stock exchange, human auctioneer is replaced with computer server.

transactions by mining is indispensable. So, firstly, we remark that the most important designing strategy is placed on removing the kinds of malicious bidder in the kind of P2P auction.

The second remark will come out from the use of blockchain. We cite an important remark from Manski (2017):

> Blockchain technologies are reconfiguring the global economy, though often in contradictory ways. Blockchain technologies are disrupting key economic and financial sectors. Some blockchain applications allow for democratization of finance, services, agriculture, and governance, yet they may also deepen inequality and weaken democracy. We need new understandings of the countervailing tendencies of blockchain technologies and the contingencies that shape their deployment.

3 Byzantine General Problem

3.1 Common Knowledge

It matters for years in social science and coordination problems that how common knowledge does exist. Common knowledge is formed only after there is a unique signal that is recognized by each other, as Schelling (1960) argued[8]. As shown in the last subsection, the existence of malicious will will disturb/prevent from forming **Common Knowledge Rationality**, which is not premised on malicious intent, because all agents should be reasonable/rational.

On the contrary, man who is born malicious can lie. Thus common knowledge in human communication is always exposed to falsification. Interestingly, this kind of danger/risk also applies to information transmission of a machinery system. In this case, accidental faulty transmission will play a role of human lier. It may be easily verified in the two party message to observe a propagation of a new message. Figure 6a indicates a unique transmission of message emanating from a node. While Fig. 6b indicates a mutual formation of a value system of the network.

3.2 Where the Problem Is

In the following, we will discuss the so-called Byzantine General problem. This problem has been already solved by Pease et al. (1980). This paper smartly "where the problem is" as follows:

> [The Byzantine General Problem] addressed a set of isolated processors, some unknown subset of which may be faulty, that communicate only by means of two-party messages. Each nonfaulty processor has a private value

[8] See Common Knowledge, Stanford Encyclopedia of Philosophy https://plato.stanford.edu/entries/common-knowledge/.

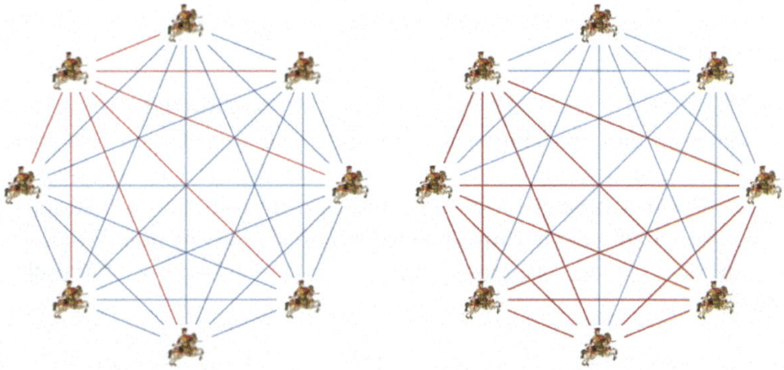

(a) A uniqe transmission (b) A mutual transmission

Fig. 6. P2P propagation of faulty and non-faulty message in the two party message system

> of information that must be communicated to each other nonfaulty processor. Nonfaulty processors always communicate honestly, whereas faulty processors my lie. The problem is to devise an algorithm in which processor to infer a value for each processor. The value inferred for a nonfaulty processor must be that processor's private value inferred for a faulty one must be consistent with the corresponding value inferred by each other nonfaulty processor.

In short, the Byzantine General Problem implies that "when a consensus is formed on a decentralized network, if there is a malicious person, it is impossible to form an agreement that can guarantee reliability". This is a problem that must be solved on a P2P system like a virtual currency[9]. In the following, we abbreviate the Byzantine General Problem as **BGP**.

3.3 Two Generals Problem

We begin with an elementary problem called Two Generals Problem. There is a fortress that cannot be surrendered by a single attack, and the attack is then designed to be performed by two generals. On topography, the method of communication between two generals is limited to dispatching messengers. Whether or not to perform an attack at the scheduled attack time depends on the content of the agreement between the generals.

We examine a consensus formation among the two generals. It is necessary to communicate with each other, agreeing on the attack time, and confirming that the other party knows the agreement. In order to avoid an infinite loop of consensus formation, we are induced to rank the two generals into the commander and the deputy commander each two.

[9] Cited from https://note.mu/rottenmarron/n/ne98508d1b958.

If we employ the oral messages due to easy sending, we will need more assumptions on them:

- The messages to be send are assumed to be transmitted properly. In other words, rebels cannot interfere with communication between two generals.
- The recipient has right knowledge of the sender of the message. In other words, rebels can't obscure messages by sending additional messages.
- It can be confirmed whether oral messages can be sent or not. We may promise the default message as "Retreat", for example. Otherwise, "Attack" will be confirmed.

The commander general then decides the attack time, transmits the message of attack to the deputy commander by dispatching a messenger, and waits for the consent from the deputy army. If a reply is not received, either a messenger's accident to send or a deputy commander's betrayal. In general, it is impossible to design algorithms that can be safely agreed.

3.4 Byzantine General Problem and Its Practical Implication

Now we generalize the Two Generals Problem above stated. According to Siddharth (2011)[10], we give a brief profile of the proof on the BGP. In our environment, there are at least the supreme commander and two lieutenant generals.

- If all the lieutenant generals are faithful, they will move with the same action plan.
- If the number of rebels is small, it will not be moved to the opposite plan.
- In order to realize the above, generals must come up with the same plan through the same method of information collection.

Thus it then follows the next statement:

Proposition 1. *If no more than 2/3 generals are faithful, no solution can be found. Even with 3 generals, the protocol will not work if there is one rebel.*

The set of adjacent nodes i categorized by lieutenant name is called *regular* when the following conditions are met. Each node is an adjacent node of node i. A node that does not belong to the rule set has a route that does not pass through the node i that originates from the node of the set. If any node has a neighbor of p-*regular*, the graph is said to be p-*regular*. Therefore, a $3m-$*regular* graph is defined with a minimum of $3m + 1$ nodes.

In general, the BGP will be stated as the next theorem.

---

[10] https://www.slideshare.net/siddharthchaudhry/byzantine-general-problem-siddharth-chaudhry?from_action=save.

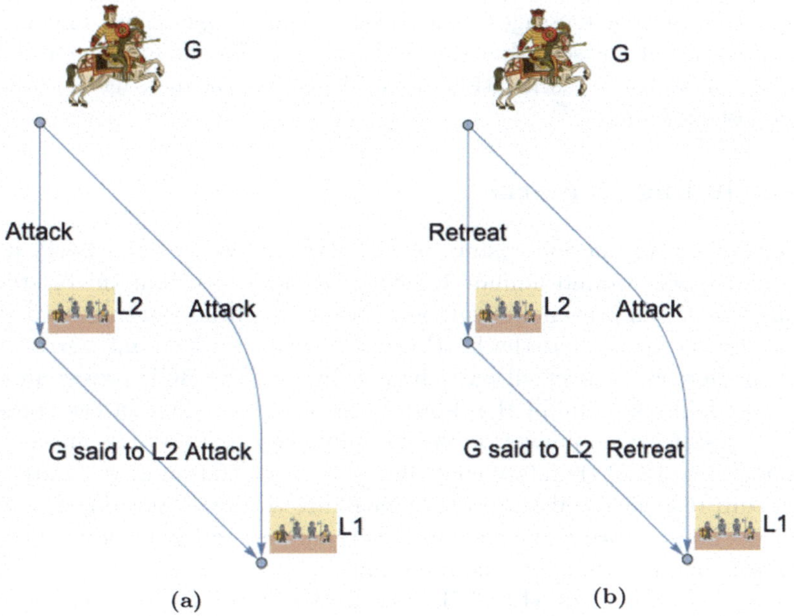

Fig. 7. BGP **G**: General; **Li**: Lieutenant *i*

Theorem 1. *If there are m rebels, a solution is not found if there is no more than* $3m + 1$ *of the number of rebels.*

Finally, the **algorithm** for detecting the arrival stream of messages will be *recursively* defined:

- Every lieutenant general forwards the received value to all other lieutenant.
- The supreme commander sends his value to all lieutenants.
- Each lieutenant broadcasts the commander's order to other lieutenants who have not had the value.
- Next, take **the majority function** of the received values.
- Index the messages occurring each round by lieutenant name i

In **majority voting**, throughout input synchronization, every input will be transferred/aggregated into the same decision value. If the input is not faulty, the non-faulty arrival stream will follow the above values. The recipient knows who sent it. A fully connected network is then formed. Given a signed message, the recipient would not need to spend time to know who sent it[11]. In order to avoid forgery of the messages, however, the message is to be signed in the form of node $i = (M, S(M))$ where M is a message, $S(M)$ is a signed message. The proof of the BGP thus has demonstrated a set of necessary conditions to fulfill

[11] To detect the absence of message transmission, **a time-out mechanism** must work for message transmission.

a transmission of true messages, using cryptographic operations and modular arithmetic. Hence it has seen that the BGP prepared for the fundamental design information, as well as a justification for verification of the message of remittance by mining blockchains.

4 Concluding Remarks

As we examined an auction system of P2P type as well as the BGP, it matters lies and rebels around human decision. We have seen that the existence of malicious bidder can prevent the auction transaction from working normal. It is quite interesting for us that the BGP theorem does not demand a strict fulfillment of the agreement over all participants. In fact, the BGP proves an extent to allowable failures to fulfill the desired (physical) decision. In the context of machinery system, communication media failure can be viewed as another form of rebellion. The BGP therefore contributed to a realization of the **fault torelant system**.It is also well known that there are as usual the sample programs of either auction or voting attached in Ethereum package. Not only auction but also majority voting will share the same idea.

In order to implement **the fail safe factor** into the network system of greater number of *nodes*, however, it takes some grand idea to keep the working of P2P network almost always fair. This grand idea is just the smart contracts on the Ethereum platform. It is noted that we require a huge but distributive infrastructure to support this idea.

Thus we temporarily sum up our discussion about P2P network here:

- Market is not a given entity but a designed artifact.
- Market is designed at basis of the existence of malicious participants.
- Blockchain is simply a possible mean to achieve a distributive ledger system to avoid risks emanating from malicious wills.
- P2P network, although it guarantees democratic, will be faced with a huge complex relationship of links and nodes beyond human intelligence to be solved by a new technology like blockchain.
- Needless to say, distributed network and distributed ledger will be applied not only to smart contracts but also to local currencies.

References

Mirowski, P.: Markets come to bits: evolution computation and markomata in economic science. J. Econ. Behav. Organ. **63**(2), 209–242 (2007)

Manski, S.: Building the blockchain world: technological commonwealth or just more of the same? Strat. Change **26**(5), 511–522 (2017)

Pease, M., Shostak, R., Lamport, L.: Reaching agreement in the presence of faults. J. Assoc. Comput. Mach. **27**(2), 228–234 (1980)

Schelling, T.: The Strategy of Conflict. Harvard University Press, Cambridge (1960)

Takago, T.: Kenro-na smart contract kaihatsu-no-tameno blockchain nyumon (Introduction of blockchain to develop a robust smart contract), Gijutsu-Hyoron-sha (2017). (in Japanese)

Automatic Collision Avoidance Using Deep Reinforcement Learning with Grid Sensor

Ryohei Sawada(✉)

National Maritime Research Institute, National Institute of Maritime, Port and Aviation Technology, Tokyo, Japan
sawada-r@nmri.go.jp

Abstract. This paper presents an automatic collision avoidance algorithm for multiple ships using reinforcement learning (RL). Obstacle zone by target (OZT) is used to grasp multiple ships' dynamic information in the form of 2-dimensional areas. OZT shows a dangerous area where collisions may happen. Then a new method using a virtual sensor which is separated in a grid is proposed to detect multiple ships simultaneously. The sensor detects OZTs efficiently and provides information about where OZTs expands as a part of a state vector with a fixed dimension as inputs for a RL algorithm. I applied a deep RL algorithm. An agent of deep RL learned manoeuvre using a set of ship encounter situations called Imazu problem. The learned model can avoid all encounter situations of up to three target ships in simulations. The proposed approach can learn manoeuvre to manage both waypoint navigation and collision avoidance.

Keywords: Collision avoidance · Multiple ships · Reinforcement learning · OZT

1 Introduction

Recently, research and development of autonomous ships is work in progress by many organizations and countries. Discussion on the regulatory of autonomous ships as Maritime Autonomous Surface Ships (MASS) including definition of MASS and degrees of autonomy is ongoing for concerns about safety, security and environment in International Maritime Organization (IMO) [1]. There are some kinds of technology to realize autonomous ships and one of most important technology is automatic collision avoidance. In particular, automatic collision avoidance in congested waterway such as around a bay is one of challenging tasks for automatic ships.

Previous studies proposed many kinds of approach for automatic collision avoidance. These approaches uses a variety of algorithms such as dynamic programming [7,8], evaluation of collision risk by fuzzy reasoning and fuzzy control

H. Sato et al. (Eds.): IES 2019, PALO 12, pp. 17–32, 2020.
https://doi.org/10.1007/978-3-030-37442-6_3

[9,10], A* algorithm [12], Nagasawa model of operational preference and collision risk [13,14], velocity obstacle method [16], reinforcement learning algorithm [17,18]. However, most existing approaches do not scale encounters of multiple target ships or take account of only one target ship at once [2].

In this paper, I propose a new automatic collision avoidance algorithm using obstacle zone by target (OZT) and deep reinforcement learning for encounters of multiple ships. OZT is a method to represent collision risk as areas where collision may happen like a predicted areas of danger (PAD). The difference between OZT and PAD is that OZT is calculated based on time and distance at the closest point of approach (CPA). The proposed approach uses proximal policy optimization algorithm (PPO) [5] to improve efficiency of learning of collision avoidance manoeuvre. PPO is an algorithm of deep reinforcement learning and considered that provides satisfying and stable results if enough data is available [21]. And a new type of virtual sensors is proposed, which extends from the center of a ship and is separated in a concentric circle grid and detects the obstacle zone by target (OZT) with collision course [3] to recognize multiple ships simultaneously. The simulator as an environment for learning was built on OpenAI Gym which is widely used in reinforcement learning and its environment has the standard interface for reinforcement learning. By numerical simulations were carried out and the system learned through Imazu problem, which is a series of ship encounter situations including some kinds of situation that is very difficult to avoid collisions. The learned models avoid up to three ships successfully. This approach can take account of multiple ships in a long-range simultaneously and enable a ship to reach a waypoint as a destination while avoiding multiple ships.

2 Related Work

In earlier work, Imazu and Koyama applied dynamic programming to avoid multiple ships in 1980s [7,8]. Discrete control variables of three levels of ship's speed and four levels of amount of heading changes were available for control in this simulation. Their approach evaluated simulation results in point of view about the distance of each ships, risk assessment using probability of collision and rules of collision avoidance they proposed based on the Convention on the International Regulations for Preventing Collisions at Sea (COLREGs).

Kouzuki and Hasegawa [9] proposed an algorithm using a fuzzy controller that reasons collision risk (CR) calculated from distance of closest point of approach (DCPA) and time of closest point of approach (TCPA) to evaluate risk of collision and determine the timing to steer. Then, Cai and Hasegawa [10] evaluated the system using Imazu problem [11] which is a series of ship encounter situations.

As another approach to evaluate the risk of collision, a computational model of operational preference proposed by Nagasawa et al. [13,14] is also widely used. Sera et al. conducted model experiments using the Nagasawa model and succeeded collision avoidance with up to three ships [15]. However, even though they had adjusted the parameters of Nagasawa model for their model experiments, they reported the collision in their experiments using four model ships.

In recent years, since techniques of machine learning is developing rapidly, reinforcement learning, which is one of machine learning is begun to be applied to automatic collision avoidance. Mitsubori et al. used Q-learning algorithm which is one of classical reinforcement learning algorithms to collision avoidance [17]. They prepared a sea-area model which was separated in a quadratic grid for reading a situation using reinforcement learning. They put the no-entry girds on the sea-area model and the agent learned maneuvering according to kinds of girds with going to the goal grid from the start grid. Shen et al. developed the automatic collision avoidance system using deep Q-network (DQN) which is an algorithm of reinforcement learning [18]. Their system used the ship bumper model to switch between two modes of waypoint navigation and collision avoidance. They also used many kind of methods such as predicted area of danger (PAD), ship domain, ship contour and motion vector of ships to avoid ships. other obstacles such as side walls of the basin were represented in the form of polygons. The system detected them by twelve detection lines. They carried out numerical simulations and model experiments to validate the developed method. However, this method using the detection lines may increase the risk of overlooking of target ships or other obstacles far away from the own ship because the detection lines are straight and one line can detect only one object at the same time.

3 System Overview

3.1 Detection of OZT by Grid Sensor

Proposed approach used OZT to process information of ships in the target area. In collision avoidance tasks in actual sea, pilots use DCPA and TCPA to know how close and when the own ship and a target ship will get at the CPA each other. When DCPA is very short to avoid a target ship, the own ship takes action to avoid the ship. This information is very useful for collision avoidance, however, indices of CPA don't provide information about which direction is safe. Thus, pilots cannot know the safe direction without changing the course practically. In addition, pilots take account safe passing distance to determine their ships' course to be changed. It is same to automatic collision avoidance system using DCPA and TCPA. To prevent collision, systems should predict motions of other ships and probability of collisions to operate ships. Therefore, OZT is adapted to the proposed approach to simplify for evaluation of traffic for collision avoidance. There are some versions of calculation and expression of OZT. In this paper, an OZT is defined as a capsule-shaped area calculated by using collision courses [3,4] as follow. An OZT is an area where collision may happen when the own ship's speed and the target ships' speed and course will not change in the future. Figure 1 shows example of an OZT of the own ship created by the target ship that is the area inside of blue line. The set of collision courses of the own ship C_o is calculated as below.

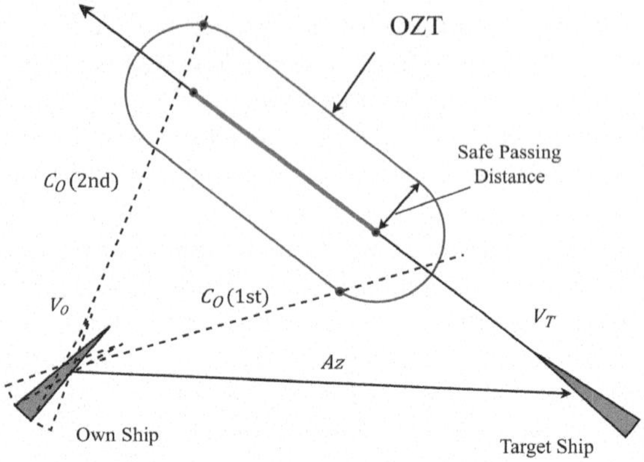

Fig. 1. Computation of OZT

$$C_O = \begin{cases} Az \pm \alpha - \arcsin\left\{\frac{V_T}{V_O} \arcsin\left(Az \pm \alpha - C_T\right)\right\}, \\ Az \pm \alpha - \pi + \arcsin\left\{\frac{V_T}{V_O} \arcsin\left(Az \pm \alpha - C_T\right)\right\}, \quad (V_T > V_O) \end{cases} \tag{1}$$

where $\alpha = \arcsin(r/d)$. r is the safe passing distance and d is a distance between the own ship and a target ship. V_O is the own ship's speed and V_T is a target ship's speed. Az is an azimuth of a target ship's position from the own ship and C_T is a course of a target ship.

Calculating relative velocity and course of a target ship using C_O calculated from Eq. (1), a set of DCPA and TCPA for each C_O can be obtained as follow.

$$DCPA = d\left|\sin\left(C_R - Az + \pi\right)\right| \tag{2}$$

$$TCPA = \frac{d\left|\sin\left(C_R - Az + \pi\right)\right|}{V_R} \tag{3}$$

where C_R and V_R are a relative course and speed of a target ship against C_O respectively.

Using this set of TCPA, a line segment (the bold part in Fig. 1) on the target course can be obtained. An OZT is an area consists of a set of points which distance from the line segment is less than the safe passing distance r. Next, it is required to process information of a set of OZTs in the target area to use it as input into the system using reinforcement learning. Because networks in reinforcement learning only accept vectors which dimension is fixed. However, to avoid multiple ships, the automatic collision avoidance system should track more than one OZT simultaneously and the number of ships may be change during navigation. One solution is, as mentioned before, detection lines by Shen et al. [18]. This approach is similar to LiDAR used to detect obstacles in the field of self-driving cars [19]. This approach has two problems. First, detection lines extend radially and are sparse so that it is difficult to recognize other ships

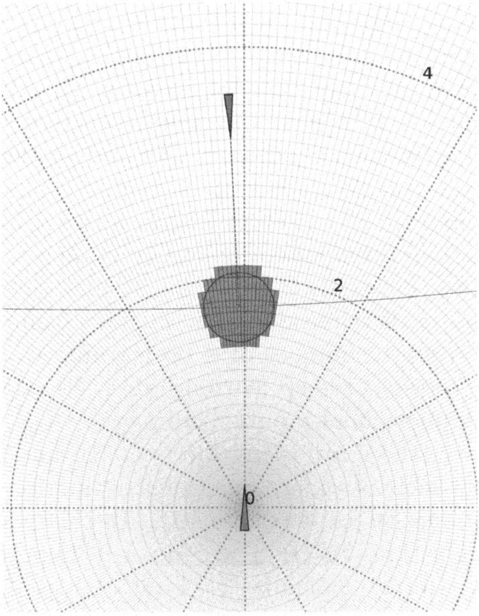

Fig. 2. An OZT in a head-on situation and detection by the grid sensor (The unit of a radius is nautical mile and the size of ships' plots is 4 times of the full-scale)

far away from the own ship. And each detection line can detect only at once, so when two ships in the same direction, the own ship can recognize only one target ship which is closer to the own ship than the other. Incidentally, ships at sea avoid a collision according to the COLREGs. The rule 8 of the COLREGs states that action to avoid a collision should be positive and made in ample time. Therefore, it is necessary to observe the situation from an ample distance and choose an appropriate action with ample time. To solve these problems, I designed the grid sensor that is a virtual sensor that extends from the center of the own ship and is separated by evenly spaced intervals of the angle direction and the radius direction in a concentric circle grid. When a grid cell overlaps with an OZT, it is judged detecting the OZT on the cell of the grid sensor. After detection, the grid sensor returns a state vector which dimension equals to the number of its grid cells. Grid cells detecting OZTs return 1.0 and the others return zero. The schematic diagram of detection of OZTs by using a grid sensor is shown in Fig. 2. Gray cells are the grid cells which are overlapping with the OZT created by the target ship on the upper side of the figure. In this way, the collision avoidance system recognizes OZTs as a vector with a fixed dimension regardless of changes of OZTs.

3.2 Deep Reinforcement Learning Algorithm: PPO

The proposed collision avoidance algorithm is based on deep reinforcement learning. Deep reinforcement learning is a kind of machine learning and a combination of reinforcement learning (RL) and deep learning. RL algorithms uses an environment and agents. An environment returns information of state of the environments by given some actions taken by agent(s). Agents can take action in the environment and aim to maximize the cumulative reward. In collision avoidance tasks, an agent corresponds to the own ship and an environment consists of some components such as target ships and waypoints. There is a wide variety of algorithm to determine next action according to an environment's state. Function approximation is to approximate a policy function or a value function with parameterized functions. For some kinds of algorithms such as on-policy actor-critic method, probability distributions of actions is given by a policy function. Deep RL algorithms apply techniques of deep learning to function approximation. Using expressive power of deep learning, deep RL algorithms can control agents based on high-dimensional inputs like images.

As mentioned above, there are many kinds of RL and deep RL algorithms depending on processes such as update functions, evaluation of a state and determination of action (stochastic or deterministic). In this paper, I used proximal policy optimization algorithm (PPO). The code of PPO in this paper was implemented using deep RL library machina [21] built on top of PyTorch [22] which is a deep learning framework. This algorithm was developed referencing trust region policy optimization algorithm (TRPO) which update a policy under a constraint expressed in terms of the Kullback-Leibler divergence. PPO in machina is an actor-critic method and uses the advantage like the asynchronous advantage actor-critic (A3C) [6]. Actor-critic method uses two networks basically. One is a policy $\pi(a_t|s_t;\theta)$ which is formulated in the form of a posterior probability distribution for a given state vector to determine an action. The other is a value function $V(s_t;\theta_v)$ to evaluate a state of an environment. Where a_t is an action and s_t is a state at step t. θ and θ_v are parameters of two networks of a policy and a value function respectively. In my implementation, these two networks are independent and don't share any part of them. A update using the advantage, if without using the state-action value function directly, performs via $\nabla_\theta \log \pi_\theta\left(a_t|s_t;\theta\right)\hat{A}\left(s_t,a_t;\theta,\theta_v\right)$ obtained from an estimate of the advantage given by $\hat{A}\left(s_t,a_t;\theta,\theta_v\right)=\sum_{i=0}^{k-1}\gamma^i r_{t+i}+\gamma^k V\left(s_{t+k};\theta_v\right)-V(s_t;\theta_v)$, where k can vary from a state to a state and is upper-bounded by max step of each episode. In the implementation of machina, a set of estimates of the advantage in a batch is normalized with its variance and mean. The distinctive feature of PPO is to use the modified objective function after the idea of TRPO to update policy with the simple implementation. A primary variant of PPO called PPO-clip stabilize policy update by letting a objective function as Eq. 4.

$$L^{CLIP}\left(\theta\right)=\hat{E}_t\left[\min\left(r_t\left(\theta\right)\hat{A}_t, clip(r_t\left(\theta\right),1-\epsilon,1+\epsilon)\hat{A}_t\right)\right] \tag{4}$$

where $r_t(\theta) = \pi(a_t|s_t;\theta)/\pi(a_t|s_t;\theta_{old})$, ϵ is a hyper-parameter and $\hat{A}_t$ is an estimate of advantage at the step t. For updates of value function, the Monte-Carlo method is adopted. Specifically, the Monte-Carlo method minimizes the mean squared error between cumulative discounted reward $G_t^T = r_{t+1}+\gamma r_{t+2}+\cdots+\gamma^{T-1}r_T$ and a current value function $V(s_t|\theta_v)$.

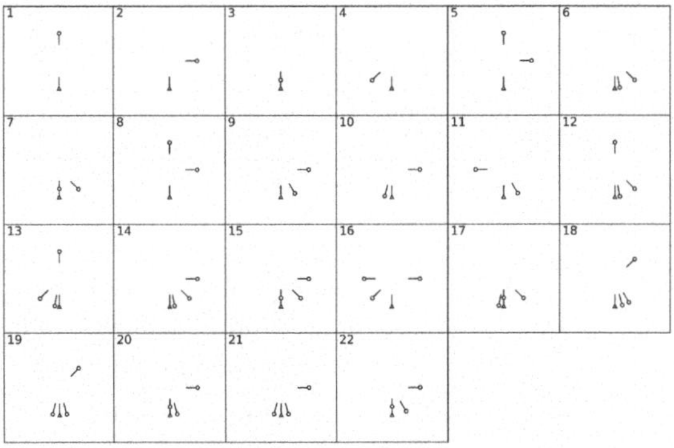

Fig. 3. Imazu problem

4 Learning Method

4.1 Environment for Learning of Collision Avoidance

The environment for learning collision avoidance consists of target ships, a waypoint and a target area. In this paper, it was assumed that the target area was an open sea with no obstacles such as coast lines, buoys. A waypoint was prepared as a destination of a own ship which is an controllable agent of this learning. The own ship must go to a given waypoint while avoiding target ships. Target ships were set in the target area according to a set of encounter situations. The encounter situations used during learning affects quality of the collision avoidance model. It is suitable that the set of situations includes from easy one like 1 on 1 encounters to difficult one like encounters of many ships. Cai and Hasegawa proposed an evaluation method of performance of automatic collision avoidance systems. In this method, they used Imazu problem [11] as a benchmark which is a set of ship encounter situations as shown in Fig. 3. Imazu problem consists of basic ship encounters of 1 on 1 and difficult situations of multiple ships. In Fig. 3, numbers on the top left in each of boxes indicate the number of cases of Imazu problem. Each of short bars from triangles (the own ship) or circles (target ships) is a velocity vector of each ship. All ships were set to arrive at the center of boxes at the same time. According to the report by Cai and Hasegawa, the problem may become easier by the elimination of the collision risk when the avoidance maneuvering by target ships is permitted. Therefore, in this paper, target ships was able to only go straight without making any changes their course by waypoint navigation or avoidance maneuvering. In addition, in order to improve the generalization performance of the learned model, one random case that 3 ships were randomly arranged was prepared, and the learning was carried out under the problem of total 23 cases including 22 cases of the Imazu problem. The position and course of each ship for every cases were set so as to collide at the origin of the space fixed coordinates. In the learning, 1 case out of 23 case was chosen randomly when it was initialized in each episode start, and target ship was arranged according to configurations of each case. For the initial

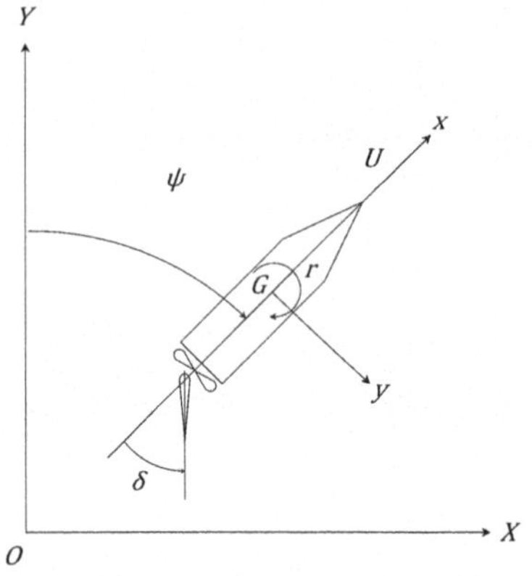

Fig. 4. Coordinate system of ship motion

Table 1. Subjects of ships for learning

Subjects	Value
K	0.10 [1/s]
T	50.0 [s]
T_E	2.5 [s]
Ship length between perpendiculars L_{PP}	106 [m]
Ship breadth B	16.2 [m]
Ship's speed U	12.0, 8.4 [kt]

heading angle of target ships in each case in Imazu problem, see [11]. This time, each ship was positioned at the set speed so that the time until the collision would be 30 min. The own ship was positioned at (0, −6) [nm] and its azimuth was randomly set in the range of −5 to + 5 deg for every episode. The motion of ships was calculated by Nomoto's equation for the heading and a primary delay equation for the rudder motion as Eq. 5 and the coordinate system of ship motion is shown in Fig. 4.

$$\begin{bmatrix} \dot{\psi} \\ \dot{r} \\ \dot{\delta} \end{bmatrix} = \begin{bmatrix} 0 & 1 & 0 \\ 0 & -1/T & K/T \\ 0 & 0 & -1/T_E \end{bmatrix} \begin{bmatrix} \psi \\ r \\ \delta \end{bmatrix} + \begin{bmatrix} 0 & 0 & 0 \\ 0 & 0 & 0 \\ 0 & 0 & \delta_E/T_E \end{bmatrix} \quad (5)$$

where ψ is a heading angle. r is a rate of turn. δ is a rudder angle of a ship and δ_E is a command rudder angle. T and T_E are time coefficients of heading and rudder motion. K is a gain. All of own and target ships had the same parameter for motion calculation assuming a kind of cargo ships. In this paper, all target ship shown as a circle had the same speed to own ship's speed except overtaken ships like in case 3, which speed was 70% of own ship's speed. The speed U of the own ship was set to 12 knots in full-scale. The own ship and target ships cannot change their speed. The Runge-Kutta method was used to integral motion equations. An action at step t : a_t was chosen from $a_t \in [-\delta_0, 0°, \delta_0]$. δ_0 is a command of a rudder angle which is a constant value set to 20°.

As a state vector from an environment, networks of a policy and a value function of PPO get inputs as follows: 1. OZT information by a grid sensor detection results, 2. normalized values of a heading angle, a rate of turn, a speed and a rudder angle of the own ship and 3. normalized values of an azimuth angle and a distance to a waypoint and a rudder angle from autopilot toward a waypoint. In this algorithm, dynamic information of target ships is grasped only from a detection result of a gird sensor. Thereby it is a feature of the proposed algorithm that the dynamic information of the target ships and position information of the own ship are not handled directory. In order to reduce computational complexity, detection of a grid sensor and updating of a command rudder angle were carried out every 5 s in the simulation time, and for the motion calculation, the interval of the integration was set to 1 s. The setting of other learning environments is shown in Tables 1 and 2. In this study, I implemented the environment using Python with OpenAI Gym which is used as a standard platform for development and evaluation for RL algorithms. By this implementation, it is easy to apply any other deep RL algorithms to this environment.

Table 2. Configurations of environment

Subjects	Value
Safe passing distance	0.3 [nm]
Grid sensor	
Angle of detection	360 [deg]
Grid space on angular direction	2.0 [deg]
Radius of sensor	6.0 [nm]
Grid space on radius direction	0.2 [nm]
Detection intervals	5.0 [s]

4.2 Design of Rewards

The rewards were designed by dividing them into two kinds of rewards: basic rewards, which were added at each step, and achievement rewards, which were given at the end of each episode. Here, an episode is a sequence from a start of a simulation to a termination of the simulation by satisfying following conditions. The end conditions of an episode were that a ship deviates from a specified area, the distance to one of other ships around the own ship becomes less than or equal to the safe passing distance (it is judged as a collision), or the distance to a waypoint becomes less than or equal to a specified distance. The basic rewards $Costs$ were defined as shown in Eqs. 6–8. The $Costs_{wp}$ calculated by Eq. 7 was a positive reward that was given more as the own ship approaches a given waypoint. In addition, for compliance with the COLREGs, a small positive reward was given as $Cost_{SB}$ to encourage the own ship to pass through the area on the right side of the line connecting the initial position of the own ship with a given waypoint, so that an agent basically learns to avoid toward the starboard side.

$$Costs = Costs_{wp} + Costs_{SB} \tag{6}$$

$$Costs_{wp} = (0.01 Costs_{Az_{wp}} + 0.1) \tanh(1/d_{wp}) \tag{7}$$

$$Costs_{SB} = \begin{cases} 0.025, & Az_{wp} \geq 0 \\ 0.0, & Az_{wp} < 0 \end{cases} \tag{8}$$

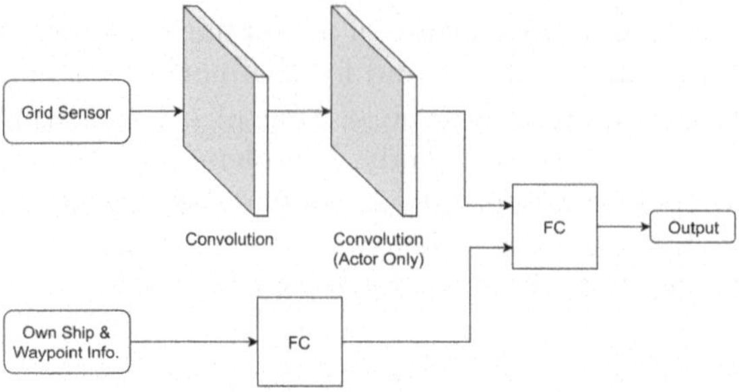

Fig. 5. Illustration of networks in PPO

where, $Costs_{Az_{wp}} = (\pi/4 - \min(|Az_{wp}|, \pi/4))/(\pi/4)$. d_{wp} and Az_{wp} are the distance and the azimuth to a given waypoint from the own ship. Results were set as follows according to the end condition of each episode: -50 for deviation from the target area, -50 for a collision, and $+50$ for reaching within the specified distance from a given waypoint without collisions. The scale of achievement rewards was decided from the result of the preliminary learning carried out beforehand as the scale for that the effect of $Costs_{Az}$ with a small value does not disappear, while encouraging the own ship to avoid collision and deviation of the course is prioritized.

4.3 Structure of Networks and Update Method

The policy and value function used in PPO are represented by neural networks as described above. At this time, among the states given as inputs, the values of the grid sensor and the information of the own ship and a given waypoint are different in properties. A state vector from the grid sensor needs to be processed by convolution layers used in deep learning using images. On the other hand, real numerical data such as dynamic information of the own ship can be processed in full connected layers because of its small number of dimensions. Therefore, as shown in Fig. 5, a state vector of the grid sensor and a state vector of other information were processed separately by the convolution layers and the full connected layer (FC), and then finally, outputs of each layer were concatenated and passed through the full connected layer. In the implementation, two networks were defined completely independently without sharing any parameters of networks. Here, only the policy network had two convolution layers, and the network of the value function had only one convolution layer. Adam [20] was used to update networks. Except for the learning rate, all Adam's hyper-parameters were default values of Pytorch. Details of any other networks' hyper-parameters are described in the appendix.

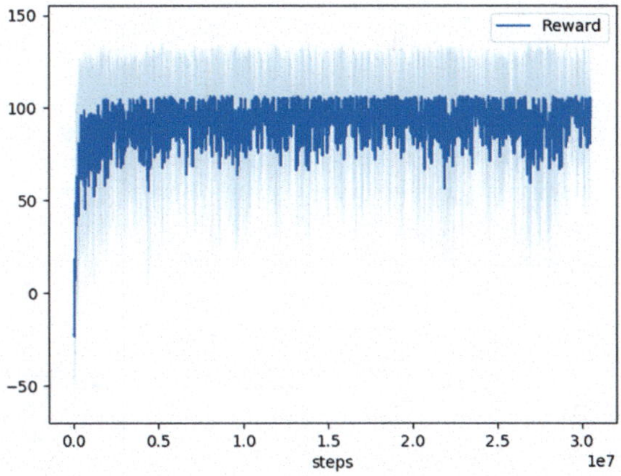

Fig. 6. Moving average reward of each iteration of the test model

5 Simulation Results

The model used in the evaluation was trained for 30,480,816 steps (43565 episodes), and its average cumulative reward per 1 iteration was 94.78, the median was 105.9, and the maximum was 106.9. Figure 6 shows the moving average of the cumulative reward per iteration during learning. In the figure, the dark line shows the moving average and the light area shows the standard deviation for each iteration. Using this model, simulation was carried out using all 22 cases of the Imazu problem, and the performance of collision avoidance was evaluated.

Figure 7 shows trajectories in all 22 cases of Imazu problem. Figure 8 shows the minimum passing distance of the own ship and target ships in each time step. The course stability was not specifically included in the design of rewards, but trajectories by the learned model was smooth. According to the minimum passing distance, the learned model performed avoidance maneuvering with keeping the safe passing distance of 0.3 nautical miles for all cases of Imazu problem. For these simulations, it was necessary for the own ship to solve danger of collisions only by turning without increasing and decreasing speed since target ships did not give way. However, in many cases such as case 18 and case 19, avoidance maneuvering which went around the stern of target ships was learned. Especially, in cases 19 and 20, it can be seen that the learned model can select its course properly even in complicated situations in which target ships came from left and right and from north and south. The lateral deviation from the original course on the line of $x = 0$ during avoidance maneuvering was within 2 nautical miles in all cases.

Considering the condition in which target ships did not give way, I consider that the learned model made an appropriate decision with following the COLREGs in all cases except case 6, 12 and 14. The reward for the COLREGs (Eq. 8) was simple, encouraging right-handed traffic, and did not specifically correspond

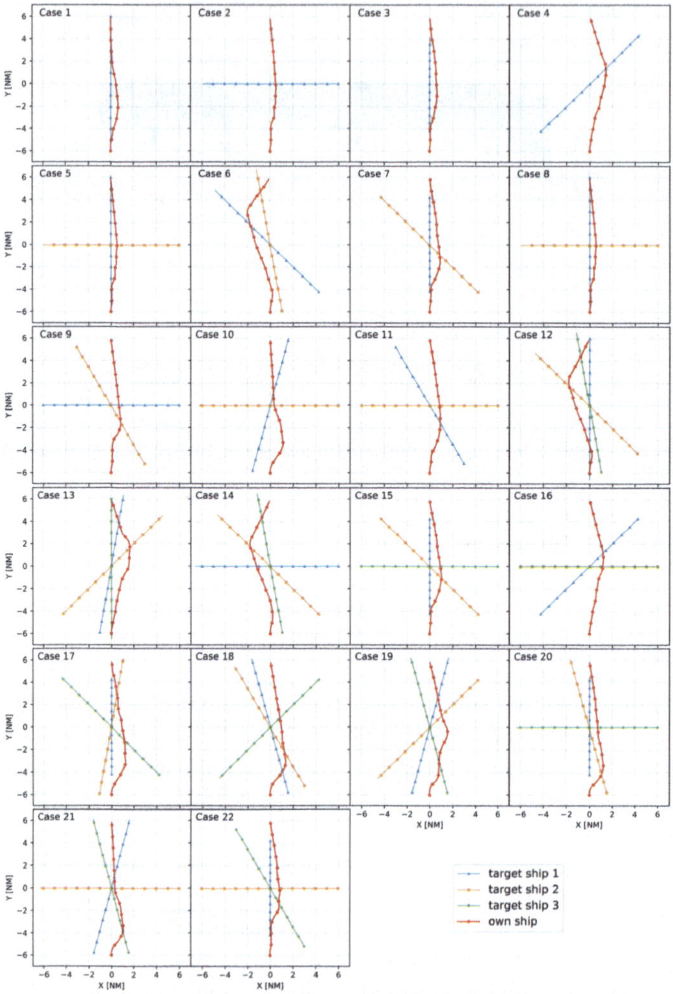

Fig. 7. Trajectories through Imazu problem

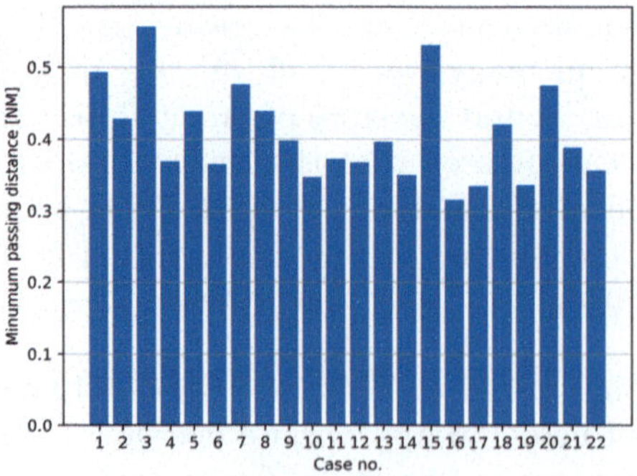

Fig. 8. Minimum passing distance in each case

to each of encounter situations between two ships specified in the COLREGs. It was difficult to follow the COLREGs for the own ship because one of the target ships in case 6, 12 and 14 was very close to the own ship at the start. Therefore this simple reward may be enough to make an agent follow the COLREGs.

Finally, robustness on maneuverability of the learned model was also confirmed. In additional simulations, each combination of $T = 25, 50, 100, 200\,[s]$, $K = 0.01, 0.05, 0.1\,[1/s]$ was verified using Imazu problem. As a result, it was confirmed that the learned model could avoid target ships for all combinations of K and T. However, if the maneuverability was remarkably different from the configurations of the learning, for example, when $T = 200\,[\mathrm{s}]$ and $K = 0.1\,[1/\mathrm{s}]$, the course of the own ship may not be stable in some cases. Therefore, in practical use, it is desirable to prepare learned models of several patterns according to representative ship types and their maneuverability, and to choose a learned model which its maneuverability is similar to a ship to be applied.

6 Conclusions

In this study, a method combining detection of OZTs by the grid sensor and deep RL algorithm was proposed for collision avoidance of multiple ships and waypoint navigation in parallel. The grid sensor can efficiently detect OZTs of multiple ships in its range, and can express the surrounding situation in detail in the form which is easy to apply to the control algorithm by converting this as a fixed dimensional state vector. By giving detection results of the grid sensor to the deep RL algorithm PPO, flexible maneuvering was learned. As a subject of the learning, I carried out the learning using Imazu problem. As a result, the model learned collision avoidance maneuvering in various cases including the difficult encounter situations involving multiple ships. The learned model is flexible because the navigation route changes according to the situation and corresponds to each encounter situation.

For future works, I would like to build more advanced collision avoidance models considering such as the bow crossing range. In addition, as a research of collision avoidance maneuvering using learned models, I would like to classify and evaluate of marine traffic in terms of the requirement for automatic avoidance system.

A Appendix

The configuration of the networks and the hyper-parameters of PPO in this paper are shown in Table 3. Machina supports only Ubuntu and doesn't supports Windows officially. Therefore, some implementations in the machina code such as the use of multiprocessing in Pytorch and Python and the log output format have been modified for Windows.

Table 3. Hyper-parameters values for PPO

Parameters	Value
Optimizer	Adam
Learning rate (Actor, Critic)	$3.0 \cdot 10^{-4}$
Discount *gamma*	0.995
1st convolutional layer for grid sensor	
Number of output channel	256
Kernel size	8
Stride	4
Padding	8
The number of channels of output	32
2nd convolutional layer for grid sensor number of output channel	128
Kernel size	4
Stride	2
Padding	4
The number of channels of output	32
Hidden units of fully connected layer for other state vector	256
Hidden units of fully connected layer for merged output	128
Size of clipping gradient norm	10.0
Nonlinearity	ReLU
Number of processes to parallel sampling	6
Batch size	256

References

1. IMO takes first steps to address autonomous ships (2018). http://www.imo.org
2. Varas, J.M., Hirdaris, S., Smith, R., Scialla, P., Caharija, W., Bhuiyan, Z., Mills, T., Naeem, W., Hu, L., Renton, I., Motson, D., Rajabally, E.: MAXCMAS project: autonomous COLREGs Compliant Ship Navigation. In: Proceedings of the 16th Conference on Computer Applications and Information Technology in the Maritime Industries 2017, pp. 454-464 (2017)
3. Imazu, H.: Computation of OZT by using collision course. Navigation **188**, 78–81 (2014)
4. Imazu, H.: Evaluation method of collision risk by using true motion. Int. J. Mar. Navig. Saf. Sea Transp. (TransNav) **11**(1), 65–70 (2017)

5. Schulman, J., Levine, S., Abbeel, P., Jordan, M., Moritz, P.: Proximal policy optimization algorithms, arXiv preprint arXiv:1707.06347 (2017)
6. Mnih, V., Badia, A.P., Mirza, M., Graves, A., Lillicrap, T., Harley, T., Silver, D., Kavukcuoglu, K.: Asynchronous methods for deep reinforcement learning. In: Proceedings of the 33rd International Conference on Machine Learning, PMLR, vol. 48, pp. 1928-1937 (2016)
7. Imazu, H., Koyama, T.: The optimization of the criterion for collision avoidance action-II. J. Jpn. Inst. Navig. **72**, 23–30 (1985)
8. Imazu, H., Koyama, T.: The optimization of the criterion for collision avoidance action-III. J. Jpn. Inst. Navig. **73**, 19–26 (1985)
9. Kouzuki, A., Hasegawa, K.: Automatic collision avoidance system for ships using fuzzy control. J. Kansai Soc. Nav. Arch. Jpn. **205**, 1–10 (1987)
10. Cai, Y., Hasegawa, K.: Evaluating of marine traffic simulation system through imazu problem. In: The Proceedings of Japan Society of Naval Architecture and Ocean Engineering, vol. 17, pp. 191–194 (2013)
11. Imazu, H.: Research on collision avoidance manoeuvre (in Japanese). Ph.D. thesis. University of Tokyo, Japan (1987)
12. Hu, L., Naeem, W., Rajabally, E., Watson, G., Mills, T., Bhuiyan, Z., Salter, I.: COLREGs-compliant path planning for autonomous surface vehicles: a multi-objective optimization approach. In: 20th IFAC World Congress, vol. 50, pp. 13662–13667 (2017)
13. Nagasawa, A., Hara, K., Inoue, K.: The subjective difficulties of the situation of collision avoidance-I: toward the rating by simulation. J. Jpn. Inst. Navig. **79**, 91–100 (1988)
14. Nagasawa, A., Hara, K., Inoue, K., Kose, K.: The subjective difficulties of the situation of collision avoidance-II: toward the rating by simulation. J. Jpn. Inst. Navig. **88**, 137–144 (1993)
15. Taniguchi, Y., Matsuda, A., Sera, W., Terada, D., Hashimoto, H.: Validation of a ship collision avoidance algorithm in congested sea area by means of model experiment. In: The Proceedings of Japan Society of Naval Architecture and Ocean Engineering, vol. 23, pp. 627–632 (2016)
16. Kuwata, Y., Wolf, M.T., Zarzhitsky, D., Huntsberger, T.L.: Safe maritime autonomous navigation with COLREGs, using velocity obstacles. IEEE J. Ocean. Eng. **39**, 110–119 (2014)
17. Mitsubori, K., Kamio, T., Tanaka, T.: Finding the course and collision avoidance based on reinforcement learning algorithm. Navigation **170**, 26–31 (2009)
18. Shen, H., Hashimoto, H., Matsuda, A., Taniguchi, Y., Terada, D., Guo, C.: Automatic collision avoidance of multiple ships based on deep Q-learning. Appl. Ocean Res. **86**, 268–288 (2019)
19. Rachman, A.S.A.: 3D-LIDAR multi object tracking for autonomous driving: multi-target detection and tracking under urban road uncertainties, Master thesis, Delft University of Technology, Netherlands (2017)
20. Kingma, D.P., Ba, J.: Adam: a method for stochastic optimization. In: Proceedings of the 3rd International Conference on Learning Representations, International Conference on Learning Representations (ICLR) (2015)

21. DeepX, Inc.: Machina a library for real-world deep reinforcement learning (2019). https://machina-rl.org/
22. Paszke, A., Gross, S., Chintala, S., Chanan, G., Yang, E., DeVito, Z., Lin, Z., Desmaison, A., Antiga, L., Lerer, A.: Automatic Differentiation in PyTorch, NIPS Autodiff Workshop (2017)

Two Way Allocation Methods in SAVS for Large Depopulated Area

Sho Iwata(✉) and Keiji Suzuki(✉)

Future University Hakodate, Hakodate, Japan
{g2119005,kjsuzuki}@fun.ac.jp

Abstract. In depopulated areas, many areas are introducing On-Demand transportation to improve local traffic. Hiyama southern region is a depopulated area located in the southern part of Hokkaido. It is necessary to secure efficient means of transportation due to the decline of the means of transportation and the aging of the area. This research aims to introduce Smart Access Vehicle Service (SAVS) for wide and depopulated area. In order to introduce SAVS in Hiyama southern region, it analyzed problems and issues with SAVS simulation. From the analysis results, an improved assignment algorithm was proposed and analyzed by simulation.

Keywords: SAVS · Large depopulated area · Allocation algorithm · Simulation

1 Introduction

1.1 Demand Based Traffic

There are several types of demand-type traffic. In particular, there are many researches on on-demand (full-demand) traffic. In a depopulated area where there are few local residents has the problem due to profitability of not being able to provide sufficient transportation means for users. In addition, in areas where has advanced in areas becoming aging seriously, the frequency of movement for residents is low and the percentage of people with traffic that can not easily move by themselves. So, it is necessary to secure transportation means in aging areas. In such depopulated areas, there are many areas that introduce on-demand transportation as an alternative to public transportation. However, the on-demand traffic introduced so far has limited locations for getting on and off, needs to have a reservation in advance, and there are remaining issues such as the operator manually creating a route plan. In recent years, the introduction of full-demand type transportation that transfers without restriction of the place of getting on and off is in progress and research and development are actively conducted. Research includes the verification of the availability of demand buses by simulation [1], on-demand bus allocation, and verification of operation methods [2,3].

H. Sato et al. (Eds.): IES 2019, PALO 12, pp. 33–45, 2020.
https://doi.org/10.1007/978-3-030-37442-6_4

1.2 Purpose of Research

In this research, purpose of introduce the full-demand type public transport service SAVS in a wide and depopulated area. However, there is no case of demand type traffic in wide and depopulated special area, and the possibility of operation is unknown. So, it is necessary to analyze and evaluate by simulation. It constructed and experimented with the SAVS simulation environment and clarified the analysis and multiple issues from the results. Furthermore, the solution method is show by comparison with the existing allocation algorithm by SAVS simulation which applied the allocation algorithm which solved the problem point.

2 SAVS

2.1 SAVS Information

Smart Access Vehicle Service (SAVS) is a type of Demand Responsive Transportation (DRT). SAVS is a full demand transportation service. The research and the development of SAVS has been started since 2013. Prior research includes a research of SAVS operation method by simulation [4] and a demonstration experiment in Hakodate City [5–7]. Currently, demonstration experiments are being conducted in various parts of Japan, and development is underway for full-scale operation. Compared with other demand-type traffic, it has door-to-door operation without having a fixed route, real-time allocation and without presupposing reservation, and it possible that operate ride sharing. In other words, the difference between taxis and SAVS is that the vehicle allocation and calculation system automatically finds the optimal vehicle and assigns it by telling the passengers where to get on and off. The system configuration of SAVS is configured as follows.

1. Application for resister of passenger's demand (Ride APP)
   The screen of application used by the passenger is shown in Fig. 1. The passenger inputs the ride on point and ride off point from the map displayed on the screen and issues a demand. If the demand is accepted and the assignment is complete, the screen will display the information such as the vehicle number and when it will arrive.
2. Application for the SAV driver to check and drive the demand (Traveling APP)
   The screen shot of the application used by driver is shown in Fig. 2. The application display the passenger's getting on/off position and the passenger information. When a new passenger is added, it is changed in real time and notified to the driver.
3. Allocation system that assigns demand to the most suitable vehicle and plans travel route
   SAVS finds the best vehicle among the running vehicles by the sequential optimal insertion method. The outline of the sequential optimal insertion method Fig. 3 is shown below.

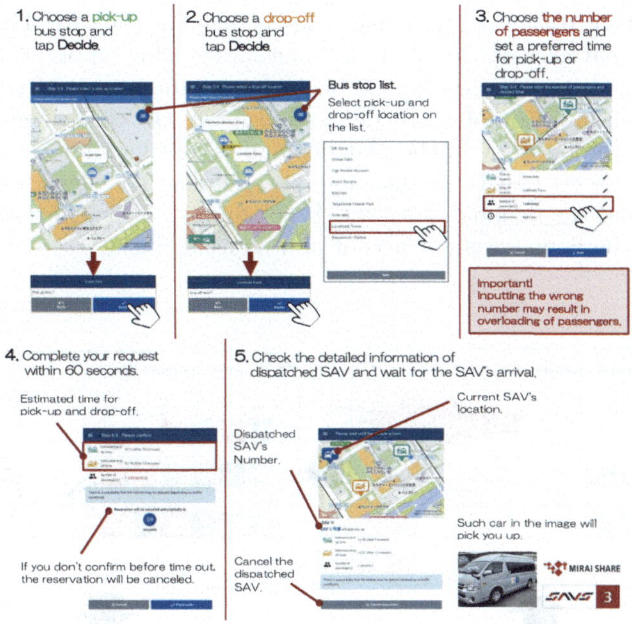

Fig. 1. Ride App

When the allocation system receives a new demand, the demand is presented to each running vehicle. Each vehicle agent inserts the get on point and the get off point of the demand into any place of the existing waypoint list respectively, and compares the delay time of the scheduled achievement time of all the demand by the insertion with the scheduled time before the insertion. Furthermore, the estimated time to achieve the new demand is obtained, and

Fig. 2. Traveling App

the total of the estimated time and delay is taken as the cost. These calculate all possible patterns of the boarding point and getting off point, and with the via pattern with the lowest cost. This is the minimum cost of vehicle agents. At this time, insertion of a new demand excludes the case where the existing demand or the deadline of the new demand is overdue. For all of the vehicle agents, the lowest cost is sought and a new demand is assigned. It change the route list and the demand is accept. If not found vehicle agent satisfying the condition, the demand is not accept and no assignment.

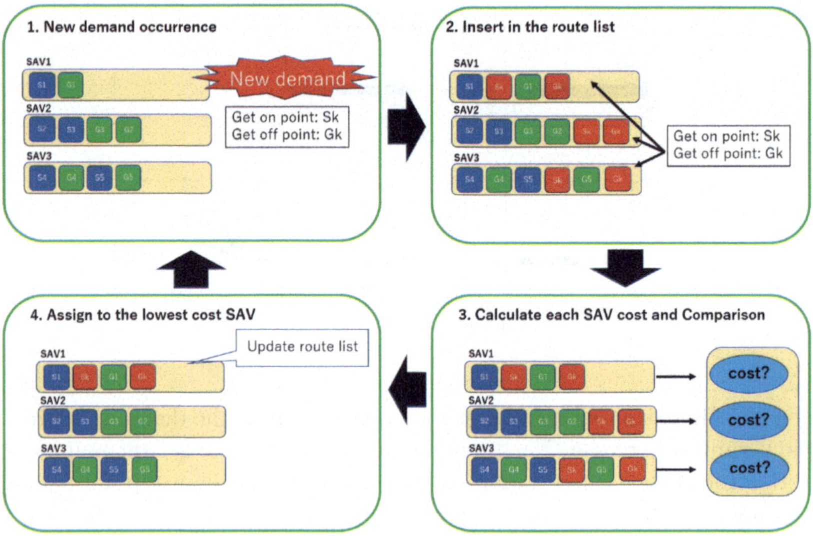

Fig. 3. Sequential optimal insertion method

2.2 SAVS Operation in Regional Areas

Hiyama southern region is a large and depopulated area. In addition, the operation with door-to-door which is a feature of SAVS considered to work effectively because the area is becoming seriously. However, looking at the past SAVS service area, the service area is at most about 12 km^2 Table 1. Therefore, it is necessary to examine the effectiveness of the operation in the special area of SAVS simulation and to investigate the issues.

Table 1. Service area and scale of SAVS

Service area	Operation scale	Number of vehicles
Tokyo Rinkai-Fukutosin area	4.42 km^2	10 cars
Yokohama-Minatomirai	8 km^2	15 cars
Aichi-Nagakute city	21.5 km^2	3 cars
Kumamoto-Arao city	25 km^2	5 cars
Nagano-Kamisuwa Town	49 km^2	9 cars
Hakodate	144 km^2	30 cars
Hiyama southern region	1278 km^2	?

3 Simulation Analysis

3.1 About Hiyama Southern Region

Hiyama southern region located in Hokkaido Donan district, consists of four towns including Assabu-town, Esashi-town, Otobe-town, and Kaminokuni-town Fig. 4. It is the depopulated area with the total population of 21,312 people, the total area of 1278 km^2 and the population density of 16.67 people. The average aging rate is as high as 38.25%. There is a problem that people can not move freely due to the decrease of the movement of local residents and due to the aging and the decline of public transportation. Currently, each town operates own pick-up bus to send and receive users via facilities and shops. Therefore, it is necessary to secure the means of transportation for the residents.

Fig. 4. Hiyama southern region

3.2 Simulation Preparation

In this section, explain the analysis of the environment construction and experimental evaluation to perform the SAVS simulation. In this research, used traffic simulation software SUMO (Simulation of Urban Mobility). SUMO is an open source road traffic flow simulator and it is possible to use free road network. By importing free road map data, it also supports construction of the same road network as in the real world. In addition, simulations with general vehicles, programs possible for moving specific vehicles, and it is also possible to respond to changes in travel route in real time such as SAVS. Based on these points, it is judged that it would be desirable to use SUMO and to carry out a simulation of SAVS assuming operation in Hiyama southern region. The simulation environment was constructed by linking SUMO with the sequential optimal insertion method and constructing a road network. Location information and route change of SAV vehicles, which are using the API provided by SUMO. The road network creation was converted from the Openstreetmap and using the tools of SUMO and road network of Hiyama southern region. The left of the figure is the road network immediately after conversion Fig. 5. However, there are roads that do not need to use SAV, such as forest roads and agricultural roads. These roads are deleted and the road network is reconstructed suitable for SAVS operation.

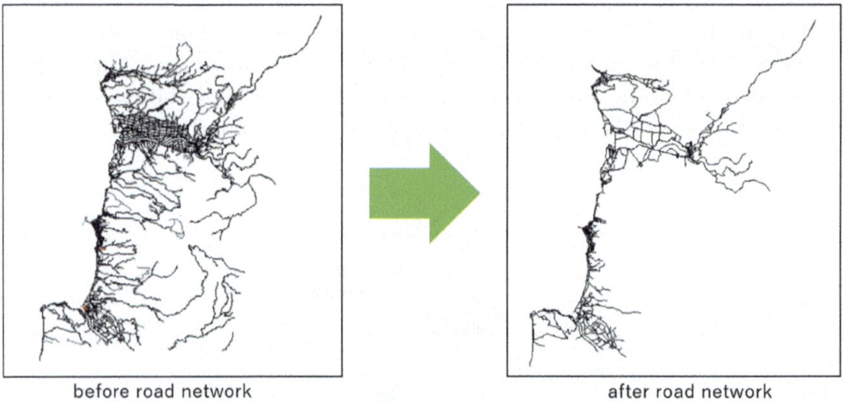

Fig. 5. Hiyama southern region network

Simulation settings

The common setting items in each simulation are defined as follows.

- The simulation time is assumed to be 12 h (43,200 s), assuming daily operation.
- The maximum speed of the vehicle is 50 km/h.
- The maximum number of passengers on the vehicle is 4.
- The number of general vehicles traveling in the town is always 400 to 500 cars.

- The time limit is set within 1800 s from the demand occurrence time, which is considering the operation in a wide range.
- The waiting point of the vehicle is set at one place each town. Vehicles that have not been transferred are returned to the waiting point.
- The demand occurrence point is narrowed to 220 places where the building exists and randomly selected.

The simulation pattern is set as follows.

The number of vehicles for SAVS to five and ten. The number of generated demand case is set to 100, 250, 500 and 2000. The simulation is performed with a total of 8 pattern settings.

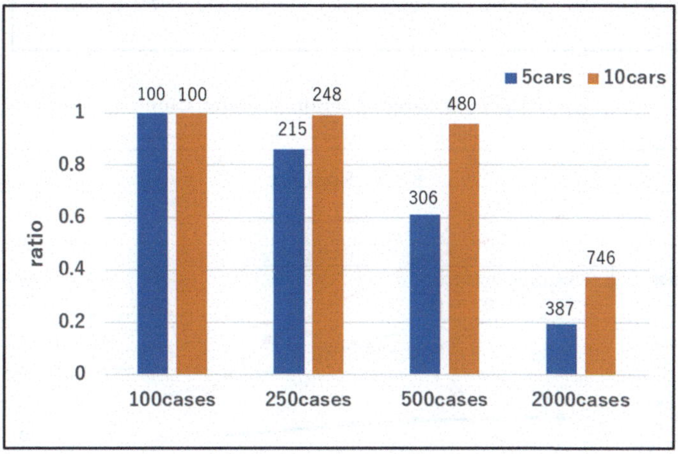

Fig. 6. Number of achievements

3.3 Experiment and Consideration

3.3.1 Results

- Number of achievements
  The number of demands achieved is shown for each number of occurrences Fig. 6. 100 cases were all achieved. 2000 was hardly achieved due to a lack of vehicles.
- Average Ride on waiting time
  For each occurrence, it represents the average waiting time until boarding Fig. 7. If the number of vehicles is sufficient, the waiting time is short.
- Demand processing results by number of vehicles
  The figures Fig. 8, Fig. 9 below show the average distance of accepted demand, the average distance of not accepted demand, the achieved demand rate, and ride-share rate, respectively. The vertical axis represents the demand distance

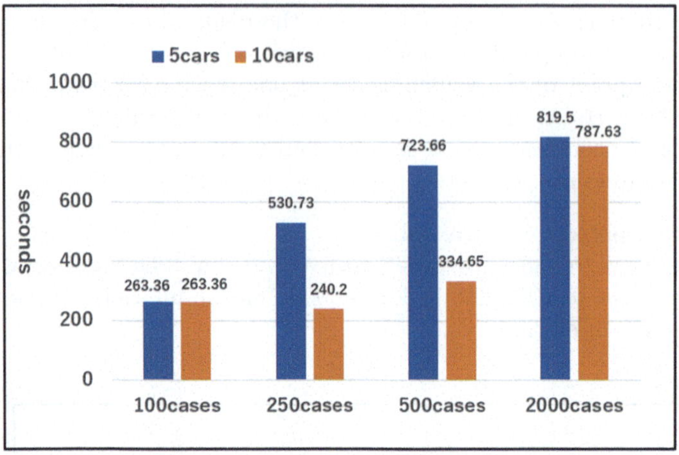

Fig. 7. Average ride on waiting time

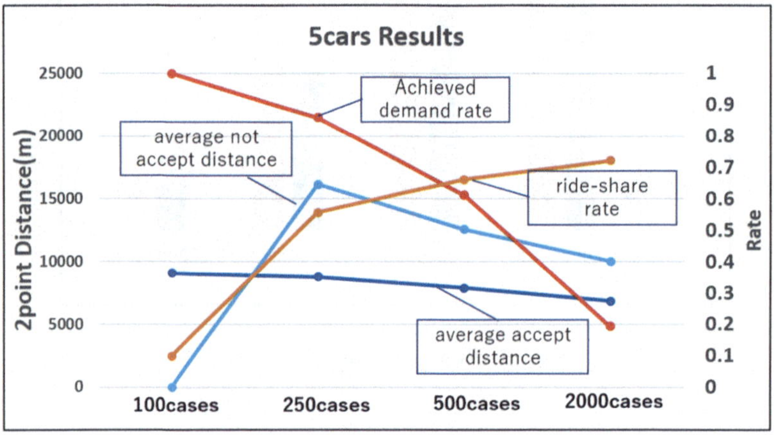

Fig. 8. Demand processing 5 cars results

(the distance between two points, the get on point and the get off point) and the rate. The horizontal axis represents the each demand cases.

At first, from the relation of the ride-share rate, it is understood that the ride-share rate in both 100 cases (5 and 10 cars) is about 0.1. There is almost no ride sharing. On the other hand, for more than 250 demand cases, operations are being carried out ride sharing.

Next, from the relationship of the average distance of not accepted demand. When the achieved demand rate exceeds 0.8, the average distance of not accepted demand exceeds 20000 m. In addition, since the demand distance that can be traveled within the time limit (1800 s) is 22000 to 23000 m. So, the "long distance" demands that can not be transport within the time limit are accepted.

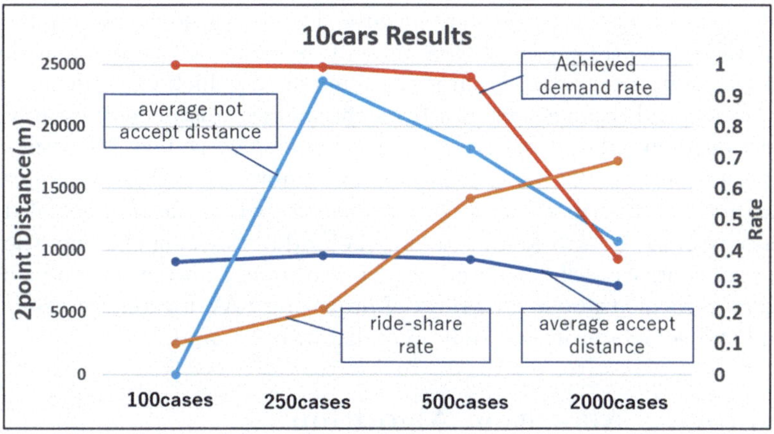

Fig. 9. Demand processing 10 cars results

Next, If the number of vehicles is sufficient for the number of generated demands. It can be understood that the average distance is 9000 to 10000 m from the relationship of the average distance of the accepted demands. In the case of 2000 cases, the average distance of accepted demand is around 12000 m, and the average distance of satisfied demand is 7500 m, so it can be seen that the number of vehicles is insufficient for the number of demand occurrence.

3.3.2 Consideration and Measures

Based on the experimental results, the discussion and the measures for the problems are described. The approximate number of vehicles required was determined as follows from the relationship between the achieved demand rate, the average distance accepted demands, and the average distance not accepted demands Table 2.

Table 2. Required number of vehicles

Number of demand occurrences	Number of vehicles
100 cases	5 cars
250 cases	5–10 cars
500 cases	More 10 cars
2000 cases	More 30 cars

According to the ride-share rate, if the number of demand occurrences is 250 or more, the operation is carried out while the boarding takes place. But, in less than 250 cases, almost not ride-share occurs. This is considered to be due to the fact that the demand generated time interval is long and the demand does not occur in the same time zone. Even if the number of vehicles is sufficient for the number of demand occurrences and from the average distance of establishment and rejection demands, in many cases the demand exceeding 20000 m will not be established due to the relationship of time limit. From these considerations, the measures are described. As a countermeasure when the number of demand occurrence can not be secured, there is a method of devising the operation time zone and concentrating the demand occurrence time. In order to cope with long distance movement, there is a method of preparing SAV for long distance movement and the vehicle processes only long distance demand.

4 Improving Allocation Algorithm

From the considerations and measures described in Sect. 3.3.2 as a problem, the "long-distance" demand that requires resulted that the problem of limited time is not accepted. In this section, in order to be able to accept "long distance" demand, propose a improve allocation algorithm. It is performed the simulations and compared these with the conventional method.

4.1 Two Way Allocation Methods

This algorithm performs allocation paying attention to demand distance. In the conventional allocation algorithm, the optimum demand is found out of all the vehicles and the allocation is performed. The operation range was about 12 km, so there was no need to separate the vehicles. However, if there long-distance demand, it takes a long time to move. Therefore, once demand is assigned, it may be considered that no other demand can be assigned to the vehicle. It is difficult to process "long distance" demand and "short- and medium" demand at the same time together. So, it is predict that efficient operation could be possible by classifying the demand according to the distance. Show the flow of allocation algorithm Fig. 10. Specifically, it prepare SAV for long distance demand and differentiate short and medium distance demand. When the demand occurs, the distance between two points (get on and get off point) is acquire. If it is more than the set value, it is allocation to the long distance SAV, and if it is less than it, the allocation short and medium distance SAV.

4.2 Simulation Analysis

Simulation settings same as Sect. 3.2 and 500 cases simulations. Here, it represent the "short middle distance" SAV as GSAV and the "long distance" SAV as LSAV. But, the time limit of the demand assigned to LSAV is 3600 s. The number of total 9 vehicles of 5 GSAV and 4 LSAV, and total of 14 vehicles of 10 GSAV and 14 LSAV. The demand distance allocated to LSAV was divided into more than 15 km and more than 20 km. A total of simulation are four patterns.

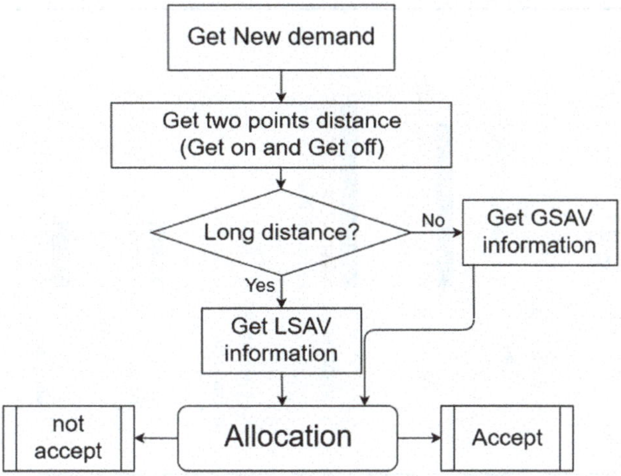

Fig. 10. Two way allocation algorithm

4.3 Results and Consideration

Shows the number of TOTAL processing and the number of GSAV and LSAV processing Fig. 11. Shows the average waiting time for all demand, and the average waiting time for GSAV and LSAV Fig. 12. The conventional allocation algorithm shows the same result as 500 cases 5 cars and 10 cars in Sect. 3.3.1.

From number of achievements Fig. 11, it was 500 cases, 14 cars and 20 km that accepted all demands. Roughly, if there are 14 or more vehicles, it is possible to process and divide 500 requests. Compared with conventional allocation algo-

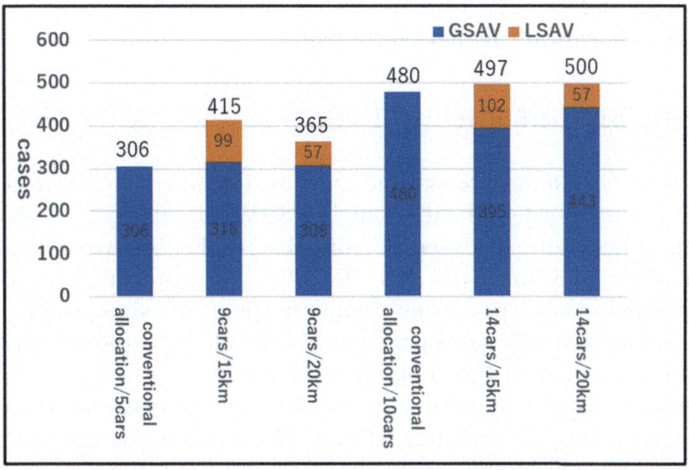

Fig. 11. Number of achievements by two way allocation algorithm

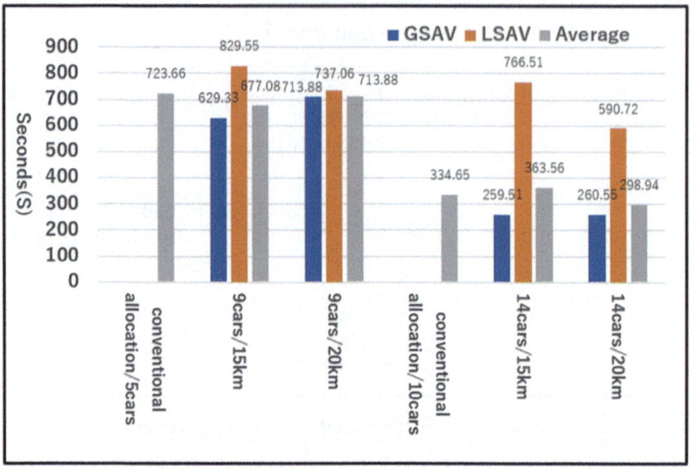

Fig. 12. Average ride on waiting time by two way allocation algorithm

rithm, it can be said that GSAV handles short and medium distance demand, and LSAV accepts long distance demand that could not be accepted by conventional allocation algorithm. From average ride on waiting time Fig. 12, GSAV is shorter than the conventional allocation algorithm. LSAV is a reasonable waiting time considering that it handles long distance demand. By setting the allocation to the "long distance" SAV to 20 km, the result was the most efficient. The number of LSAVs processed was almost the same for both 9 and 14 cars. In other words, by dividing the number of vehicles between GSAV and LSAV and optimizing according to the number of demand. It is considered possible to operate efficiently with the number of vehicles without waste. From the results of ride on waiting time and demand processing, it can be said that two way algorithm is the effective allocation calculation method.

5 Conclusion and Future Tasks

In this research, an analysis was conducted to introduce SAVS to a wide and depopulated area. In order to carry out the SAVS simulation, construct a simulation environment and analyzed it from the results. As a results, there was the problem of the movement distance. To solve this problem, proposed two way allocation algorithm and performed demand 500 cases simulations. From the results, it is show that efficient operation is possible by separating the vehicles to be allocate according to the demand distance.

In the future tasks, it is necessary to carry out simulations with the same demand as the real world demand. Also, in this simulation, the time not takes get on and off into consideration, but in order to obtain more accurate results, it is necessary to carry out a simulation that takes the time to get on and get off time.

References

1. Noda, I., Shinoda, K., Ohta, M., Nakashima, H.: Evaluation of usability of dial-a-ride system using simulation. Inf. Process. Soc. J. **49**(1), 242–252 (2008)
2. Tsubouchi, K., Yamato, H., Hiekata, K.: On the advantage of the on-demand bus system in less populated area. J. Robot. S. Jpn. **27**(1), 115–121 (2009)
3. Yamato, H., Tsubouchi, K., Hiketa, K.: A research on real-time scheduling algorithm for on-demand bus and evaluation by simulation. Transp. Policy Stud. Rev. **10**(4), 02–10 (2008)
4. Koshiba, H., Noda, I., Yamashita, T., Nakashima, H.: Evaluation of Demand Bus for Real Operation by Bus Simulator SAVSQUID Considering Real Environment. Computer software journal **31**(3), 141–155 (2014)
5. Nakashima, H., Noda, I., Matsubara, H., Hirata, K., Tayanagi, E., Shiraishi, Y., Sano, S., Koshiba, H., Kanamori, R.: Concept and implementation of a new transportation system that unifies the bus and taxi services. In: Proceedings of JSCE, vol. 71, no. 5, pp. I875–I888 (2015)
6. NakaShima, H., Koshiba, H., Sano, S., Ochiai, J., Shiraishi, Y., Hirata, K., Noda, I., Matsubara, H.: Smart access vehicle system: implementation and evaluation of a vehicle operation system for demand responsive public transportation. Inf. Process. Soc. J. **57**(4), 1290–1302 (2016)
7. Hirata, K., Suzuki, K., Noda, I., Ochiai, J., Kanamori, R., Matsudate, W., Nakashima, H., Sano, S., Shiraishi, Y., Matsubara, H.: Design and implementation of a platform for fully automatic real-time full-demand traffic system SAVS. IPSJ-Technical reports, vol. 2017-ITS-68, no. 1, pp. 1–6 (2017)

Proposal of Mathematical Model of Prediction of Human Movement Using Position Information

Shunsuke Hatadani[1](✉), Yuki Ohira[2], Akira Ishii[3], and Yasuko Kawahata[4]

[1] Graduate School of Sustainability Science, Tottori University, 4-101, Koyama Minami, Tottori 680-8522, Japan
hassium502@gmail.com

[2] Department of Management of Social Systems and Civil Engineering, Tottori University, 4-101, Koyama Minami, Tottori 680-8522, Japan

[3] Department of Applied Mathematics and Physics, Tottori University, 4-101, Koyama Minami, Tottori 680-8522, Japan

[4] Faculty of Social and Information Studies, Gunma University, 4-2 Aramaki-machi, Maebashi, Gunma 371-0044, Japan
kawahata@gunma-u.ac.jp

Abstract. Smartphones, which are essential to our lives, contain various types of information. Among them, GPS data is being used for behavior analysis. It is becoming clear that the routine behavior of city dwellers follows a certain pattern, based on preferences and life rhythms, and it is possible to predict traffic congestion by reading these behavior patterns.

Keywords: GPS · Foreigner · Simulation

1 Introduction

In recent years, the number of foreign tourists has increased, and the number of those tourists who use Japanese travel guides is also increasing. Such a guide is often a volunteer who is familiar with an area [1]. A traveler books the guide service in advance, communicates with the guide by e-mail, and makes any requests before the day of the trip. The guide receives the request and recommends sightseeing spots and shops as well as convenient modes of travel among other things. In this way, tourists and guides create sightseeing plans together. On that day, they take trips according to the sightseeing plan. This travel plan is a combination of the detailed needs of the travelers and advance knowledge of the local guides in the area, and thus is a kind of optimized travel plan. NEC Solution Innovators Ltd., provides volunteer guides with a system for reporting the details of activities of foreign travelers such as sightseeing, dining, travelling, and shopping. The daily reports include information on the trip's start time, the order of the locations they visit, travel mode, how much money was spent as well as what was bought, the purpose of the visit, and the trip end time. Using this

H. Sato et al. (Eds.): IES 2019, PALO 12, pp. 46–51, 2020.
https://doi.org/10.1007/978-3-030-37442-6_5

sequentially ordered trip data, it is possible to analyze the diverse behaviors of foreign tourists. Therefore, as a first step toward understanding the behavior of foreign tourists, the aim of this study is to calculate the frequency of movements between areas and predict the typical movements of tourists. In this paper, we analyzed data at a coarse resolution (municipality). Applying similar methods to smaller spatial units may reveal needs of foreign tourists that are not obvious to Japanese people.

2 Proposed Calculation Model

2.1 Calculation Model

The use of several straight lines is a simple way to model movement. Consider a case in which a person is at position i at time t0, at position r at times t1 and t2, and at position j at time t3, as shown in Fig. 1. We define the position of a person at time t as I(t), as in the following example.

$$\begin{aligned} I(t0) &= i \\ I(t1) = I(t2) &= r. \\ I(t3) &= j \end{aligned} \tag{1}$$

To simplify the calculation, the position is replaced by a numerical value (Fig. 1). When positions i, r, and j are defined as i = 0, r = 1, and j = 2, (1) is rewritten as follows:

$$\begin{aligned} I(t0|i) &= 0 \\ I(t1|r) = I(t2|r) &= 1. \\ I(t3|j) &= 2 \end{aligned} \tag{2}$$

Generalizing (2) yields

$$I(tn|\beta) = I(tn-1|\alpha) + M(\alpha, \beta). \tag{3}$$

Equation (3) indicates that the position at one time can be represented by a numerical value.

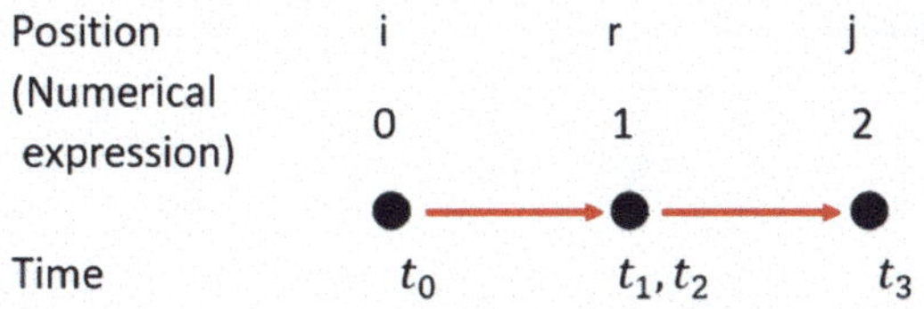

Fig. 1. Position movement model.

2.2 Function M(α, β)

Function M(α, β) is a function given by the probability P$\alpha\beta$ (tn-1, tn) of the movement of a person (discussed later). When a person decides to move from position α to position β, the movement is given by

$$M = \begin{cases} \beta - \alpha(\alpha < \beta) \\ 0(\alpha = \beta) \\ \alpha - \beta(\alpha > \beta) \end{cases}, \quad (4)$$

where α and β are obtained by replacing the position with a numerical value and M(α, β) is represented by a numerical value.

Suppose that the movement from α to β is determined by probability P$\alpha\beta$ (tn-1, tn). Then, M = 0 indicates that there is no movement and M $=\mid \beta - \alpha \mid$ indicates that movement occurs.

2.3 Probability

Probability P$\alpha\beta$ (tn-1, tn) represents the probability of movement from α to β during the period tn-1 to tn and is calculated using the following procedure. Consider the movement of a person of a certain nationality. For each city, we count the number of movements from α to β. Here, only one day of movements is considered. However, several dates are considered and added to the existing number of movements for each nationality. After calculating the number of movements from α to β, the total number of movements from α is calculated. For each city, the number of movements is calculated as follows:

$$P\alpha\beta(t(n-1), t_n) = \frac{Totalnumberofmovementsfrom\alpha to\beta}{Totalmovefrom\alpha} \quad (5)$$

$$P\alpha\beta(t(n-1), t_n) = \frac{Totalnumberofmovementsfrom\alpha to\beta}{Totalmovefrom\alpha} \quad (6)$$

This probability of movement in a city is calculated for each nationality. In this calculation, the probability is assumed to be constant over time.

3 Tourist Data

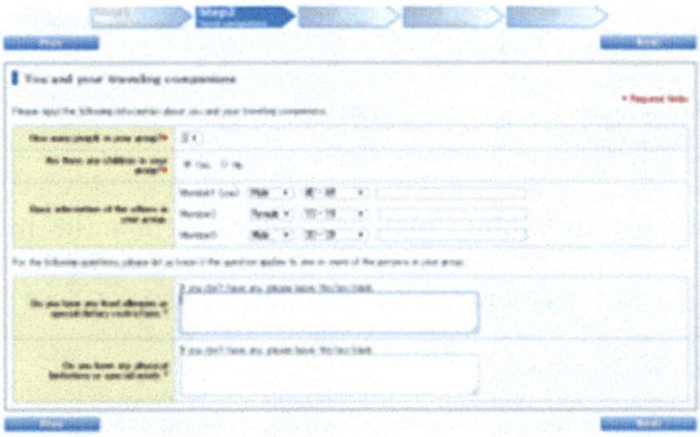

Fig. 2. Online registration in which tourists can enter various preferences and requirements, enabling geographic and chronological information to be added to this data.

3.1 Data Acquisition

Registrations in the tourist information journal were used to collect the data used in this study. Because it is possible to register in this journal online, any type of device, such as a PC, tablet, or smartphone, can be used for electronic registration. The registration information is then put into a standard format so that a person's itinerary can be systematically registered. The registered tourist areas and sightseeing spots are included in a list of choices that are displayed to enable users to input their destinations, reducing the burden on the user during registration. This also reduces the possibility of spelling inconsistencies caused by manual input. The Fig. 2 shows the screenshot of the registration system.

3.2 Aims of Data Analysis

The data in this study are not used in the same way as other data (e.g., the location update history of roaming cellular phones, GPS data obtained from smartphone tourism applications, or data obtained from comments on social networking sites and virtual forums). These data of previous studies are used for understanding traveler attributes (nationality, language, age, gender, group size, traveling with or without children, and dietary restrictions). Additionally, they are used for understanding the types of areas, facilities, and shops visited by travelers with various attributes and the amount spent on goods and services at these locations [2].

4 Calculation Result Using Tourist Data

Four municipalities are included in this study: Shizuoka City, Fuji City, Fujinomiya City, and Fujieda City. The transitive probability matrix for the cities was calculated using (5) and is shown in Fig. 3. The results of prediction using this matrix are close to the actual data (Fig. 4).

	Shizuoka City	Fuji City	Fujinomiya City	Fujieda City
Shizuoka City	0.971	0	0.029	0
Fuji City	0.5	0.5	0	0
Fujinomiya City	0	0.333	0.667	0
Fujieda City	0	0	0	0

Fig. 3. Transitive probability matrix results for the cities in this study

Actual data	Calculation result 1	Actual data	Calculation result 2
shizuoka	shizuoka	fuji	shizuoka
shizuoka	shizuoka	fuji	shizuoka
shizuoka	shizuoka	shizuoka	shizuoka
shizuoka	shizuoka	shizuoka	shizuoka
shizuoka	shizuoka	fujinomiya	fujinomiya
shizuoka	shizuoka	fujinomiya	fuji
shizuoka	shizuoka	fujinomiya	fuji
shizuoka	shizuoka	fuji	shizuoka
shizuoka	shizuoka		shizuoka
shizuoka	shizuoka		shizuoka

Fig. 4. Comparison of the results calculated using the matrix in Fig. 3 with the actual data

The results in Fig. 3 establish that the probability of trips that do not leave Shizuoka City is very high. Hence, we repeated 100 simulations consisting 10,000 trials of nine movements. In these results, 76.38% of all trips remain in Shizuoka City on average.

4.1 Prediction Using Clusters

We predicted movements according to city, but only used movements at a coarse granularity. Therefore, we thought it might be possible to predict detailed movements by grouping clusters. Here 12 clusters of movements around the following locations were identified: Shizuoka station, Shinfuji station, Fujinomiya station, Shimizu station, Nihondaira, Higashishizuoka station, Mihonomatsubara, Marikoinari, the zoo, Mt. Fuji, Fujieda station, and other locations. The transitive probability matrix for the clusters for Germans is shown in Fig. 5. This transitive probability matrix also predicts the data well (Fig. 6).

	Shizuoka	Shinfuji	Fujinomiya	Shimizu	Nihon daira	Higashishi zuoka	Mihonoma tsubara	Mariko inari	Zoo	Mt.Fuji	Fujieda	Others
Shizuoka	0.9048	0	0.0476	0	0	0.0476	0	0	0	0	0	0
Shinfuji	0.5	0.5	0	0	0	0	0	0	0	0	0	0
Fujinomiya	0	0.3333	0.6667	0	0	0	0	0	0	0	0	0
Shimizu	0.1667	0	0	0.3333	0	0.3333	0.1667	0	0	0	0	0
Nihondaira	0	0	0	0	0	0	0	0	0	0	0	0
Higashishizuoka	0	0	0	0.6667	0	0.3333	0	0	0	0	0	0
Mihonomatsubara	0	0	0	0.5	0	0	0.5	0	0	0	0	0
Marikoinari	0	0	0	0	0	0	0	0	0	0	0	0
Zoo	0	0	0	0	0	0	0	0	0	0	0	0
Mt.Fuji	0	0	0	0	0	0	0	0	0	0	0	0
Fujieda	0	0	0	0	0	0	0	0	0	0	0	0
Others	0	0	0	0	0	0	0	0	0	0	0	0

Fig. 5. Transitive probability matrix for clusters (for Germans)

Actual data	Calculation result 1	Actual data	Calculation result 2
shizuoka	shizuoka	mihomatsubara	shizuoka
shizuoka	shizuoka	shimizu st.	shizuoka
shizuoka	shizuoka	higashishizuoka	shizuoka
shizuoka	shizuoka	shimizu st.	higashishizuoka
shizuoka	shizuoka	mihomatsubara	higashishizuoka
shizuoka	shizuoka	mihomatsubara	higashishizuoka
shizuoka	shizuoka	mihomatsubara	higashishizuoka
shizuoka	shizuoka	shimizu st.	shimizu
shizuoka	shizuoka	shimizu st.	mihonomatsubara

Fig. 6. Comparison of the results calculated using the matrix in Fig. 5 with actual data

Figure 5 shows that the probability of remaining around Shizuoka station is higher than it is for other locations, although it is not as high as the probability of staying in Shizuoka City shown in Fig. 3. In the same way as we did for cities, we found that approximately 40.65% trips remained in the area around Shizuoka station.

5 Summary and Future Prospects

We determined that it is possible to predict the movement of people using the model in (3). In this study, we found that more detailed predictions are possible using clusters instead of cities. However, this method is still inadequate because we did not consider the time and purpose of the stay at each location. Hence, we plan to improve the computational model with these details, and we would like to be able to predict finer-grained information. In addition, our prediction considered the simple case of Shizuoka prefecture in Japan. It will hence be necessary to verify whether similar predictions can be performed in more complicated cases such as Tokyo and Kyoto.

Acknowledgements. We would like to thank in particular Yuki Moriyama, Shinichirou Yamada, Mingyi Sun, and Taketo Kawamura from NEC Solution Innovators, Ltd., for access to this valuable data.

References

1. Pentland, A.: Social Physics: How Good Ideas Spread—The Lessons from a New Science. Penguin Press, New York (2014)
2. Kawahata, Y., Moriyama, Y., Yamada, S., Sun, M., Kawamura, T.: Analytical the large-scale collection of data on the results of the guides for Foreigners visiting Japan. In: 2017 IEEE International Conference on Big Data (Big Data), Boston, MA, pp. 4760–4764 (2017)

Public Transport Smart Card Data Analysis Using Tucker Decomposition

Mio Hosoe(✉) and Masashi Kuwano(✉)

Department of Management of Social Systems and Civil Engineering,
Tottori University, 4-101 Koyama-Minami, Tottori, Japan
d19t4003b@edu.tottori-u.ac.jp, kuwano@tottori-u.ac.jp

Abstract. With the development of information and communication technology, various kinds of data have been collected and accumulated in large quantities. Along with this, many methods for analyzing large-scale data have been proposed. This study extracts travel characteristics from smart card data of a private rail system using Tucker decomposition, a valid method for analyzing high order data. Applying Tucker decomposition to a 6th-order tensor dataset (weatherday of the weektime of daypassenger typeorigin stationdestination station), the results reveal what kind of users move from which station to which station, in what weather, on which days of the week, and at what time of day. Moreover, the dataset of 6,489,600 elements was compressed to 1,440 elements with the suggested approach, and the main travel patterns of passengers could be confirmed.

Keywords: Tensor factorization · Smart card · High order data · Travel pattern

1 Introduction

With the rapid development of information and communication technology, a large variety of data have been generated and accumulated. In the field of transportation research, interest in large-scale data such as global positioning system (GPS) data [1–3], probe vehicle data [4,5] and mobile phone data [6–8] has been increasing. These large-scale traffic data record individual human activities at both temporal and spatial scales. With the emergence of such data, it has become possible to understand detailed travel behavior which could not be obtained with conventional paper-based surveys.

In addition to the above-mentioned data, interest in smart card data is also increasing. Smart cards were originally developed for toll collecting. However, they also allow us to track the relationship between origin and destination because smart card data contain information about which people passed the ticket gate at which station at what time of day, as well as at which station and at what time of day they ended their trips. In addition, the great advantage of

H. Sato et al. (Eds.): IES 2019, PALO 12, pp. 52–66, 2020.
https://doi.org/10.1007/978-3-030-37442-6_6

smart card data is that they can be collected on a long-term basis and often include personal attributes of card users such as age, address, and occupation.

Taking advantage of smart card data, many researchers have been working to understand smart card users behavior [9–12]. Medina [13] presented a method to identify temporal (weekly) patterns of primary activities performed by public transport users in Singapore. Using trips reported in a household travel survey, the activities traveled to and from by public transport were extracted from this dataset and used to develop models for workers and students. According to the start time and duration of activity, these models estimated the likelihood of such activity being HOME, WORK/STUDY, or OTHER. Then, consistent activities were extracted from smart card data for a one-week period and these extracted activities were classified by type. By applying the DBSCAN clustering algorithm to the 14-dimensional vector which is composed of the start time and the duration of work or study activities during 7 consecutive days, the author recognized the most common primary activity patterns. Using smart card data from Beijing, Zhang et al. [14] developed a method to identify the group travel behavior which they defined as two or more persons intentionally traveling together from a single origin to a single destination. Their method did not allow a reliable estimate of the group travel behavior share of overall travel, but did enable a description of the characteristics and the spatial and temporal patterns of group travel behavior. Concretely, their results showed that group trips tended to occur on weekends, afternoons and public holidays. In addition, they showed that stations and lines serving leisure destinations had the relationship with group travel behavior. The authors concluded that group travel patterns were different from patterns of individual travel in terms of both time and space. Zhao et al. [15] presented a method for predicting individual mobility represented as a chain of trips, each defined as a combination of the trip start time, origin and destination. They first predicted whether the user will travel on a given day, and then, if so, predicted the attributes of the next trip (trip start time, origin and destination). The authors confirmed the prediction accuracy of the model using smart card data from London. Their trip prediction model achieved a high level of accuracy. On the other hand, the prediction accuracy was different for each attribute: in particular, trip start time was less predictable than origin and destination, as temporal choice showed higher intra-personal variability than spatial choice.

By using smart card data in the provincial city of Japan, this study reveals the travel behavior patterns. This study thinks that attributes in smart card data should be considered as many as possible and uses six attributes in the analysis: weather, day of the week, time of day or time period, passenger type, origin station, and destination station. When considering multiple variables, there is a method which characterizes multivariate data by summarizing it into a vector or a matrix. However, when the data has third order or higher relations, it cannot be represented with this method without loss of the original structure [16]. To solve this problem, tensor decomposition was proposed by Hitchcock [17]. A tensor representation permits us to summarize multivariate data into a multi-dimensional array [18,19] and tensor decomposition enables us to analyze it while maintaining the original data structure. Hence, this can be an effective analysis method for high-order data.

Tensor decomposition makes it possible to use simple structures with relatively few parameters to reproduce the interactions of the complex multivariate data structure. It has two main decomposition models: CANDECOMP/PARAFAC (CP) decomposition and Tucker decomposition [20]. Han and Moutarde [21] applied CP decomposition to a 3rd-order tensor (linktime sampling steptemporal traffic sequence) constructed from simulated traffic state data. They identified spatial and temporal patterns of traffic states by characterizing the three attributes. Sun and Axhausen [22] applied probabilistic tensor factorization to smart card data collected in Singapore. Their model was based on the concept of Tucker decomposition and probabilistic latent semantic analysis (PLSA). By analyzing a 4th-order tensor (time of daypassenger typeorigin zonedestination zone), they revealed the interactions and dependencies of the four attributes.

This study represents smart card data using a 6th-order tensor (weatherday of the weektime periodpassenger typeorigin stationdestination station) and applies Tucker decomposition. The results reveal the correlations between weather, day of the week, time of day, passenger type, origin station and destination station by analyzing each attribute while maintaining the high-order data structure.

2 Data Descriptions

To understand the travel patterns of smart card users, this study uses the smart card data for a private rail system in Takamatsu City, Kagawa Prefecture, Japan (Fig. 1). Takamatsu is located in the center of Kagawa Prefecture and is its capital. As of October 2017, the population of the city was 428,872, of which 115,228 (about 27%) were 65 years old or older. The aging of the population is increasing in the suburbs of Takamatsu and the number of elderly persons in Takamatsu is also expected to keep increasing. Therefore, it is important to maintain transportation which is available to people who cannot drive a car, such as senior citizens. Moreover, Takamatsu is well known for its many traffic accidents. The overall number of traffic accidents in Kagawa Prefecture is the highest in Japan. To address these problems, it is important to improve the rail system, and in order to plan such improvements, understanding the travel behavior of current users of the rail system is necessary.

This study focuses on Kotoden, the rail system in Takamatsu. Kotoden is managed by the Takamatsu-Kotohira Electric Railroad Company. Figure 2 shows the Kotoden route map. In this figure, it can be seen that there are 52 stations in total, on three lines (Kotohira line, Nagao line, and Shido line). Kawaramachi station is connected to all three lines, so Kotoden passengers can transfer to any line at this station.

In this study, we examine the usage history data from IruCa, which is a smart card that can be used for traveling on Kotoden. Although IruCa can also be used for riding on a Kotoden bus or other additional bus routes, this study uses only the usage history data for the three rail lines.

The first step in this study was to extract data pertaining to 9,008,709 valid Kotoden trips using the following criteria: (1) collected during the 15 months from December 1, 2013 to February 28, 2015, (2) card used by passenger for any train ride during the hours of operation between 5 a.m. and 12 p.m. on any of the three lines, and (3) more than 60 s taken to move between stations. As a second step, users were divided into two ticket type groups: commuter pass users and non-commuter pass users. Because the travel patterns of users having the commuter pass shows little variation, this study only focuses on the records related to use of one of the five types of non-commuter pass tickets (adults, students, children, senior citizens, persons with a disability). As a result, this study uses the 2,821,281 data points related to this group of users.

Figure 3 shows usage by the time of day the trip was taken (riding time period). The time periods with the largest average total numbers of users per day are from 07:00 to 07:59 and from 08:00 to 08:59. After that, the average total number of users decreases. There is an increasing trend from 15:00 onward and the usage starts to decrease again after 18:00. Figure 4 shows usage by passenger type. Of the five passenger types, the group with the largest average total number of users per day is Adults.

In addition to the usage history data of IruCa, this study uses weather data from Yahoo! Japan which shows the daily weather of specific areas in Japan for the past 15 years. It is likely that the weather strongly influences the decisions of train users with non-commuter tickets, therefore this study included the weather data in the analysis. For example, a person who usually walks or uses a bicycle may instead use the train on a rainy day. Figure 5 shows the summary of the Takamatsu weather data for the 15-month study period.

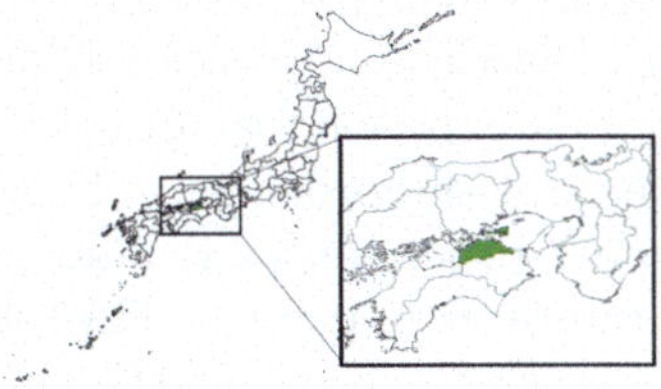

Fig. 1. Kagawa Prefecture

Fig. 2. Kotoden route map

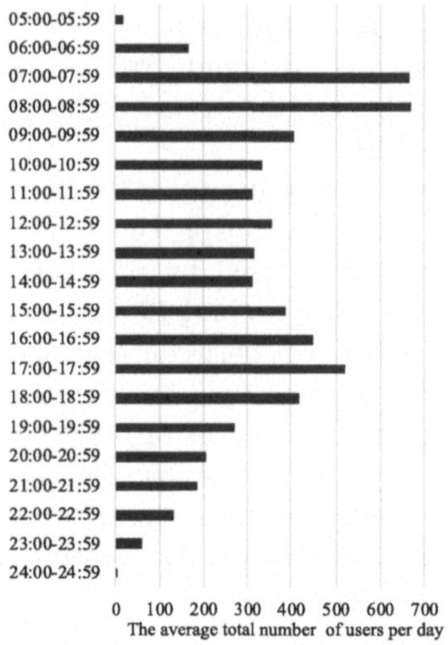

Fig. 3. Train usage by time of day

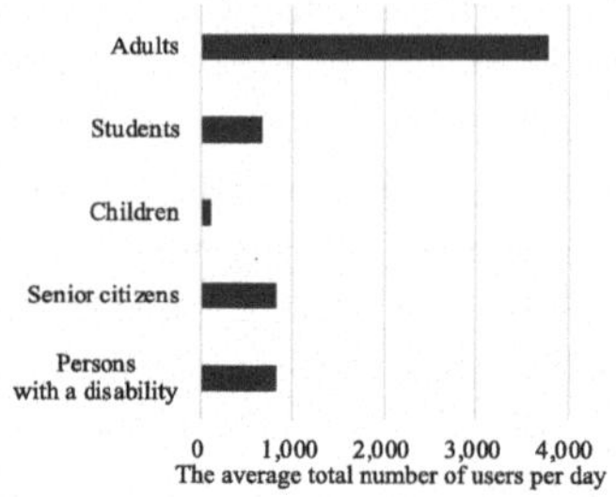

Fig. 4. Train usage by passenger type

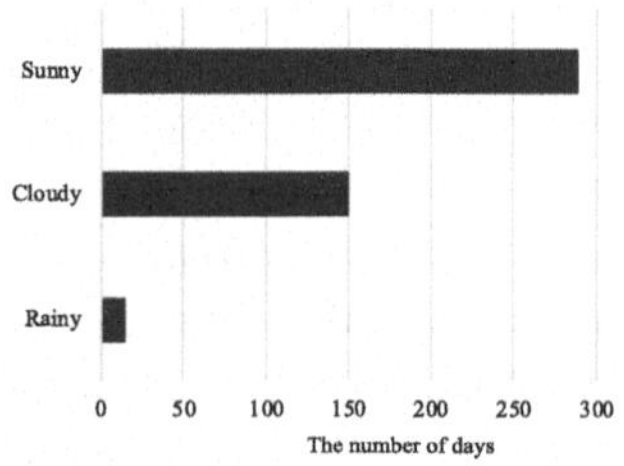

Fig. 5. Takamatsu weather (December 2013–February 2015)

3 Tucker Decomposition

Tensor factorization is an analytical method that approximates high-order data (degree 3 or more) with the product of low-order data to make complex structured data easy to interpret. A tensor of degree n is expressed as an nth-order tensor. For example, it is called a 3rd-order tensor if the order is 3. The lowest order tensors have specific common names: a 0-order tensor is a scalar, a 1st-order tensor is a vector, and a 2nd-order tensor is a matrix. This study decomposes high-order data into matrices, each of which represent the characteristics of an attribute (hereafter a factor matrix), and a core tensor. To do this, Tucker decomposition one of the available tensor factorization methods is used.

For the case of a 3rd-order tensor, the conceptual diagram of Tucker decomposition is shown in Fig. 6. $\boldsymbol{X}$ is a 3rd-order tensor of $I \times J \times K$. Tucker decomposition is defined as follows: for $\boldsymbol{X}$ with three attributes whose dimensions are given by I, J, and K, it compresses dimension I to Q $(Q \leq I)$, dimension J to R $(J \leq R)$, and dimension K to S $(K \leq S)$. That is, X is separated into matrix

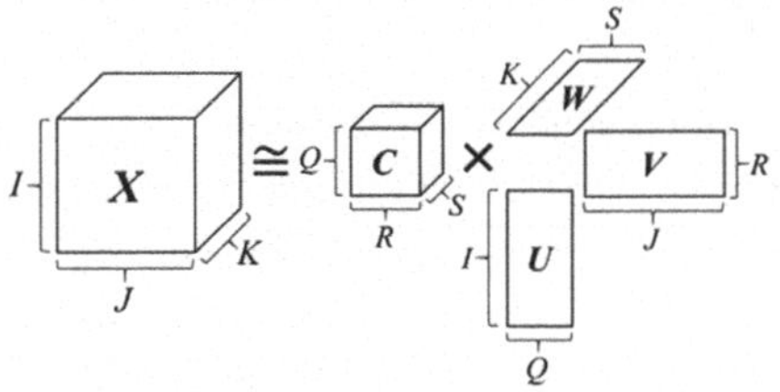

Fig. 6. The conceptual diagram of Tucker decomposition (3rd-order tensor)

$\boldsymbol{U}$ $(I \times Q)$ with information on the axes of I and Q, matrix $\boldsymbol{V}$ $(J \times R)$ with information on the axes of J and R, and matrix $\boldsymbol{W}$ $(K \times S)$ with information on the axes of K and S. This decomposition is called rank-Tucker decomposition, where the matrices $\boldsymbol{U}$, $\boldsymbol{V}$, and $\boldsymbol{W}$ are factor matrices, and $\boldsymbol{C}$ is the core tensor made up of the three orders of $Q \times R \times S$. The size of Q, R, and S is the number of factors for each attribute. This operation is expressed by the Eq. (1).

$$\boldsymbol{X} \cong \boldsymbol{C} \times \boldsymbol{U}^T \times \boldsymbol{V}^T \times \boldsymbol{W}^T \tag{1}$$

Equation (1) for each element is as shown in Eq. (2).

$$x_{ijk} \cong \sum_{q=1}^{Q} \sum_{r=1}^{R} \sum_{s=1}^{S} c_{qrs} u_{iq} v_{jr} w_{ks} \tag{2}$$

In Tucker decomposition, compression is performed to different dimensions (QRS) for each attribute.

Tucker decomposition has $\boldsymbol{C}$, $\boldsymbol{U}$, $\boldsymbol{V}$, and $\boldsymbol{W}$ as parameters, and these parameters are estimated by giving a non-negative value constraint so as to minimize the square error. In other words, rank-Tucker decomposition of a 3rd-order tensor $\boldsymbol{X}$ is formulated as the following problem:

$$\begin{aligned} &min_{\boldsymbol{C},\boldsymbol{U},\boldsymbol{V},\boldsymbol{W}} f(\boldsymbol{C},\boldsymbol{U},\boldsymbol{V},\boldsymbol{W}) \\ &\qquad f(\boldsymbol{C},\boldsymbol{U},\boldsymbol{V},\boldsymbol{W}) = \left\| \boldsymbol{X} - \boldsymbol{C} \times \boldsymbol{U}^T \times \boldsymbol{V}^T \times \boldsymbol{W}^T \right\|_F^2 \\ &s.t\ c_{qrs} \geq 0, u_{iq} \geq 0, v_{jr} \geq 0, w_{ks} \geq 0 \end{aligned} \tag{3}$$

where $\|\|_F$ is the Frobenius norm.

First, the core tensor $\boldsymbol{C}$ and factor matrices $\boldsymbol{U}$, $\boldsymbol{V}$, and $\boldsymbol{W}$ decomposed by a 3rd-order tensor $\boldsymbol{X}$ using high-order singular value decomposition (HOSVD) are treated as an optimal initialization. Then, this study calculates optimal values by the gradient descent method. For detail, refer to the paper written by Chen et al. [23].

In Tucker decomposition, we can understand the characteristics of data with the combination of the factor matrices $\boldsymbol{U}$, $\boldsymbol{V}$, and $\boldsymbol{W}$, and the core tensor $\boldsymbol{C}$, optimized by Eq. (3). In addition, because the elements of the factor matrices and the core tensor are conditioned on being non-negative, it is possible to intuitively capture the characteristics of data from the estimated result.

Table 1. Six attributes in smart card data

Attribute	No. of categories	Description
Weather	3	(1) Sunny, (2) Cloudy, (3) Rainy
Day of the week	8	(1) Monday, (2) Tuesday, (3) Wednesday, (4) Thursday, (5) Friday, (6) Saturday, (7) Sunday, (8) Public holiday
Time of day	20	(1) 5:00–5:59, (2) 6:00–6:59, (20)24:00–24:59
Passenger type	5	(1) Adults, (2) Students, (3) Children, (4) Senior citizens, (5) Persons with a disability
Origin station	52	52 stations
Destination station	52	52 stations

4 Results

To understand the travel patterns of Kotoden passengers, a 6th-order tensor is used in the analysis. This simultaneously reveals what kind of people move from which station to which station, in what weather, on which days of the week, and at what time of day using six attributes: weather, day of the week, time of day (time period), passenger type, origin station, and destination station. Table 1 shows details of the six attributes.

The 6th-order tensor has $3 \times 8 \times 20 \times 5 \times 52 \times 52 = 6,489,600$ elements. This study applies Tucker decomposition to this 6th-order tensor and extracts a factor matrix for each attribute (weather, day of the week, time of day, passenger type, origin station, and destination station) and the core tensor. Through trial and error, this study determined the number of factors for each attribute so that the result is easily interpreted. The weather factor matrix contains 2 factors, day of the week factor matrix has 3 factors, time of day factor matrix has 4 factors, passenger type factor matrix has 3 factors, origin station factor matrix has 5 factors, and the destination station factor matrix has 4 factors. The order of the core tensor equals the order of the original tensor and the number of elements corresponds to the product of the number of factors of each attribute. Thus, it becomes a 6th-order tensor with $2 \times 3 \times 4 \times 3 \times 5 \times 4 = 1,440$ elements. From the above, it can be said that a dataset having 6,489,600 elements was compressed to 1,440 elements of data. Although the factor matrices of each attribute and the core tensor are estimated at the same time, the results for each factor matrix and core tensor are considered in order below.

4.1 Factor Matrices

4.1.1 Weather Factor Matrix

Weather factor matrix results are shown in Fig. 7. The highest numerical value of the weather category in the first factor is Sunny. The second factor has higher numerical values of the weather categories Cloudy and Rainy than the first factor does. Therefore, it can be said that the first factor is Sunny and the second factor is CloudRainy. This result is intuitively reasonable.

The following other factor matrices are also considered in the same way used to determine the weather factors; that is, we compared the highest values of the categories comprising each factor matrices and considered what each factor matrix represents.

4.1.2 Day of the Week Factor Matrix

Day of the week factor matrix results are shown in Fig. 8. The first factor in the day of the week factor matrix is MondayThursday and the second factor is TuesdayWednesdayFriday. These results show that on weekdays passengers behavior is similar on Monday and Thursday as well as on Tuesday and Wednesday and Friday. The third factor is SaturdaySunday Public holiday, which means that behavior on weekends and holidays is different from that on weekdays.

4.1.3 Time of Day Factor Matrix

Time of day (time period) factor matrix results are shown in Fig. 9. The 20 time periods (each hour, from 05:00–05:59 to 24:00–24:59) resulted in four time of day factors: the first factor is 05:00–06:5909:00–10:59, the second factor is 07:00–08:59, the third factor is 11:00–17:59, and the fourth factor is 18:00–24:59. Only the first factor included two non-consecutive time periods. It can be said that there are similar travel behaviors at 05:00–06:59 and 09:00–10:59.

4.1.4 Passenger Type Factor Matrix

Passenger type factor matrix results are shown in Fig. 10. In this factor matrix, the five passenger types of Adults, Students, Children, Senior citizens, and Persons with a disability were compressed into three passenger type factors. The first factor is Adults, the second factor is Students, and the third factor is ChildrenSenior citizensPersons with a disability. From the above, it can be said that the passenger types Children, Senior citizens, and Persons with a disability (i.e., people who are generally referred to as vulnerable road users) have similar travel patterns whereas the passenger types Adults and Students move in ways that are independent of other users patterns.

4.1.5 Origin Station Factor Matrix

Because of the large number of origin stations (52), it is complicated to compare and consider all the numerical values of its factor matrix. Therefore, to determine similarities between origin stations, this study adopts the approach of including only stations showing numerical values of 50% or more of the maximum value in a given factor. This approach results in 11 origin stations being identified in the five origin station factors; the results of the factor matrix for these 11 stations are shown in Fig. 11.

Focusing on the first factor, the numerical value of Takamatsu-Chikko is 4.14, and it is the largest value in this factor. Because 50% of 4.14 is 2.07, all origin stations with values of 2.07 or higher would be identified in the first factor. However, as there are no other origin stations showing 2.07 or more in this factor, the first factor only expresses Takamatsu-Chikko.

As a result of considering the other factors using the same procedure, we identified the second factor as KataharamachiKawaramachi, the third factor as Ritsurin-KoenSanjoOtaHayashimichi, the fourth factor as OtaBusshozan, and the fifth factor as SanjoEnzaHayashimichiKatamotoFuru-Takamatsu.

There are some duplications in the stations that represent each factor. Ota was extracted with the third factor and the fourth factor, and both Sanjo and Hayashimichi were extracted with the third factor and the fifth factor. Based on this, although Ota, Sanjo, and Hayashimichi are related, it becomes clear that each station has similarities with other stations, too.

Along the three lines of Kotoden, there are many facilities such as commercial facilities, educational institutions, and other transportation hubs around Takamatsu-Chikko, Kataharamachi, and Kawaramachi, the stations located nearest the central area of Takamatsu, and there are houses and fields in the periphery. Ritsurin-Koen, Sanjo, Ota, Busshoza, Enza, Hayashimichi, Katamoto, and Furu-Takamatsu are surrounded by residential areas. Although these stations are located on different lines, it can be guessed that they play similar roles.

4.1.6 Destination Station Factor Matrix

The results of the destination station factor matrix were determined in the same way as the origin station factor matrix. As a result, eight stations were extracted as shown in Fig. 12. The results show that the first factor is Takamatsu-ChikkoKataharamachi, the second factor is KataharamachiKawaramachiRitsurin-koen, the third factor is Kawaramachi, and the fourth factor is SanjoOtaBusshozanHayashimichi.

As can be seen by comparing Figs. 11 and 12, the stations extracted in the destination station factor matrix are all also extracted in the origin station factor matrix. However, the combinations of stations constituting each factor are different. Although Sanjo, Ota, and Hayashimichi appeared in several factors in the origin station factor matrix, these stations were extracted only from the second factor in the destination station factor matrix. In addition, Ritsurin-Koen, which has similarities with Sanjo, Ota, and Hayashimichi in the origin station factor matrix, is similar with Kataharamachi and Kawaramachi in the destination station factor. Then, although Kawaramachi was related to other stations (the origin station second factor and the destination station second factor), it appears that it also has its own role as a destination station (the third destination station factor).

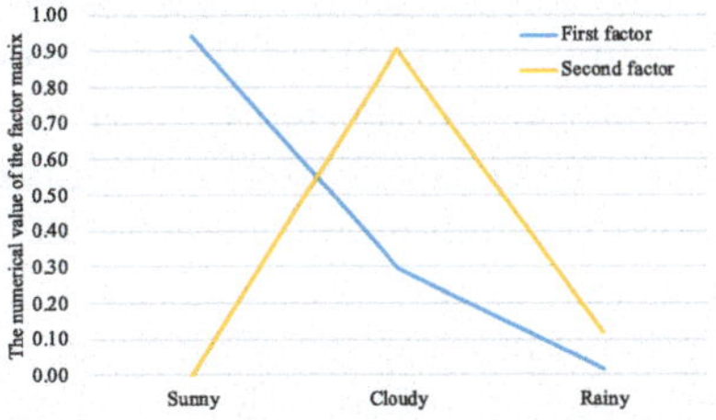

Fig. 7. Weather factor matrix

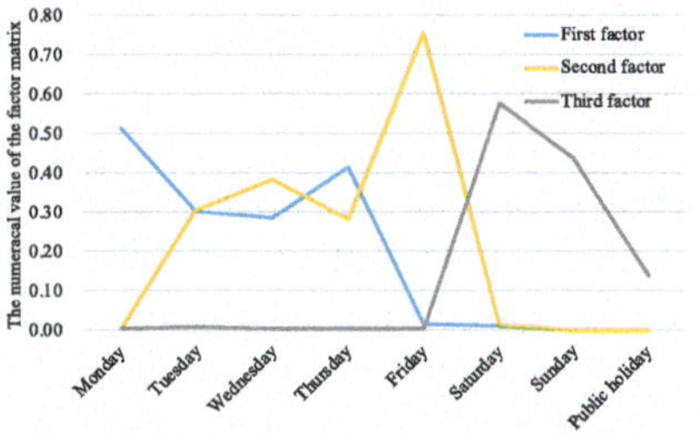

Fig. 8. Day of the week factor matrix

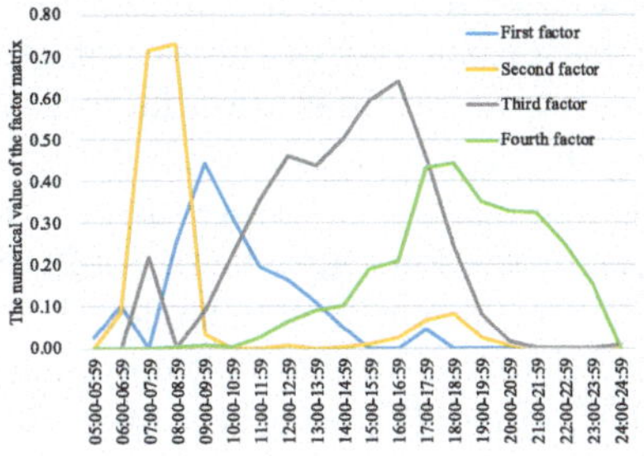

Fig. 9. Time of day factor matrix

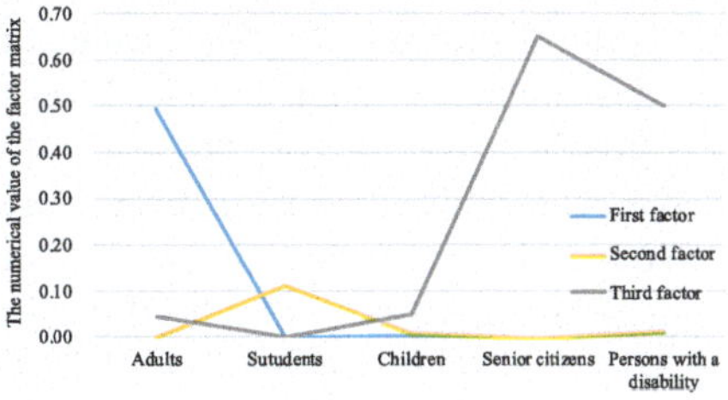

Fig. 10. Passenger type factor matrix

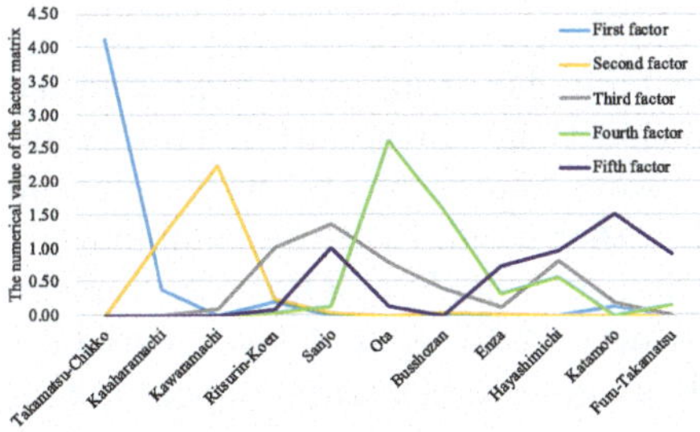

Fig. 11. Origin station factor matrix

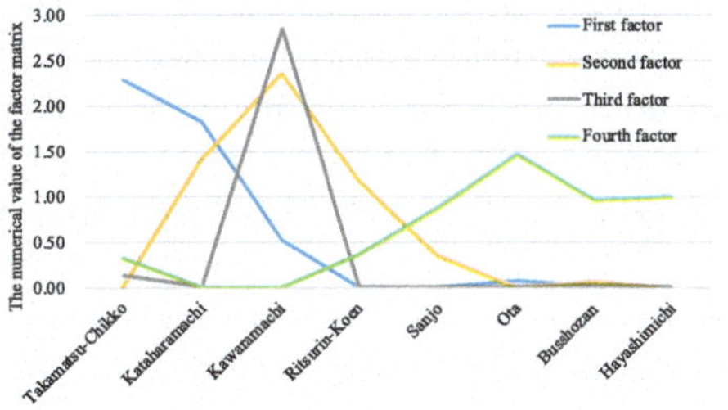

Fig. 12. Destination station factor matrix

4.2 Core Tensor

The core tensor represents a combination of the weather factors, the day of the week factors, the time of day factors, the passenger type factors, the origin station factors, and the destination station factors. This combination allows us to understand the travel patterns of Kotoden users; that is, what kind of people move from which station to which station in what weather, on which days of the week, and at what time of day. The numerical value of the core tensor is regarded as the amount of usage of Kotoden. The number of core tensor elements equals the product of each attribute, thus the core tensor has 1,440 elements ($2 \times 3 \times 4 \times 3 \times 5 \times 4$).

In the 1,440 combinations, 725 elements with a value of 0 are excluded because an element of the core tensor with a numerical value of 0 is regarded as a travel pattern that Kotoden passengers never use. Furthermore, in order to extract more characteristic travel patterns from the 715 elements showing non-zero values, this study extracted elements showing a minimum value of 53.02, that is, 5% or more of the maximum value (1060.48) of the core tensor. The 5% level was selected as a standard statistical level used in a variety of applications. Hereafter, this study describes the 124 elements extracted from the core tensor and uses these to understand the movement patterns of Kotoden users. The core tensor results for these 124 elements are shown in Fig. 13.

The number shown on the vertical axis of Fig. 13 indicates a combination of factors: it is arranged in order of the weather factor, the day of the week factor, the time of day factor, the passenger type factor, the origin station factor, and the destination station factor. For example, [1, 1, 1, 1, 1, 3] is the combination of the weather first factor, the day of the week first factor, the time of day first factor, the passenger type first factor, the origin station third factor, and the destination station first factor. This combination represents the behavior that is from Ritsurin-KoenSanjoOtaHayashimichi to Takamatsu-Chikk Kataharamachi by Adults at 05:00–05:5909:00–10:59 of MondayThursday in weather categorized as Sunny. Focusing on the top two combinations [1, 1, 2, 2, 4, 4] and [1, 1, 1, 2, 4, 4]we note that these represent the behaviors that are from KataharamachiKawaramachi to SanjoOtaBusshozanHayashimichi by Adults at 18:00–24:00 when Sunny and differ only in the day of the week factor. One travel pattern is TuesdayWednesdayFriday and the other is MondayThursday, thus both representing weekday behaviors. Moreover, we considered that these travel patterns represent return-home behavior on weekdays because Katarahamachi and Kawaramachi are near the city center of Takamatsu. On the other hand, the third combination [1, 1, 1, 2, 2, 3]is the behavior that is from Ritsurin-KoenSanjoOtaHayashimichi to Takamatsu-ChikkoKataharamachi by Adults at 07:00–08:59 on TuesdayWednesdayFriday when Sunny, and probably shows the behavior of shopping or recreation.

Focusing on the combination of the origin station factor and the destination station factor, it is found that the combination of KataharamachiKawaramachi to SanjoOtaBusshozanHayashimichi appears 24 times in the 124 elements. These travel patterns can be considered return-home behaviors from the commercial area to the residential area based on the geography of Takamatsu. Also, many of these 24 elements are behaviors by Adults or Students at 11:00–17:59 or 18:00–23:59. In addition, there are almost no differences in the number of elements of the day of the week factor. It can be said that Adults or Students frequently use Kotoden to move from KataharamachiKawaramachi to SanjoOtaBusshozanHayashimichi between 11:00 and 23:59 during the week.

Focusing on the combination of ChildrenSenior citizensPeople with disability, this combination appears in six elements. Three of these elements represent the behavior of KataharamachiKawaramachi to SanjoOtaBusshozanHayashimichi at 11:00–17:59 and the other three elements represent the

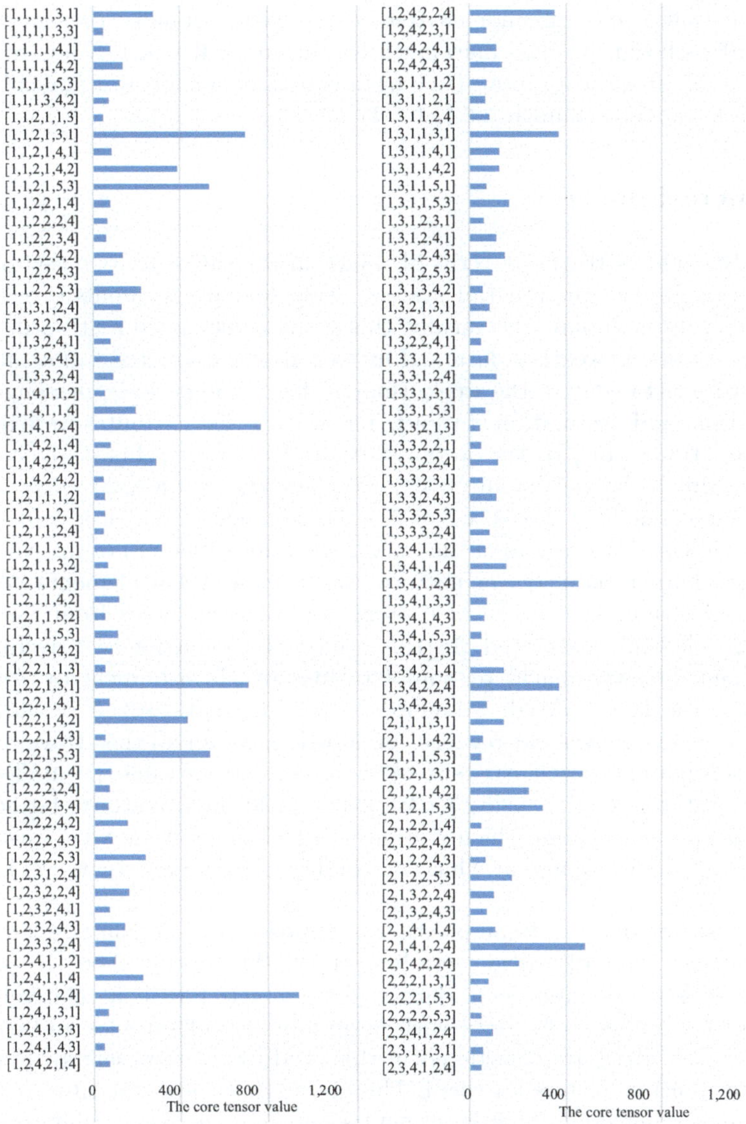

Fig. 13. Core tensor

behavior of OtaBusshozan to KataharamachiKawaramachiRitsurin-Koen at 05:00–06:5909:00–10:59. Since the combinations of origin station and destination station are similar, it can be said that the station sets representing the behaviors of OtaBusshozan to KataharamachiKawaramachiRitsurin-Koen correspond to the station sets representing the behaviors of KataharamachiKawaramachi to SanjoOtaBusshozanHayashimichi. From the above, it is considered that ChildrenSenior citizensPeople with disability passengers move from OtaBusshozan to KataharamachiKawaramachiRitsurin-Koen at 05:00–06:5909:00–10:59 and they move back from KataharamachiKawaramachi to SanjoOtaBusshozanHayashimichi at 11:00–17:59.

5 Conclusions

This study explores travel patterns of smart card (IruCa) users based on usage history data for the private rail system, Kotoden, in Takamatsu, one of the provincial cities in Japan. Specifically, this study constructed a 6th-order tensor using six attributes: weather data and five attributes extracted from smart card usage history data (day of the week, time of day (time period), passenger type, origin station, and destination station). Then, these six attributes were analyzed at the same time using Tucker decomposition. As a result, this study succeeded in compressing a dataset having 6,489,600 elements ($3 \times 8 \times 20 \times 5 \times 52 \times 52$) to 1,440 elements ($2 \times 3 \times 4 \times 3 \times 5 \times 4$). Moreover, this study focused on elements showing 5% or more of the maximum value (1060.48) of the core tensor to extract characteristic travel patterns from these 1,440 elements. We were able to understand salient travel patterns of Kotoden users from the results of the 124 elements extracted. In this study, the behavior of traveling from KataharamachiKawaramachi to SanjoOtaBusshozaHayashimichi by Adults at 18:00–24:59 on TuesdayWednesdayFriday on a day categorized as Sunny was the most frequent movement pattern. In addition, we found that travel patterns could be compared for each attribute from the core tensor value. By analyzing six attributes at the same time using Tucker decomposition, this study clarified what kinds of people move from which station to which station, in what weather, on which days of the week, and at what time of day as characteristic travel behavior patterns.

In this study, Tucker decomposition is applied to more complex and large-scale data than those of previous studies [21, 22]. The results show the potential for characteristics of large-scale data to be extracted with a relatively small number of factors, even for data with large numbers of samples and elements. Moreover, each attribute considered in this study were represented by not one factor, but multiple factors for them. This shows that the six attributes affect to Kotoden user s behavior. That is, it can be said that these six attributes should be considered to understand travel patterns.

Tucker decomposition has already used in image processing field and signal processing field [24–26]. However, there are few studies applying the method in transportation field. Therefore, this study show effectiveness of the analysis method for large-scale data in transportation field.

It is likely that many of the IruCa users are residents of Kagawa Prefecture because IruCa can only be used within this prefecture. That is, the IruCa data may not be the best source for analysis to understand the sightseeing and traveling movement patterns of tourists who live outside of Kagawa Prefecture.

However, analysis including travelers movements is necessary in order to comprehensively understand Kotoden rail passenger demand, and we must consider how to collect tourists movement data.

In addition, it is necessary to analyze the six attributes used for this study with additional attributes that indicate conditions around the origin or destination stations such as the number of commercial facilities and houses. If this analysis is successful, it will be possible to make a transportation plan to contribute to increasing the number of Kotoden users. Tucker decomposition can be a useful method, even in more complex analyses with more comparison attributes.

Acknowledgements. The authors especially thank Takamastu-Kotohira Electric Railroad Co. in Takamatsu City, Kagawa Prefecture, Japan for providing smart card data.

References

1. Hofleitner, A., Herring, R., Bayen, A.: Arterial travel time forecast with streaming data: a hybrid approach of flow modeling and machine learning. Transp. Res. Part B **46**, 1097–1122 (2012)
2. Vij, A., Shankari, K.: When is big data enough? Implications of using GPS-based surveys for travel demand analysis. Transp. Res. Part C **56**, 446–462 (2015)
3. Simoncini, M., Taccari, L., Sambo, F., Bravi, L., Salti, S., Lori, A.: Vehicle classification from low-frequency GPS data with recurrent neural networks. Transp. Res. Part C **91**, 176–191 (2018)
4. Jenelius, E., Koutsopoulos, H.N.: Travel time estimation for urban road networks using low frequency probe vehicle data. Transp. Res. Part B **53**, 64–81 (2013)
5. Yang, H., Wang, Z., Xie, K., Dai, D.: Use of ubiquitous probe vehicle data for identifying secondary crashes. Transp. Res. Part C **82**, 138–160 (2017)
6. White, K.M., Hyde, M.K., Walsh, S.P., Watson, B.: Mobile phone use while driving: an investigation of the beliefs influencing drivers' hands-free and hand-held mobile phone use. Transp. Res. Part F **13**, 9–20 (2010)
7. Jarv, O., Ahas, R., Witlox, F.: Understanding monthly variability in human activity spaces: a twelve-month study using mobile phone call detail records. Transp. Res. Part C **38**, 122–135 (2014)
8. Choudhary, P., Velaga, N.R.: Modelling driver distraction effects due to mobile phone use on reaction time. Transp. Res. Part C **77**, 351–365 (2017)
9. Ma, X., Wu, Y.J., Wang, Y., Chen, F., Liu, J.: Mining smart card data for transit riders' travel patterns. Transp. Res. Part C **36**, 1–12 (2013)
10. Zhong, C., Manley, E., Arisona, S.M., Batty, M., Schmitt, G.: Measuring variability of mobility patterns from multiday smart-card data. J. Comput. Sci. **9**, 125–130 (2015)
11. Alsger, A., Tavassoli, A., Mesbah, M., Ferreira, L., Hickman, M.: Public transport trip purpose inference using smart card fare data. Transp. Res. Part C **87**, 123–137 (2018)
12. Wang, Z., Hu, Y., Zhu, P., Qin, Y., Jia, L.: Ring aggregation pattern of metro passenger trips: a study using smart card data. Phys. A **491**, 471–479 (2018)
13. Medina, S.A.O.: Inferring weekly primary activity patterns using public transport smatr card data and a household travel survey. Travel Behav. Soc. **12**, 93–101 (2018)

14. Zhang, Y., Martens, K., Long, Y.: Revealing group travel behavior patterns with public transit smart card data. Travel Behav. Soc. **10**, 42–52 (2018)
15. Zhao, Z., Koutsopoulos, H.N., Zhao, J.: Individual mobility prediction using transit smart card data. Transp. Res. Part C **89**, 19–34 (2018)
16. Simsekli, U., Virtanen, T., Cemgil, A.T.: Non-negative tensor factorization models for Bayesian audio processing. Digit. Sig. Proc. **47**, 178–191 (2015)
17. Vazifehdan, M., Moattar, M.H., Jalali, M.: A hybrid Bayesian network and tensor factorization approach for missing value imputation to improve breast cancer recurrence prediction. J. King Saud Univ. Comput. Inf. Sci. **31**, 175–184 (2018)
18. Fan, H., Kuang, G., Qiao, L.: Fast tensor principal component analysis via proximal alternating direction method with vectorized technique. Appl. Math. **8**, 77–86 (2017)
19. Taneja, A., Arora, A.: Cross domain recommendation using multidimensional tensor factorization. Expert Syst. Appl. **92**, 304–316 (2018)
20. Yao, D., Yu, C., Jin, H., Ding, Q.: Human mobility synthesis using matrix and tensor factorizations. Inf. Fusion **23**, 25–32 (2015)
21. Han, Y., Moutarde, F.: Analysis of large-scale traffic dynamics in an urban transportation network using non-negative tensor factorization. Int. J. Intell. Transp. Syst. Res. **14**, 36–49 (2014)
22. Sun, L., Axhausen, K.: Understanding urban mobility patterns with a probabilistic tensor factorization framework. Transp. Res. Part B **91**, 511–524 (2016)
23. Chen, X., He, Z., Wang, J.: Spatial-temporal traffic speed patterns discovery and incomplete data recovery via SVD-combined tensor decomposition. Transp. Res. Part C **86**, 59–77 (2018)
24. Feng, B., Lu, W., Sun, W., Huang, J., Shi, Y.Q.: Robust image watermarking based on tucker decomposition and adaptive-lattice quantization index modulation. Sig. Process. Image Commun. **41**, 1–14 (2016)
25. Wang, L., Bai, J., Wu, J., Jeon, G.: Hyperspectral image compression based on lapped transform and tucker decomposition. Sig. Process. Image Commun. **36**, 63–69 (2015)
26. Yin, W., Ma, Z.: LE & LLE regularized nonnegative tucker decomposition for clustering of high dimensional datasets. Neurocomputing **364**, 77–94 (2019)

Estimating Impressions for Clothing, Landscape, and Indoor Images Using CNN

Mizuki Kambe[1(✉)], Soichiro Yokoyama[2], Tomohisa Yamashita[2], and Hidenori Kawamura[2]

[1] Graduate School of Information Science and Technology, Hokkaido University, Sapporo, Hokkaido, Japan
kambe@complex.ist.hokudai.ac.jp
[2] Faculty of Information Science and Technology, Hokkaido University, Sapporo, Hokkaido, Japan
{yokoyama,tomohisa,kawamura}@complex.ist.hokudai.ac.jp

Abstract. In the fashion industry, development and sale of goods is being carried out based on the brand image and impressions such as "girly" or "elegant." However, the impressions and brand image vary depending on the recipient; therefore, there is a need for a system to quantitatively evaluate the impressions of people. In this study, we use two datasets that include clothing images with impression attributes, and landscape and indoor images according to the brand image. We developed a system to evaluate the impression using convolutional neural network and discuss the results.

Keywords: Fashion · Impression · CNN

1 Introduction

In the fashion industry, goods are developed and sold based on impression words such as "girly" and "elegant." There is no clear definition to the impression words, and it may vary depending on the recipient. However, there is a certain common recognition in the fashion industry, which cannot be quantitatively analyzed. It is the designers who decide to what kind of goods should be made. For this reason, a system capable of quantitatively evaluating the impressions is desired. In this study, we define quantitative evaluation of impression based on how many people out of 100 feel the same impression. We prepared two datasets for this. One dataset included clothing images with impressions, and the other included landscape and indoor images according to the brand image.

H. Sato et al. (Eds.): IES 2019, PALO 12, pp. 67–78, 2020.
https://doi.org/10.1007/978-3-030-37442-6_7

2 Related Works

Fashion Dataset. Fashion is one of the major areas in computer vision, and many datasets for fashion are available. Some datasets include a particular fashion category, while other datasets include more types of fashion items [4,5]. Some datasets include supplemental information such as attributes and landmarks for more fine-grained analysis. These datasets are mostly designed for image retrieval of the same product. For example, DeepFashion dataset is the largest dataset and includes 800,000 images [8]. Moreover, some datasets aim at analyzing individual fashion items from a street photo image [6,7,17,18,20,22]. Datasets for image retrieval have only image-level annotations, *i.e.*, they have pixel-level annotations. On the other hand, few datasets deal with impressions such as "cute" or "girly." The dataset made by Sirion Vittayakorn *et al.* includes attributes that represent an impression; however, most attributes are visual ones [15].

Attribute Learning. Research to learn attributes have been made in various fields [1,2,10,19]. In fashion, learned attributes are used for image retrieval of the same product [8] and for clothing search with certain attributes [21]. In addition, there are studies to learn from the noisy data on the Web [11,12,15]. Although most attributes considered in these studies are visual ones, some studies handle impression attributes. Xiong *et al.* learned the impression from the shape of a mobile phone using a support vector regression [16]. Vaccaro *et al.* extracted the attributes from the text given to a fashion image, and learned the relationship between the visual attributes and impression attributes using a multi-language topic model [14]. Talebi *et al.* learned the image quality using a convolutional neural network (CNN) [13]. However, these studies to learn impression attributes in terms of number of data and attributes are few. In this study, we prepare datasets that include many attributes and data in order to learn it.

3 Datasets

We use the Fashion Impression Dataset that includes clothing images with impression attributes and a Brand Image Dataset that includes landscape and indoor images according to a brand image.

3.1 Fashion Impression Dataset

This dataset includes clothing images under categories such as blouse, skirt; impression attributes such as girly, sweet; and visual attributes such as plain, floral. The impression attributes have four-stage evaluation (1. Strongly disagree, 2. Disagree, 3. Agree, 4. Strongly agree) and binary evaluation (the keywords are true or not). The visual attributes have only binary evaluation. There are six four-stage evaluation attributes, and 142 binary evaluation attributes. These attributes are often used in fashion magazines and are chosen based on the designers' opinions. This dataset was made by 52 people in a fashion professional

Table 1. List of categories and number of products.

Category	Number of products
Cutsew	4014
Blouse	4448
Knit	5949
Coat	1127
Jacket	1022
Blouson	558
Skirt	4162
Pants	3013
One piece	4094
Suit	91

Table 2. Number of products with four-stage evaluation attributes.

Attribute	Strongly disagree	Disagree	Agree	Strongly agree
Cute	2137	7014	10473	8854
Beautiful	1540	5572	11656	9710
Cool	7018	9753	7743	3964
Popular with opposite sex	3670	11405	10652	2751
Sexy	8819	11593	6060	2006
Fashionable	1487	9710	12812	4469

school. Since we define quantitative evaluation of impression as how many people out of 100 feel the same impression, a clothing image tagged by multiple people is desired. However, considering the tagging cost, we prioritize the number of data, and this dataset includes the images tagged by a single person. We think that even if the quality resolution of a clothing image is not high, we can grasp the trend if the number of images is large. Tables 1, 2 and 3 show the number of products that have categories and attributes. Binary evaluation attributes are part of all. In addition, all the products have different colors. There are 28478 products and 71781 images. We add the same attributes to different colors because of tagging cost. These images are published in the EC site and most images are of clothing alone, while some images are of people or mannequins wearing the clothing.

3.2 Brand Image Dataset

The Brand Image Dataset includes landscape and indoor images according to the brand image. Each image has visual attributes such as sunflowers or stars, and impression attributes such as attractive or friendly. We deal with Brand A,

Table 3. Number of products with binary evaluation attributes.

Attribute	Number of products
Refreshing	2523
Company entrance ceremony	1031
Girly	2873
Elegant	3747
Floral	1599
No collar	970
Flare	1467
Calm atmosphere	13753

Fig. 1. Image example of the Brand Image Dataset. The first line is Brand A, which is avant-garde; and the second line is Brand B, which is cute, sweet, and sexy

which is avant-garde; and Brand B, which is cute, sweet, and sexy. Because it is easy to understand the difference between the two brands, it is assumed that the difference in the brand image is clear. On pinterest.jp, 90 people from a fashion professional school collected images according to the brand image. Moreover, they tagged the images with free-words. The images were only landscape and indoor, and we consolidated the attributes later. Finally, we gathered 5562 images—, the images corresponding to Brand A comprised 2906 sheets, and Brand B comprised 2656 sheets. The total number of attributes were 1051. Table 4 shows the top 20 occurring attributes. Figure 1 shows the image examples of the Brand Image Dataset.

4 Learning Datasets

The optimization method was used SGDR [9]. This is simple warm restart technique for stochastic gradient decent. We set $momentum = 0.9$, $batch\_size = 50$, $weight\_decay = 0.0001$, $epoch = 70$, $initial\_learning\_rate = 0.05$.

Table 4. Top 20 occurring attributes in the Brand Image Dataset.

Attribute	Number of occurrence	Attribute	Number of occurrence
Friendly	1497	Impactful	1047
Nice	1446	An elegant moment	1029
Pure white	1370	Green	1027
Girls seems to like	1341	Seems to appear in magazine	998
Pink	1287	Clean	972
Stylish	1213	Lovely	963
Beautiful	1171	Charmed	940
Pretty	1166	Want to take a picture	934
Neat	1115	Cute	931
Plant	1111	One scene of a movie	926

4.1 Learning Fashion Impression Dataset

We use the ResNet-50 [3] model and finetune the pretrained model to learn. This dataset has categories, four-stage evaluation attributes, and binary evaluation attributes. Since these contain data in different formats, we learn them by branching in the final layer of ResNet (Fig. 2). The category is represented by a one-hot label, and cross-entropy is used as the loss function. The four-stage evaluation attributes are represented by $(\text{label} - 1)/3$, strongly disagree was 0.0, disagree was 0.333, agree was 0.667, and strongly agree was 1.0. Mean Squared Error was used as the loss function. Mean Absolute Error (MAE) is used as the evaluation value. The binary evaluation attributes are represented by multi-label, and weighted binary cross entropy is used as the loss function. We used 25626 products as traindata and 2852 products as testdata. Since all the products have different colors, we use 64589 sheets as the traindata and 7192 sheets as the testdata. If we evaluate the binary evaluation attributes by usual accuracy, despite all the outputs being 0, the accuracy exceeds the 80%. This is why, we evaluate it using the average of true positive rate and true negative rate. Figure 3 shows the confusion matrix. True positive rate is $TP/(TP+FN)$, and true negative rate is $TN/(TN+FP)$. Table 5 shows the learning results. However, the evaluation value of the four-stage evaluation attributes and binary evaluation attributes are the average of the evaluation values of each attribute. The results indicate that the evaluation value of the binary evaluation attributes is 70.0%. This does not mean that it did not learn well, but it is because we evaluate using the data tagged by a single person against the data based on opinions of various persons. From Table 6, only the suit's accuracy is low; we assumed that it could be because the number of suits is less than the other products. From Table 7, it can be seen that there is a large difference in the value of the MAE for each value of training labels. This is because of the difference in the number of data (Fig. 5). However, we could not solve it by simply weighting, it is necessary to

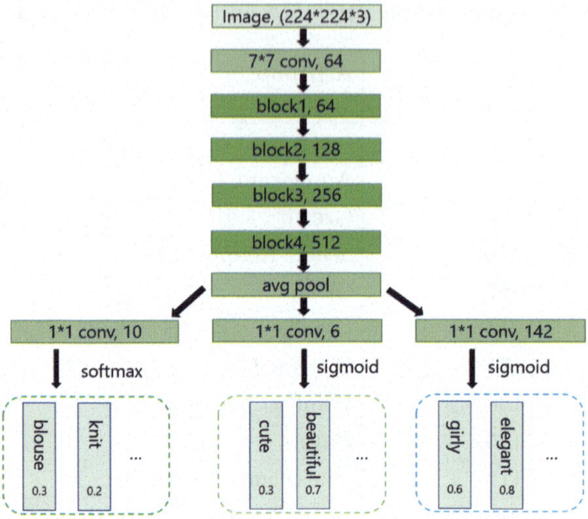

Fig. 2. Structure of ResNet.

		Predicted condition	
		positive	negative
True condition	positive	True Positive (TP)	False Negative (FN)
	negative	False Positive (FP)	True Negative (TN)

Fig. 3. Confusion matrix.

Table 5. Learning result of Fashion Impression Dataset.

MAE (four-stage)	Accuracy (binary)	Accuracy (category)
0.236	70.0%	91.0%

use another method for imbalanced data. From Table 8, the evaluation value of the visual attributes tends to be high, and the evaluation value of the impression attributes tends to be low. This is considered to be because the opinions of visual attributes do not depend on the people. In addition, we confirmed the difference in the outputs of cute and mature between the cute brand and mature brand, respectively. Figure 5 shows the examples of the outputs of the CNN for the input images and attributes. The number written in the image is the output of the CNN, and the images are sorted by the output.

Table 6. Correct rates (%) of category.

Category	Correct rate
Cutsew	82.0
Blouse	91.7
Knit	93.3
Coat	88.3
Jacket	75.4
Blouson	62.1
Skirt	98.4
Pants	94.5
One piece	95.7
Suit	3.7
All	91.0

Table 7. MAE of four-stage evaluation attributes.

Attribute	0.0 (Strongly disagree)	0.333 (Disagree)	0.667 (Agree)	1.0 (Strongly agree)	All
Cute	0.550	0.276	0.143	0.290	0.247
Beautiful	0.590	0.308	0.127	0.278	0.236
Cool	0.338	0.162	0.224	0.464	0.258
Popular with opposite sex	0.396	0.166	0.186	0.445	0.224
Sexy	0.281	0.131	0.291	0.547	0.235
Fashionable	0.512	0.223	0.126	0.389	0.217

Table 8. Evaluation values (%) of binary evaluation attributes.

Attribute	Evaluation value
Refreshing	65.1
Company entrance ceremony	72.1
Girly	68.3
Elegant	64.8
Floral	89.7
No collar	71.0
Flare	81.1
Calm atmosphere	60.7

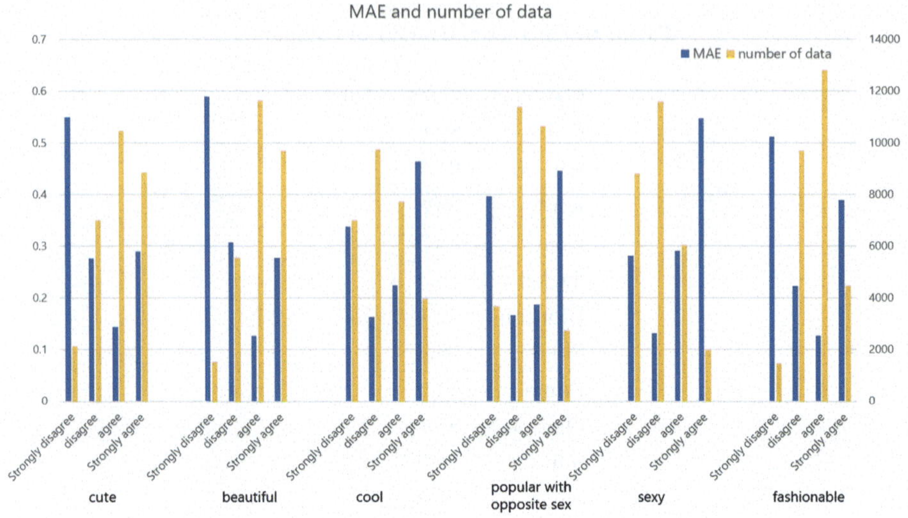

Fig. 4. MAE and number of data of four-stage evaluation attributes.

4.2 Learning Brand Image Dataset

This dataset includes images according to the brand images that are tagged. Therefore, learning is performed using only attributes, only images, both attributes and images, and predicted attributes and images. In the learning using only attributes, we used multi-layer perceptron (MLP). In the learning using only images, we finetuned pretrained ResNet-50. In the learning using both attributes and images, we extracted the features from images using pretrained ResNet-50, and we input the features and attributes to the MLP. In the learning using both predicted attributes and images, we extracted the features and predicted attributes using ResNet-50 that learned the attributes, and we input them to the MLP. We used 4449 images as the traindata and 1113 images as the testdata. Table 9 suggests that the amount of information obtained from the attributes is high, and that CNN cannot extract the information well. Further, the accuracy is not improved even when predicted attributes are used as input. We obtained the same amount of information as that from the pretrained CNN because we used finetuned CNN to predict the attributes. Figure 6 shows the input image and output value. The first line is Brand A, and the second line is Brand B. It is considered that the closer the line is to 1, the closer it is to Brand A, and the closer that line is to 0, the closer it is to Brand B. Table 10 shows the top 20 attributes in Brand Image Dataset.

Fig. 5. Examples of outputs and images for some of the attributes.

Table 9. Learning result of Brand Image Dataset.

Attributes only	Images only	Attributes and images	Predicted_attributes and images
89.1%	72.3%	90.1%	72.1%

Table 10. Correct Rates (%) of top 20 occurring attributes in Brand Image Dataset.

Attribute	Accuracy	Attribute	Accuracy
Friendly	64.4	Impactful	57.9
Nice	59.9	An elegant moment	59.2
Pure white	67.1	Green	67.5
Girls seems to like	62.2	Seems to appear in magazine	51.8
Pink	71.0	Clean	62.0
Stylish	55.9	Lovely	58.0
Beautiful	62.4	Charmed	62.2
Pretty	70.2	Want to take a picture	66.7
Neat	60.8	Cute	67.1
Plant	73.8	One scene of a movie	57.7

Fig. 6. Example of input image and output value. The first line is Brand A, and the second line is Brand B. It is considered that the closer the line is to 1, the closer it is to Brand A, and the closer that line is to 0, the closer it is to Brand B.

5 Conclusion

In this study, we developed a system to quantitatively evaluate the impression of people toward fashion, by learning through datasets. However, we could not sufficiently evaluate the impressions because the data used was tagged by a single person. Therefore, we think that it is necessary to compare the output of the CNN and the average tagged by multiple people in order to evaluate this system. Moreover, we used the impression data in the same way as normal object recognition. However, we are still to determine if we can use with the impression data depending on the recipient in the same way. For this reason, we need to address the challenge of dealing with impression data and design loss functions. In the future, this system can be applied for tying sales and impressions, and would help in analyzing the correlation between the sales and impressions.

References

1. Di, W., Wah, C., Bhardwaj, A., Piramuthu, R., Sundaresan, N.: Style finder: fine-grained clothing style detection and retrieval. In: 2013 IEEE Conference on Computer Vision and Pattern Recognition Workshops, pp. 8–13. IEEE (2013)
2. Escorcia, V., Niebles, J.C., Ghanem, B.: On the relationship between visual attributes and convolutional networks. In: 2015 IEEE Conference on Computer Vision and Pattern Recognition (CVPR), pp. 1256–1264. IEEE (2015)
3. He, K., Zhang, X., Ren, S., Sun, J.: Deep residual learning for image recognition. In: 2016 IEEE Conference on Computer Vision and Pattern Recognition (CVPR), pp. 770–778. IEEE (2016)
4. Huang, J., Feris, R., Chen, Q., Yan, S.: Cross-domain image retrieval with a dual attribute-aware ranking network. In: 2015 IEEE International Conference on Computer Vision (ICCV), pp. 1062–1070. IEEE (2015)
5. Kiapour, M.H., Han, X., Lazebnik, S., Berg, A.C., Berg, T.L.: Where to buy it: matching street clothing photos in online shops. In: 2015 IEEE International Conference on Computer Vision (ICCV), pp. 3343–3351. IEEE (2015)
6. Liang, X., et al.: Deep human parsing with active template regression. IEEE Trans. Pattern Anal. Mach. Intell. **37**(12), 2402–2414 (2015)
7. Liu, S., et al.: Fashion parsing with weak color-category labels. IEEE Trans. Multimed. **16**(1), 253–265 (2014)
8. Liu, Z., Luo, P., Qiu, S., Wang, X., Tang, X.: Deepfashion: powering robust clothes recognition and retrieval with rich annotations. In: 2016 IEEE Conference on Computer Vision and Pattern Recognition (CVPR), pp. 1096–1104. IEEE (2016)
9. Loshchilov, I., Hutter, F.: SGDR: stochastic gradient descent with warm restarts. ICLR (2017)
10. Matsukawa, T., Suzuki, E.: Person re-identification using CNN features learned from combination of attributes. In: 2016 23rd International Conference on Pattern Recognition (ICPR), pp. 2428–2433. IEEE (2016)
11. Shankar, S., Garg, V.K., Cipolla, R.: Deep-carving: discovering visual attributes by carving deep neural nets. In: 2015 IEEE Conference on Computer Vision and Pattern Recognition (CVPR), pp. 3403–3412. IEEE (2015)
12. Sun, C., Gan, C., Nevatia, R.: Automatic concept discovery from parallel text and visual corpora. In: 2015 IEEE International Conference on Computer Vision (ICCV), pp. 2596–2604. IEEE (2015)
13. Talebi, H., Milanfar, P.: NIMA: neural image assessment. IEEE Trans. Image Process. **27**(8), 3998–4011 (2018)
14. Vaccaro, K., Shivakumar, S., Ding, Z., Karahalios, K., Kumar, R.: The elements of fashion style. In: Proceedings of the 29th Annual Symposium on User Interface Software and Technology - UIST 2016, pp. 777–785. ACM Press, New York (2016)
15. Vittayakorn, S., Umeda, T., Murasaki, K., Sudo, K., Okatani, T., Yamaguchi, K.: Automatic attribute discovery with neural activations, pp. 252–268 (2016)
16. Xiong, Y., Li, Y., Pan, P., Chen, Y.: A regression-based Kansei engineering system based on form feature lines for product form design. Adv. Mech. Eng. **8**(7) (2016). https://doi.org/10.1177/1687814016656107
17. Yamaguchi, K., Kiapour, M.H., Ortiz, L.E., Berg, T.L.: Parsing clothing in fashion photographs. In: 2012 IEEE Conference on Computer Vision and Pattern Recognition, pp. 3570–3577. IEEE (2012)
18. Yamaguchi, K., Kiapour, M.H., Berg, T.L.: Paper doll parsing: retrieving similar styles to parse clothing items. In: 2013 IEEE International Conference on Computer Vision, pp. 3519–3526. IEEE (2013)

19. Yamaguchi, K., Okatani, T., Sudo, K., Murasaki, K., Taniguchi, Y.: Mix and match: joint model for clothing and attribute recognition. In: Proceedings of the British Machine Vision Conference 2015, pp. 51.1–51.12. British Machine Vision Association (2015)
20. Yang, W., Luo, P., Lin, L.: Clothing co-parsing by joint image segmentation and labeling. In: 2014 IEEE Conference on Computer Vision and Pattern Recognition, pp. 3182–3189. IEEE (2014)
21. Zhao, B., Feng, J., Wu, X., Yan, S.: Memory-augmented attribute manipulation networks for interactive fashion search. In: 2017 IEEE Conference on Computer Vision and Pattern Recognition (CVPR), pp. 6156–6164. IEEE (2017)
22. Zheng, S., Yang, F., Kiapour, M.H., Piramuthu, R.: ModaNet: a large-scale street fashion dataset with polygon annotations. In: 2018 ACM Multimedia Conference on Multimedia Conference - MM 2018, pp. 1670–1678. ACM Press, New York (2018)

Block-Based Neural Network High Speed Optimization

Kundo Lee[1,2](✉) and Tomoki Hamagami[1]

[1] Faculty of Engineering, Division of Intelligent Systems Engineering, Yokohama National University, 79-5 Tokiwadai, Hodogaya, Yokohama 240-8501, Japan
lee-kundo-yv@ynu.jp, hamagami@ynu.ac.jp
[2] Mentor Graphics Japan Co., Ltd., Tokyo, Japan

Abstract. Block-Based Neural Network (BBNN) consists of a 2-D array of memory-based modular component NNs with flexible structures and internal configuration that can be implemented in reconfigurable hardware such as a field programmable gate array (FPGA). The network structure and the weights are encoded in bit strings and globally optimized using the genetic operators. An asynchronous BBNN (ABBNN), which is a new model of BBNN, enables higher performance for BBNN by utilizing the parallel computation and the pipeline architecture. An ABBNN's operating frequency is kept stable for all scales of the network, while conventional BBNN's decreases accordingly. The architecture of ABBNN provides the capabilities to process and analyze high sample rate data at the same time. However, optimization by the genetic algorithm is a high-cost task, and the memory access is one of the causes which degrade the training performance. In this paper, we introduce a new algorithm to reduce the memory access in BBNN optimization. ABBNN, optimized with the proposed evolutionary algorithm, is applied to general classifiers to verify the effectiveness with regards to the reduction of memory access.

Keywords: FPGA · Evolvable hardware · Genetic Algorithm · Block-Based Neural Network

1 Introduction

Artificial neural network models have been successfully applied to diverse engineering problems due to their model-free approximation capacity to complex decision-making processes [1,2]. As a result, the neural network model has been dramatically improved and involved in various situations where predictions and recommendations are required. During the early stages of artificial intelligence, the analytical accuracy was the more important factor than the performance in using the algorithm. Most researchers, who were focusing on improving the algorithm, used CPU and GPU to run and train the neural network [3–5]. Although

H. Sato et al. (Eds.): IES 2019, PALO 12, pp. 79–92, 2020.
https://doi.org/10.1007/978-3-030-37442-6_8

the researchers successfully enhanced the neural network algorithm, it is still a big challenge to achieve lossless data analytics in the cloud-based system. To quickly react to continuous input, the edge devices are recommended to process and analyze the data by themselves without communication through the cloud network.

Training a neural network requires a continuous iteration to train its structure and connection weights [6]. Moreover, dynamic reconfiguration in hardware and online training are challenging tasks due to many uses of nonlinear activation function and real-value connection weights of artificial neural network [7,8]. To solve this problem, the structure and integer weights of Block-Based Neural Network (BBNN) [1] are encoded as a 2-D chromosome to enable simultaneous optimization. The BBNN's structure and weights, obtained by Genetic Algorithm (GA) [9–12] with 2-D chromosome, successfully solves various classification problems [13–15]. However, the optimization by GA requires a repetition of memory access to update the population for the next generation.

In this paper, a simple method extended on the genetic algorithm is proposed to obtain the solution with Asynchronous Block-Based Neural Network (ABBNN). ABBNN [16] is a high-performance neural network model which can be easily implemented as evolutionary hardware. The simple method will replace the conventional GA in ABBNN and show how it reduces the memory access in GA operation. This paper is organized as follows: Sect. 2 introduces the basic idea of Block-Based Neural Network and Asynchronous Block-Based Neural Network Model. Section 3 discusses the proposed evolutionary optimization method obtaining the solution with the less memory access than the conventional GA's. Section 4 will present the experiment process and the result. The suggested method in different perspective will be discussed in Sect. 5. Section 6 will conclude this paper and mention the direction for the future research.

2 Block-Based Neural Network

2.1 Basic Idea

The Block-Based Neural Network, proposed by Moon and Kong, is represented by 2-D array of blocks. Each block works as a memory-based processing unit, composed of a neural network, having four variable input/output nodes. As shown in Fig. 1, the Block-Based Neural Network consists of m stages and n columns of basic blocks. The four nodes in a basic block can be an input or output to connect a block to one other. The leftmost and rightmost blocks are connected. The output of a block becomes the input of their neighboring block. The outputs of blocks in the same stage are calculated concurrently to reflect current inputs simultaneously. Figure 2 shows the four types of a basic block. The arrow indicates the direction of the data flow. The input nodes use a linear activation function, while the output nodes use symmetric saturated linear activation function with the sum of each input's activated values.

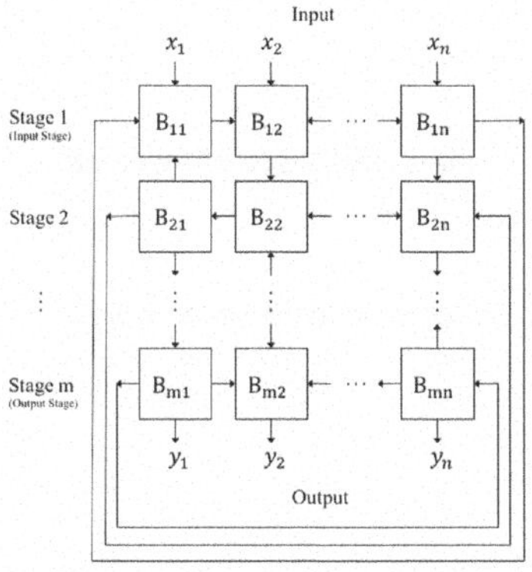

Fig. 1. Structure of BBNN [1]

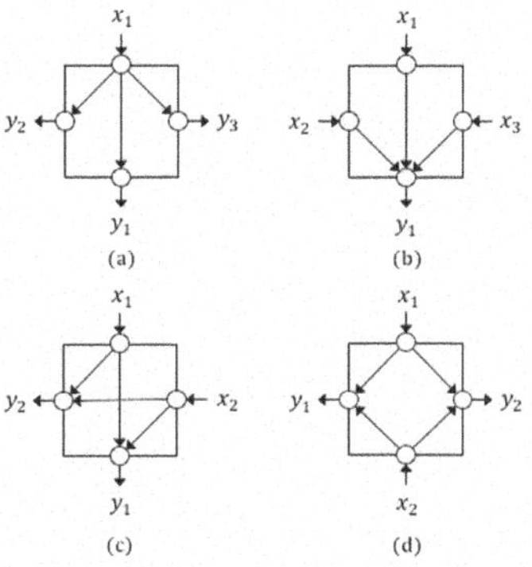

Fig. 2. Four configuration of basic block [1]

The structure and weights of BBNN are encoded into a chromosome. The genetic algorithm is used to optimize BBNN to obtain optimal structure and weights, encoded in the chromosome, according to the fitness function which represents how accurate the BBNN is in solving a problem.

2.2 ABBNN

An asynchronous block consists of six nodes which provide three pairs of data flow. The top and the bottom node only work as the input and the output, respectively. The two pairs of side nodes, which operate independently, provide parallel computation in each direction, left and right. The asynchronous Block-based Neural Network model consists of the asynchronous basic blocks and register layer to achieve high-performance BBNN.

Figure 3 shows the four types of asynchronous block. The block extends side nodes, and the nodes horizontally provide bidirectional parallel computation. All types of asynchronous block, regardless of the configuration, provide four diagonal and a downstream signal flow. These are referred to as the five common signal flows. In addition to the five common signal flows which all types support, (a) mode provides both right and left concurrent flow, (b) and (c) provide right and left flow respectively. (d) mode only contains the five common flows.

Each stage's outputs in ABBNN are connected to the register layer. The outputs of the proceeding stage are stored in the register layer while the following stage is executing. Figure 4 shows 4 × 3 pipeline BBNN example. The register layer shortens the critical path, which affects BBNN's operating frequency and hence, the high-performance BBNN becomes available.

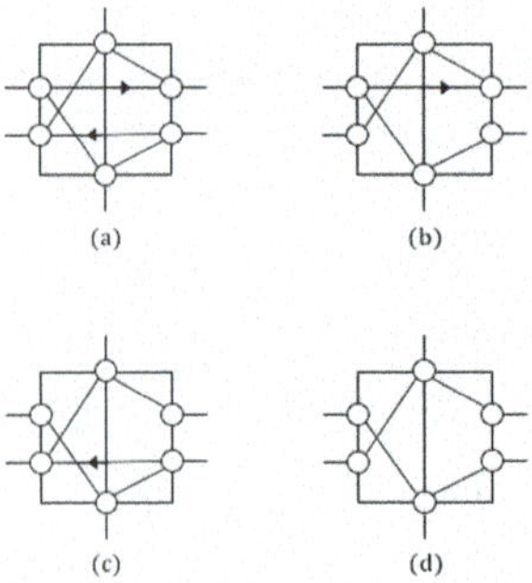

Fig. 3. Four configuration of asynchronous block [16]

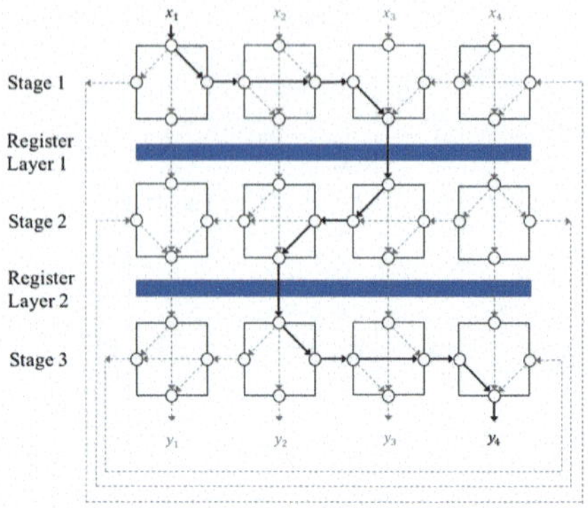

Fig. 4. 4×3 pipeline BBNN architecture

3 Evolutionary Optimization Reducing Memory Access

The genetic algorithm is a powerful method in finding the optimal solution. In ABBNN, online training by GA, which does not require the hardware reconfiguration, is available with the extended side nodes. However, ABBNN still requires an enormous amount of memory access to update the populations during the GA operation. To suppress the memory update, we suggest the approach to reduce the problem space with ABBNN.

3.1 Encoding

There are seven weights and three biases in an asynchronous block. Figure 5 shows the chromosome representing the block. The six nodes inside of a block

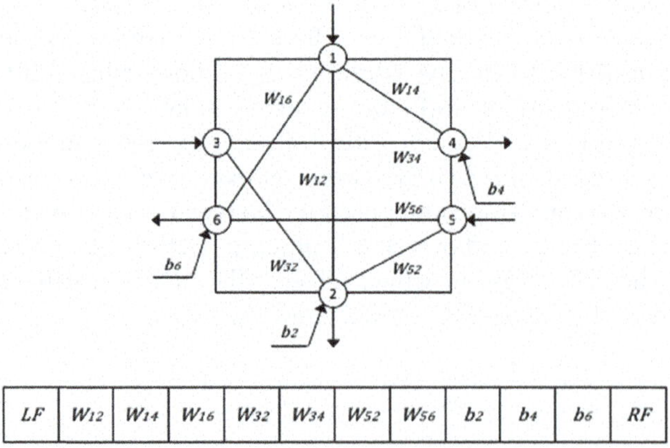

Fig. 5. Encoding for an asynchronous block

are connected through the weights. A weight W_{ij} denotes a connection from node i to j and b_k is a bias of node k. The configuration of a block is determined by the combination of $\{LF, RF\}$ in the chromosome. LF and RF are both 1-bit values, denoting left and right signal flow respectively. For example, the block becomes (a) mode which supports both left and right signal flow, when $\{LF, RF\}$ is set to $\{1, 1\}$ and (d) mode, which only supports the five common signal flows, when $\{LF, RF\}$ is set to $\{0, 0\}$.

3.2 Genetic Operators

A standard genetic algorithm has four significant steps, namely, initialization, selection, reproduction, and replacement. The parameters, such as the crossover rate, mutation rate, population size, are obtained by trial and error, preceding researches [1,14].

In the initialization stage, the program generates an initial population of chromosomes with random values. ABBNN is formed with a 2-D array of blocks internally configured by chromosome. Then the fitness of each chromosome is evaluated as below.

$$f = \frac{1}{N}\sum_{i=0}^{N} e_i^2 \tag{1}$$

$$e_i = d_i - y_i \tag{2}$$

Where N is the total number of training patterns and e_i is an ith error. The two vectors d_i and y_i are the target and actual output respectively, for the ith training pattern.

Two individuals in the population are randomly selected as parents at a time. Crossover and mutation are applied to the chosen parents. The number of the offspring is randomly decided between 1 to 5. Then, mutation is applied to each output of the crossover operation. The fitness values of offspring which are produced in the crossover operation, are measured. Then, Two individuals from the family are selected. One is the elitist, and the other is chosen by the roulette wheel method. The individuals from the family replace the parents and become a part of the next generation.

3.3 Optimization of ABBNN

ABBNN is optimized with the GA described in Sect. 3.2 and the optimization generally requires more iterations to get the solution with increasing size of problem space. To deal with the growing problem space, the divide-and-conquer approach can be one of the solutions generally used. Firstly, one divides the problem into subproblems of manageable size, then solves the subproblems. The solutions to the subproblems must be patched back together. To produce the correct solutions in this method, the subproblems must be naturally disjointed [17].

Each block in ABBNN is connected through the weight and bias. The output in the last stage is an actual output and it is compared with an expected one to measure the fitness. GA applied to ABBNN targets the whole network structure and weights. The increasing problem space becomes a penalty in the GA performance, and the memory access for updating the population is one of the causes which degrade the performance. To improve the GA performance, we suggest reducing the problem space step by step in the learning procedure to decrease the memory access for the population update. However, partitioning the problem space into a small one is not easy in ABBNN because the solution of preceding subproblem space always behaves as an offset for following subproblems when the network is divided into small problem space.

Algorithm 1. GA with Reducing Problem Space

```
CurrentStage ← 0
LastStage ← (Number of ABBNN Stages − 1)
Initialize Population
while !StopCondition do
    Evaluate Population
    GA Problem Space ← ABBNN [ CurrentStage : LastStage ]
    Randomly select 2 individuals for reproduction
    Crossover operation with (GA Problem Space) to create offspring
    Mutation operation for Offspring
    Evaluate Offspring
    Find the elitist and one by roulette wheel selection from the family
    Replace 2 individuals with 2 candidates found in previous step
    if ReduceCondition then
        CurrentStage ← CurrentStage + 1
    end if
end while
Return theBestSolution
```

Algorithm 1 shows a problem space specific method applied to the new ABBNN implementation. To reduce problem space, a limited range of chromosomes between *CurrentStage* and *LastStage* becomes a target for the crossover operation. The rest of the chromosomes are succeeded by the offspring from parents. In the family, two individuals are chosen for the next generation. One is an elitist, and the other is determined by a roulette wheel method. The two individuals replace the parents. The proposed GA only updates the limited range of chromosomes generated by the crossover operation.

3.4 Problem Space Reduction

The amount of the memory update depends on the number of individuals which GA replaces to create each generation. As shown in Fig. 5, the encoding of the asynchronous block, which forms the chromosome, is represented with 42 bits when the weight and bias are 4-bit variables.

Let GA optimize 5×7 ABBNN. It is composed of 35 asynchronous blocks, and 1,470 bits are needed to represent the ABBNN. The amount of memory access to update one individual is calculated as below.

$$BU_{conv} = \sum_{i=0}^{G} (N_{blks} \times N_{crms}) \tag{3}$$

BU_{conv} is the total bits updated in the conventional GA operation and N_{blks} in the total number of blocks in ABBNN. N_{crms} is the number of bits for the chromosome to represent each block of ABBNN and G is the number of generations which GA repeats.

The proposed GA reduces the problem space step by step, and the number of steps is determined by the number of generations and stages of ABBNN. In this paper, we set the number of generations as 1,000 for GA operation. With five stages ABBNN, the number of steps will be 200, which means that the stages in ABBNN for the crossover operation will decrease step by step every 200 generations. The whole stages in ABBNN will be a target for the crossover operation during the first 200 generations. Stages between the second and the last will be the target for the operation during the second 200 generations. The last 200 generations only targets the last stage in ABBNN. N_{stgs} is the number of stages in ABBNN. Then the amount of the reduction step is defined as,

$$N_{rstp} = G \div N_{stgs} \tag{4}$$

$$BU_{prop} = \sum_{i=0}^{S} i \times (N_{clms} \times N_{crms} \times N_{rstp}) \tag{5}$$

Equation 5 shows the amount of the memory update for 1 individual in the proposed GA. N_{clms} is the number of columns in ABBNN and N_{rstp} is the number of steps determined by Eq. 4. S is the number of stages in ABBNN.

4 Experiments

4.1 Setting

To verify the proposed method, we implemented the new approach with SystemVerilog, which is the latest version of Verilog-HDL extension.

We used 100 [1] population for GA. The mutation probabilities were set 0.05 for structure and 0.01 for weight. We randomly decided the number of children whenever crossover operation happens, between 1 to 5. In the family, two individuals are selected and replace the parents for the next generation. The problem space reduction occurs step by step whenever the specific amount of generations increases. In this experiment, the particular amount was the quotient of the number of the final generation (1,000) divided by the number of ABBNN stages. We used five datasets below to verify the new algorithm with a hardware simulator.

- **Airfoil Self-Noise Data Set** [18]
  The NASA data set comprises different size NACA 0012 airfoils at various wind tunnel speeds and angles of attack. The span of the airfoil and the observer position were the same in all of the experiments. The data set has five inputs and one output, which is a scaled sound pressure level, in decibels.
- **Concrete Slump Test Data Set** [19]
  Concrete is a highly complex material. The slump flow of concrete is not only determined by the water content, but that is also influenced by other concrete ingredients. The data set includes 103 data points. There are seven input variables, and three output variables in the data set. The initial data set included 78 data. After several years, 25 new data points are added, hence 103 data exits.
- **Forest Fires Data Set** [20]
  This is a problematic regression task, where the aim is to predict the burned area of forest fires, in the northeast region of Portugal, by using meteorological and other data. The data used to predict the burned area of forest fires by a Data Mining (DM) approach. The data consists of 12 input attributes and one output to predict the burned area of the forest (in ha), 0.00 to 1090.84.
- **Image Segmentation Data Set** [21]
  Image data described by high-level numeric-valued attributes, seven classes. The instance was drawn randomly from a database of 7 outdoor images. The images were hand segmented to create a classification for every pixel. Each instance is a 3×3 region.
- **Yacht Hydrodynamics Data Set** [22]
  Prediction of residuary resistance of sailing yachts at the initial design stage is of excellent value for evaluating the ship's performance and for estimating the required propulsive power. Essential inputs include the basic hull dimensions and the boat velocity. The Delft data set comprises 308 full-scale experiments, which were performed at the Delft Ship Hydromechanics Laboratory for the purpose. These experiments include 22 difficult hull forms, derived from a parent from closely related to the Standfast 43 designed by Frans Maas.

4.2 Results

The datasets we used, are composed of various number of attributes and instances. The diversity of datasets will support that the proposed algorithm can be an effective approach with a reduced problem space in multiple situations. The proposed algorithm shows the competitive results in finding the solution with less memory access to update the individuals in each generation.

Table 1 shows the amount of the memory update in GA operation, which repeats 1,000 generations with two individual updates in each generation. The updated bytes shows the amount of the memory written and the reduction rate indicates the proposed GA's efficiency in the aspect of the memory write operation. Comparing to the conventional GA, the proposed GA finds the solution with a 40% reduced memory update or less.

Table 1. The number of bytes updated in GA operation

ABBNN size (*Stage*×*Column*)	Updated bytes		Reduction rate
	Conventional GA	Proposed GA	
5 × 7	367,500 bytes	220,500 bytes	40%
10 × 11	1,155,000 bytes	635,250 bytes	45%
20 × 19	3,990,000 bytes	2,094,750 bytes	47.5%
50 × 22	11,550,000 bytes	5,890,500 bytes	49%

Figures 6, 7, 8, 9, 10, 11, 12, 13, 14 and 15 show the processes of convergence in the conventional and proposed GA. To ensure the convergence trend, we optimized ABBNNs with five different random seeds for each dataset. The random seeds, which we used in the experiments, are randomly selected from 1 to 70,000. Figures 6, 7, 8, 9 and 10 show the convergence processes of five-stage ABBNN and Figs. 11, 12, 13, 14 and 15 show the fifteen stage's. In both cases, the convergence processes show similar trends, and the optimization result of the proposed GA is competitive against conventional optimization, although less memory update has happened.

The fitness values in the figures are the average fitness of individuals in each generation. The scale of the fitness value in each result differs depending on the dataset because we did not rescale the expected value to focus on the convergence trend of the proposed approach.

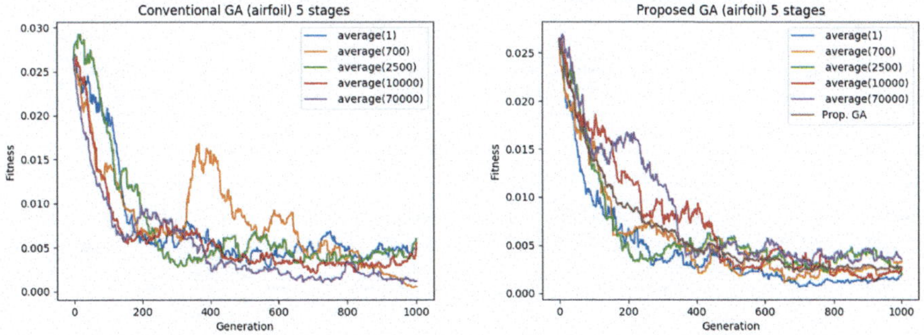

Fig. 6. Airfoil self-noise dataset (5 stages)

5 Discussion

BBNN is an algorithm that allows more straightforward implementation with reconfigurable hardware such as FPGA. The structure and weights of the BBNN are encoded with bit strings that correspond to the configuration bits of FPGA. However, the reconfiguration cost is very high because the FPGA must stop

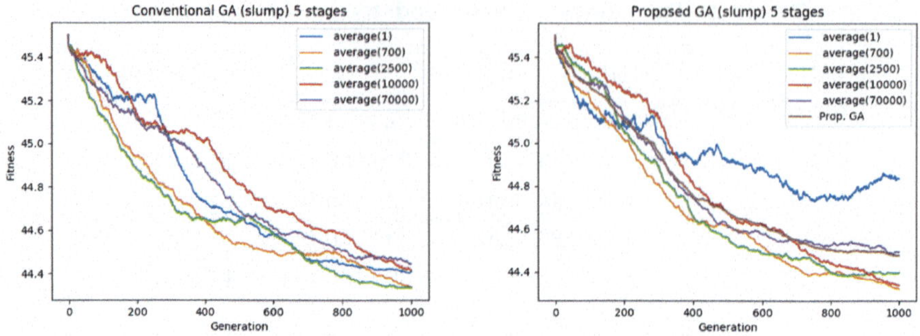

Fig. 7. Concrete slump test dataset (5 stages)

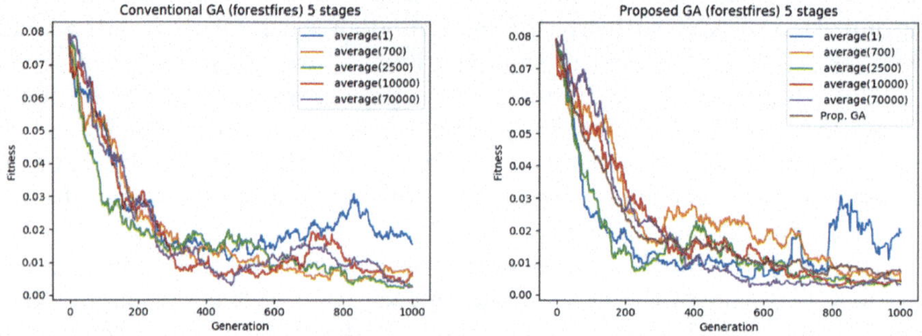

Fig. 8. Forest fires dataset (5 stages)

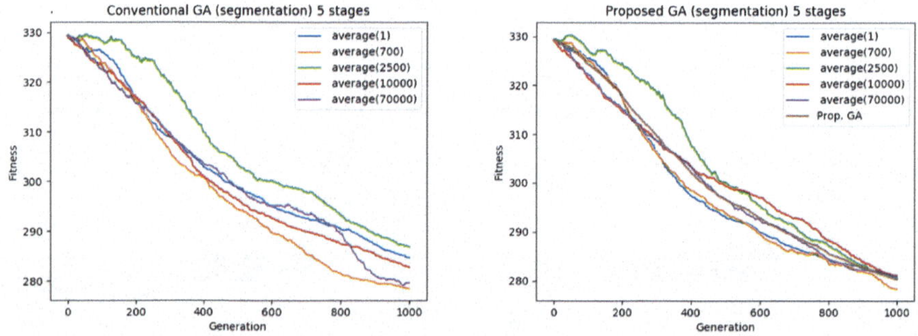

Fig. 9. Image segmentation dataset (5 stages)

running whenever the reconfiguration happens. Online training without reconfiguration is essential in finding the solution by GA, and ABBNN provides the online training capacity. Even though the online training is available in ABBNN, the optimization to evolve ABBNN is still a requirement with the size of ABBNN increasing. Reducing problem space can be an approach for high-speed GA as the experiments show. However, we need a more robust solution to improve the

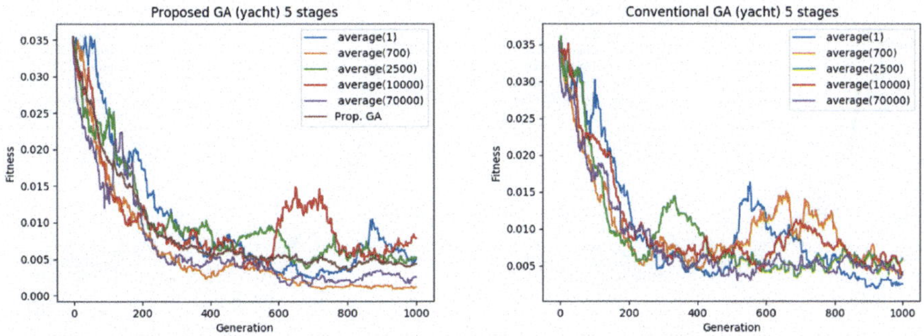

Fig. 10. Yacht hydrodynamics dataset (5 stages)

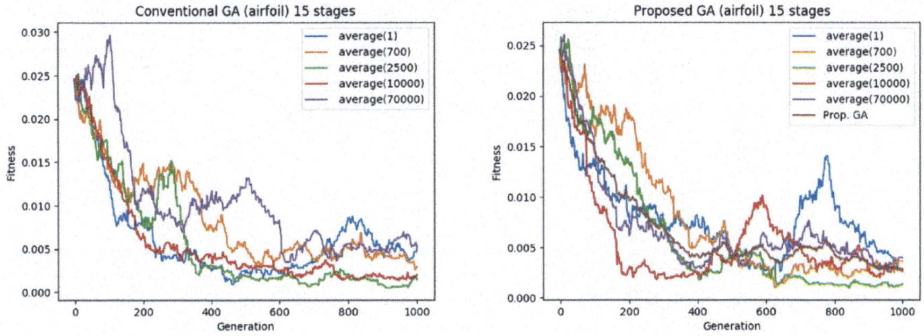

Fig. 11. Airfoil self-noise dataset (15 stages)

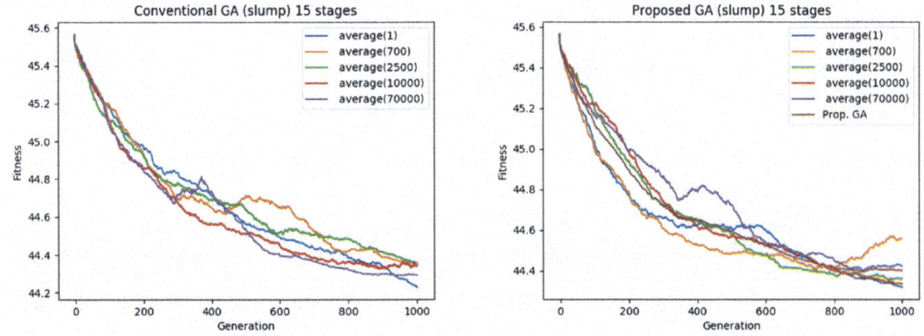

Fig. 12. Concrete slump test dataset (15 stages)

performance of optimization. The best solution to reduce problem space is to minimize the size of the neural network, which directly affects the NN's accuracy, not modifying the algorithm itself. Finding the optimal size of the neural network for a specific problem can be a better approach in reducing the problem space and designing the high-performance neural network model as well.

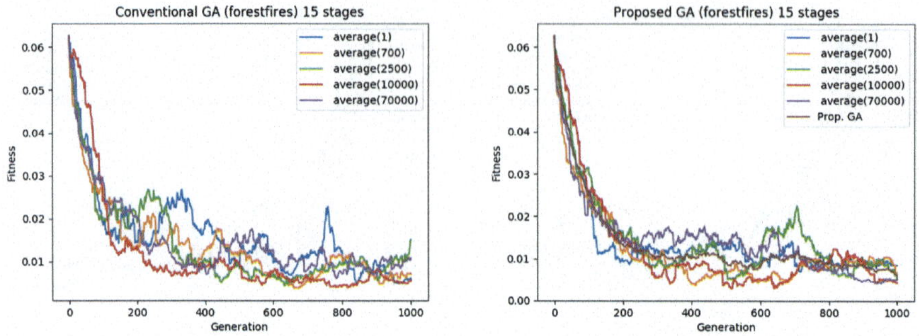

Fig. 13. Forest fires dataset (15 stages)

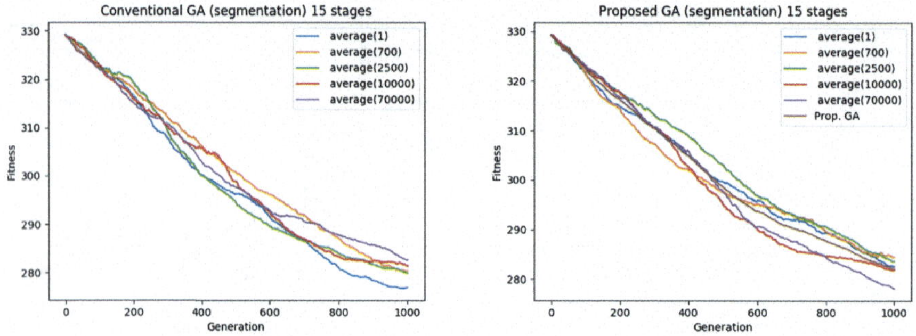

Fig. 14. Image segmentation dataset (15 stages)

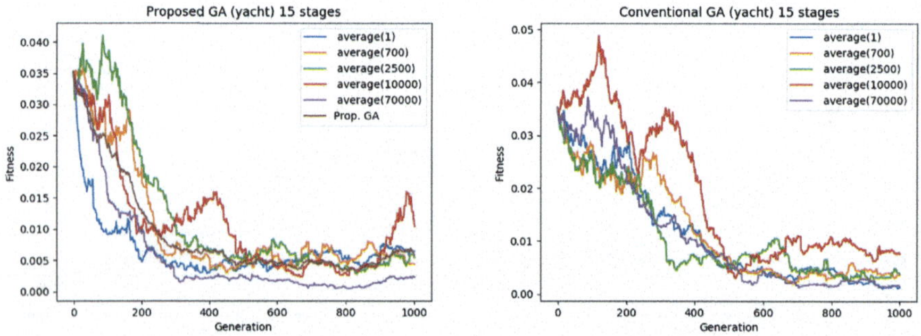

Fig. 15. Yacht hydrodynamics dataset (15 stages)

6 Conclusion

In this paper, we used an approach to reduce the problem space step by step while GA is running. The approach decreases the memory access in GA operation, and the result is competitive against conventional GA. The divide-and-conquer approach, which is generally used to deal with substantial problem space to obtain the solution early, is not easily applied on ABBNN because the connection between the blocks is very close and the proceeding stage always works as an offset for the following stages. The proposed GA reduces the problem space while GA is running and finds the solution in the reduced problem space. The problem space reduction contributes to memory access reduction and makes the GA process faster. Due to the nature of the subsequent problem space reduction, the proposed approach can apply to find the optimal network size. We will focus on a causality analysis between the network size and the NN's accuracy in the future research.

References

1. Moon, S.W., Kong, S.G.: Block-based neural networks. IEEE Trans. Neural Netw. **12**(2), 307–317 (2001)
2. Sahin, S., Becerikli, Y., Yazici, S.: Neural network implementation in hardware using FPGAs. In: Proceedings of the 13th International Conference on Neural Information Processing, vol. 3, pp. 1105–1112 (2006)
3. Krizhevsky, A., Sutskever, I., Hinton, G.E.: ImageNet classification with deep convolutional neural networks. In: Proceedings of NIPS, pp. 1106–1114 (2012)
4. Mohammed, E.Z., Ali, H.K.: Hardware implementation of artificial neural network using field programmable gate array. Int. J. Comput. Theor. Eng. **5**, 795 (2013). https://doi.org/10.7763/IJCTE.2013.V5.795
5. Sze, V., Chen, Y.-H., Yang, T.-J., Emer, J.: Efficient processing of deep neural networks: a tutorial and survey. https://arxiv.org/abs/1703.09039
6. Merchant, S., Peterson, G.D., Park, S.K., Kong, S.G.: FPGA implementation of evolvable block-based neural networks. In: Proceedings of IEEE Congress on Evolutionary Computation, pp. 3129–3136 (2006)
7. Kothandaraman, S.K.: Implementation of block-based neural networks on reconfigurable computing platforms. Master's thesis, University of Tennessee (2004). http://trace.tennessee.edu/utk_gradthes/2268
8. Merchant, S., Peterson, G.D., Kong, S.G.: Intrinsic embedded hardware evolution of block-based neural networks. In: Proceedings of Engineering of Reconfigurable Systems and Algorithms (ERSA) (2006)
9. Niknam, A., Hoseini, P., Mashoufi, B., Khoei, A.: A novel evolutionary algorithm for block-based neural network training. In: Proceedings of the First Iranian Conference on Pattern Recognition and Image Analysis (PRIA), pp. 1-6 (2013)
10. Higuchi, T., et al.: Evolvable hardware with genetic learning. In: Proceedings of IEEE International Symposium on Circuits and Systems, vol. 4, pp. 29–32 (1997)
11. Satoh, H., Yamamura, M., Kobayashi, S.: Minimal generation gap model for GAs considering both exploration and exploitation. In: Proceedings of 4th International Conference on Soft Computing, Iizuka, 30 September–5 October 1996, pp. 494–497 (1996)
12. Goldberg, D., Thierens, D.: Elitist recombination: an integrated selection recombination GA. In: First IEEE World Congress on Computational Intelligence, vol. 1, pp. 508–512 (1994)
13. Kong, S.G.: Time series prediction with evolvable block-based neural networks. In: Proceedings of IEEE International Joint Conference on Neural Networks (IJCNN-2004) (2004)

14. Jiang, W., Kong, S.G., Peterson, G.D.: ECG signal classification using block-based neural networks. In: Proceedings of International Joint Conference on Neural Networks, Montreal, Canada (2005)
15. Tran, Q.A., Jiang, F., Ha, Q.M.: Evolving block-based neural network and field programmable gate arrays for host-based intrusion detection system. In: 2012 Fourth International Conference on Knowledge and Systems Engineering (KSE). IEEE (2012)
16. Lee, K., Hamagami, T.: High performance block-based neural network model by pipelined parallel communication. IEEE J. Trans. Electron. Inf. Syst. **139**, 1059–1065 (2019). (in Japanese)
17. Kirkpatrick, S., Gelatt, C.D., Vecchi, M.P.: Optimization by simulated annealing. Science **220**(4598), 671–680 (1983)
18. Roberto, L.: UCI Machine Learning Repository. Intelnics (2014). http://archive.ics.uci.edu/ml
19. Yeh, I.-C.: Modeling slump flow of concrete using second-order regressions and artificial neural networks. Cem. Concr. Compos. **29**(6), 474–480 (2007)
20. Cortez, P., Morais, A.: A data mining approach to predict forest fires using meteorological data. In: Neves, J., Santos, M.F., Machado, J. (eds.) Proceedings of the 13th EPIA, New Trends in Artificial Intelligence (2007)
21. Carla, B.: UCI Machine Learning Repository. Vision Group, University of Massachusetts (1990). http://archive.ics.uci.edu/ml
22. Roberto, L.: Yacht hydrodynamics data set. UCI Machine Learning Repository (2013). http://archive.ics.uci.edu/ml

An Improved Auto-encoder Based on 2-Level Prioritized Experience Replay for High Dimension Skewed Data

Xin Li(✉) and Tomoki Hamagami

Graduate School of Engineering, Yokohama National University,
79-5 Tokiwadai, Hodogaya-ku, Yokohama, Kanagawa 240-8501, Japan
li-xin-jk@ynu.jp, hamagami@ynu.ac.jp

Abstract. Auto-encoder as the representative method for data dimensionality reduction and feature extraction, plays a very important role on machine learning. However, the data in the actual research work or industrial production are not always normalized data, at this time, it will lead high reconstruction error and slow convergence speed. This study proposed an improved auto-encoder and a denoising auto-encoder based on 2-level prioritized experience replay, which can improve accuracy and reduce loss, while processing a dimensionality reduction or feature extraction problem on high dimension skewed data. In order to evaluate the effectiveness of the proposed method, three models of high dimension simulation dataset which on different skewed degrees are generated. The results of evaluation experiments show that the proposed method can get lower reconstruction error than conventional method for high dimension skewed simulation data.

Keywords: Prioritized experience replay · Priority sampling · Auto-encoder · Denoising auto-encoder · Neural network · Cross-validation

1 Introduction

In the rapid development of machine learning [1], deep learning [2], and the popularity of big data technology, the intelligence and accuracy of data processing is becoming more and more important, especially for production industry and research institutions, the accuracy of algorithms directly affects productivity.

While processing data, dimensionality reduction [3] and feature extraction [4] are necessary works, also, if these works are not effective, they will greatly affect the accuracy of machine learning algorithms. As representative methods for the works, there are principal component analysis (PCA) [5] and auto-encoder (AE) [6]. PCA mainly solve linear problems, simultaneously AE mostly solve nonlinear problems. Because of AE has the structure of the neural network (NN) [7], it can be used in the expanding applicable scenes, as representative methods, there are AE, and its subspecies denoising auto-encoder (DAE) [8].

H. Sato et al. (Eds.): IES 2019, PALO 12, pp. 93–105, 2020.
https://doi.org/10.1007/978-3-030-37442-6_9

However, regarding real data, there are many data are skewed in actual production and life. Most of the production which involves the shape data, when considering the optimal solution for a particular standard, the shapes are often irregular, this brings about data skew.

For example, in the case of the shipbuilding industry, if there is such a problem, we hope to use deep learning to generate a regression model for ships to estimate the wave-making resistance [9], which is an indicator to evaluate whether the built ships have good shape. At this time, it is necessary to compress the shipshape data, however, the length between perpendiculars (LBP) [10] of ships always be huge, meanwhile, the range of forward and aft parts of ships always be short. We can know that the parts of forwarding and aft will be skewed, which have a low percentage on the amount of data and follow a different distribution, at this time the reconstruction error [11] of compressed data or the extracted feature will be high due to the influence of the skewed data, in this case, deep learning is hard to work.

For this problem, any variant of AE does not reduce the error very well, at this time, although the shipshape data are skewed, precise accuracy and convergence speed are still relevant. Based on these legacy issues, in this study, we aim at a low loss, and fast convergence speed for high dimension skewed data.

In the following, Sect. 2 describes the original AE and DAE; Sect. 3 describes how to reduce the loss rate by using the proposed method with sampling; Sect. 4 shows the experimental results through a comparison between the conventional method with the proposed method on high dimension skewed simulation data; Sect. 5 is a summary.

2 Auto-encoder

Auto-encoder (AE) is an unsupervised learning [12] algorithm mainly used for data dimensionality reduction and feature extraction.

In deep learning, AE can be used to determine the initial value of the weight matrix before the training phase begins. The weight matrix in the NN can be regarded as the feature conversion of the input data, that is, the data firstly encoded into another expression, moreover, then a series of learning performed base on this.

However, when initializing the weights, we don't know what role of initial weights plays in training, nor how the weights will change during the training. Therefore, a better idea is that when encoding with the weight matrix generated by initialization, we hope that the encoded data can better retain the main features of the original data.

About how to measure whether the encoded data retain complete information, from previous research, if the encoded data can be easily restored to the

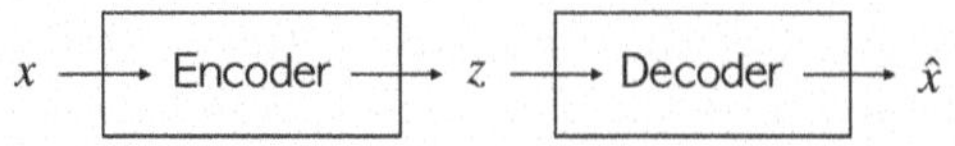

Fig. 1. The structure of AE

original data by decoding, in other words, the lower reconstruction error, the features of the data are preserved well. AE has such a NN structure, which mainly constituted by an encoder and a decoder, and perform backpropagation to solve a nonlinear reconstruction optimization problem. Figure 1 shows the structure, and Eq. 1 shows a function definition of AE.

$$\hat{x} = f(x; \theta) \tag{1}$$

Here, x is input data, and $\hat{x}$ is reconstruction data. z is an under-complete representation generated from the intermediate layer, which contains a feature of input data. θ is a weight matrix that contains parameters of AE. Through calculating the reconstruction error $L(x, \hat{x})$ between x and $\hat{x}$, the effectiveness of AE can be evaluated.

$$L(x, \hat{x}) = \sqrt{||x - f(x; \theta)||^2} \tag{2}$$

$$\theta = \arg\min \frac{1}{N} \sum_{n=1}^{N} L(x_n, \hat{x}_n) \tag{3}$$

Equation 2 shows the definition of error function, and Eq. 3 shows the target θ which we want to get, where data number is $n \in 1 \cdots N$.

AE also have some variants, to take a simple example, it contains denoising auto-encoder (DAE), variational auto-encoder (VAE) [13], or beta-VAE [14] etc.

DAE mainly used to generate generalized features by adding noise to input. VAE mainly used to get a generation model which can control output through adjust latent variable. Moreover, if a disentangled representation, which differ from the distributed representation and mapping one is hidden variable to one meaning, is necessary beta-VAE is a better choice.

However, the common problem with these methods is that, when dealing with the problem with high dimension skewed dataset, they can not reduce the error very well, moreover, the learning results are biased. That is, the problem of decrease in generalization is shown.

3 Proposed Method

In order to solve the problems described in Sect. 2, in this chapter, a sampling method for AE and DAE on the high dimension, the skewed dataset is proposed.

Figure 2 shows the AE, and Fig. 3 shows the DAE used in the chapter of experiment.

From previous research, in recent years, prioritized experience replay (PER) [15] which can enhance generalization performance through preferentially reproducing experiences on the data with few opportunities, is proposed using reinforcement learning. From examining the findings, in this research, we use priority sampling [16] to enhance opportunities for high dimension skewed data, moreover, enhance generalization performance for AE and DAE.

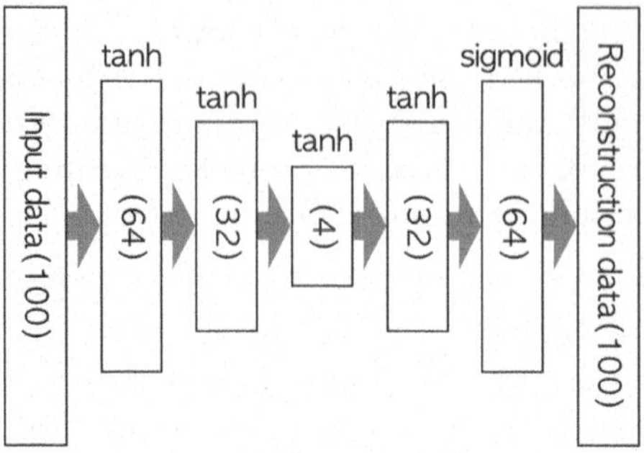

Fig. 2. The AE structure used in the experiments

Equation 4 shows the definition of PER.

$$\mathrm{PER}_i(\alpha) = \frac{p_i^\alpha}{\sum_k p_k^\alpha} \tag{4}$$

Here, p_i is the reference sampling probability based on error for sample i, and this probability can be use to perform a roulette selection. $\mathrm{PER}_i(\alpha)$ is selection probability for sample i, α is weight parameter, within the range of $0 < \alpha < 1$, $\mathrm{PER}_i(1.0) = p(i)$, and $\mathrm{PER}_i(0.0)$ makes the learning become a random sampling.

In this research, considering the idea of that using samples with significant errors preferentially for learning, we are setting the reference sampling probability according to the reconstruction error of AE and DAE.

For example, (1) firstly, starting learning progress from random sampling ($\alpha = 0$); (2) secondly, according to learning progress and variation of reconstruction error, update α; (3) finally, taking $\alpha = 1.0$ and finish learning progress.

However, when PER is performed for all samples, the sampling priority strongly depends on the distribution of reconstruction error.

For example, even if there are a small number of samples with significant reconstruction error, if there are a large number of samples with small reconstruction error, there is a possibility that the sampling does not work.

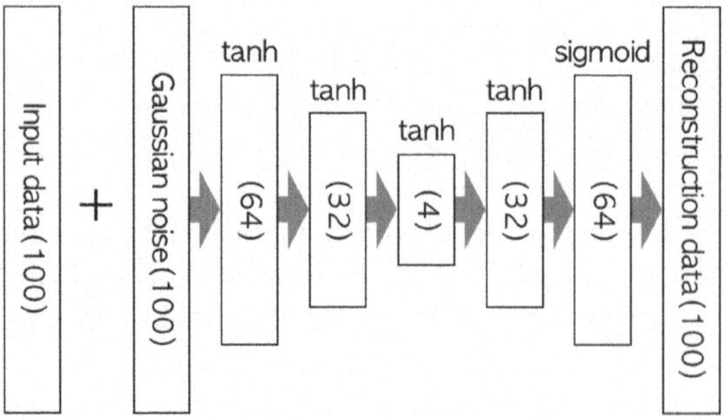

Fig. 3. The DAE structure used in the experiments

Here, we set an arithmetic progression with element number of 100 as an example of probability array, the sum of all elements in the array is 1.0, Fig. 4 shows the relationship changes between $p(i)$ with $\mathrm{PER}_i(\alpha)$ on the different α settings for this array. It can be known from this that even in the case of 100 data, if the sample of high-error data is tiny, the effect of PER will be weakened.

Moreover, the problem is that the actual application does not have only 100 data, usually with larger samples. Therefore, we proposed a method based on 2-level PER sampling, and make samples with significant reconstruction error can be preferentially selected anytime without depending on the sample number. Figure 5 shows an improved AE or DAE based on 2-level PER, and the procedure of the proposed method is shown below.

1. Let $\alpha = 0$ in first iteration, and learn to a certain number of epochs. When finish learning, using the obtained AE/DAE, calculate the reconstruction error $L(x_n, \hat{x}_n)$ for all data $n = 1 \cdots N$ (Fig. 5(a)).
2. Divide $L(x_n, \hat{x}_n)$ into groups $m = 1 \cdots M$ in units of ΔL, calculate the average value avg_m of reconstruction error for each group (Fig. 5(b)).
3. Let avg_m as reference probability, select a group based on $\mathrm{PER}_g(\alpha_1)$ (Fig. 5(c)).
4. For samples belonging to the selected group, according to the reconstruction error of these samples, select a sample based on $\mathrm{PER}_i(\alpha_2)$ (Fig. 5(d)).
5. Extract βN samples as described above, and select $(1-\beta)N$ samples randomly from all samples (Fig. 5(e)).
6. Use the selected samples as training data and learn to a certain number of epochs.
7. Go back to 2.

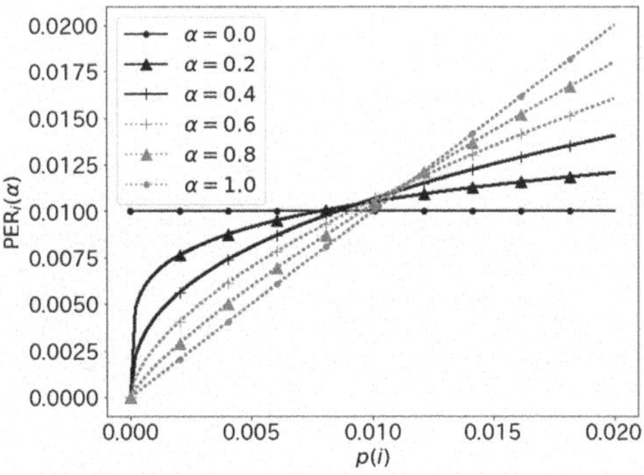

Fig. 4. The relationship changes between $p(i)$ with $\mathrm{PER}_i(\alpha)$ on the different α settings for an example of probability array

4 Experiment

4.1 Dataset

In order to evaluate the effectiveness of the proposed method, we generated three models of high dimension simulation dataset which on different skewed degrees.

For example, (1) firstly, generate two different normal distributions, according to distribution 1 and distribution 2, let μ_1, μ_2 be their expectation, and σ_1, σ_2 be their standard deviation; (2) secondly, generate 95% samples from distribution 1, moreover, generate 5% samples from distribution 2 using as skewed data; (3) finally, generate d dimension samples as described above.

Table 1 shows the settings for simulation data.

Table 1. The skewed degree control for three models of simulation data

	A model	B model	C model
μ_1	0.0	0.0	0.0
μ_2	1.0	3.0	7.0
σ_1, σ_2	0.1	0.5	1.2

In the experiment, we generated 50000 samples of dimension $d = 100$. In order to control the correlation between each dimension, the variance-covariance matrix is the right choice. From previous research, when extracting features, it is more prone to skewed issues for the data with a low correlation between dimensions. Here we let the covariance to be 0 on each two dimensions of the three models of simulation data to reduce correlation.

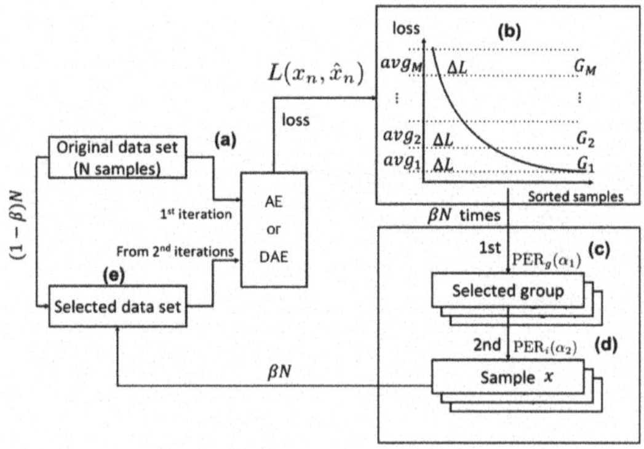

Fig. 5. Improved AE or DAE based on 2-level PER

Figure 6 shows the three models simulation data, (a)(c)(e) show the three models of simulation data on three coordinate axes, however the data have 100 dimensions, therefore considering any three dimensions in 100 dimensions data have similar relative relationships. (b)(d)(f) show the heatmap of three models simulation data, through which we can know the situation of skewed degree for the data of each model.

4.2 Experiment Environment

We used Keras [17] as an experiment tool for each experiment below. Figure 7 shows the model structure in Keras of AE or DAE for training, and when DAE, the input is corrupted by Gaussian noise [18], with 3% of data range as σ to input data.

We use the values in Table 2 as common parameter settings for improved AE and DAE, the parameters were already adjusted to appropriate values through preliminary experiments. About the parameters adjustment for α_1 and α_2, in most cases, we hope the parameter values to be higher, however when the skewed degree increased, the learning process becomes difficult, if increase parameter values blindly, it may lead to the training data changes intensely during sampling,

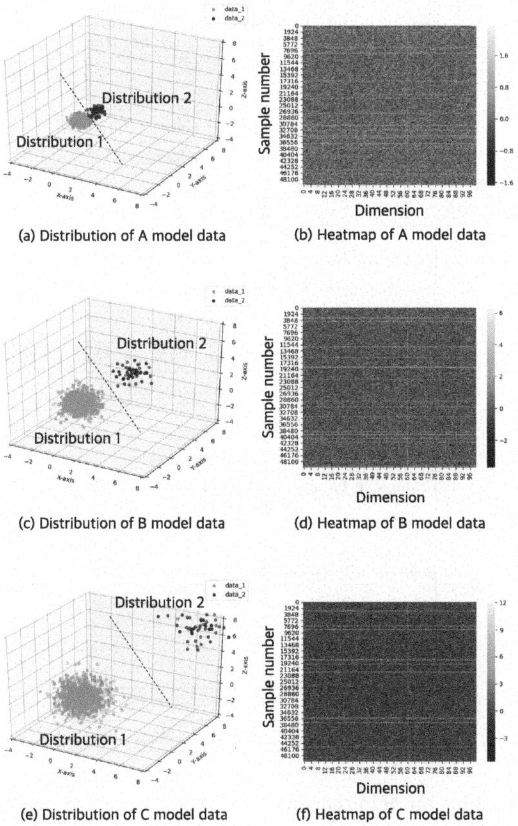

(a) Distribution of A model data (b) Heatmap of A model data

(c) Distribution of B model data (d) Heatmap of B model data

(e) Distribution of C model data (f) Heatmap of C model data

Fig. 6. The high dimension skewed simulation data of three models

this will bring a lot of instability, it is usually embodied as the learning curve up and down cyclically, therefore, this is a trade off relationship. A feasible solution is to increase the value under the premise that learning can proceed normally, and the preliminary experiment was also carried out in this way (Table 3).

However the standard parameters are not always valid for all experiments, They need to be adjusted for some specific situations. About the experiment which the parameters need to be changed, we adjusted some values additionally. In this experiment, when the data of the C model has been used, Because of the distribution of the C model perhaps lead to the learning got high reconstruction error relatively, the sampling group stride must be adjusted to a new appropriate value, in general, because of the error range becomes wider, the sampling group stride also follows the larger adjustment. Moreover, according to high reconstruction errors due to a highly skewed degree, in order to get more efficient features from learning, it also requires an increase in sampling parameter through an adjustment on α_1 and α_2. Therefore, when the processing target is C model data, some of the changes in the parameter settings are shown in Table 4.

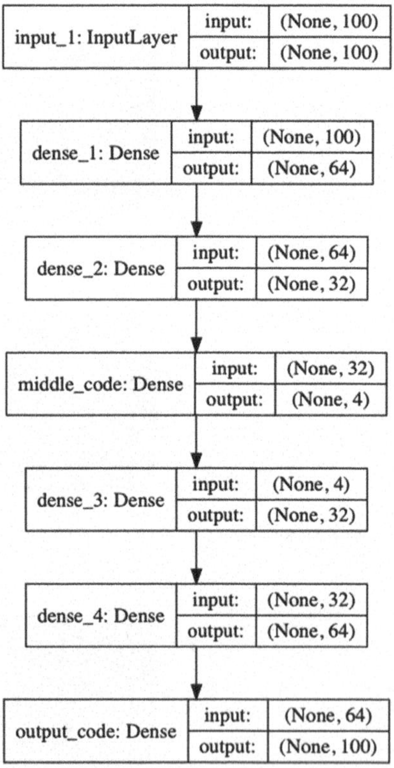

Fig. 7. The model structure of AE or DAE for training

Here, the parameters settings are based on some preliminary experiments. Regarding the difference in adjustment for improved AE and improved DAE, we consider that is because the DAE already has some generalization performance due to the addition of Gaussian noise, therefore compare to the AE, the value of α_1 which controlling the group selection can be lower, about the value of α_2 can be almost the same level as the AE.

In order to compare the proposed method with AE and DAE, 5-fold cross-validation [19] has been performed. Figure 8 shows the splits of samples through cross-validation on the experiment, here, for the simulation data of each model, 10% are generated for test data through random selection, which should be fixed

Table 2. The common parameter settings of improved AE and DAE

Simulation data number	50000
Training data number	36000
Validation data number	9000
Cross-validation data number	$45000 = 36000 + 9000$
Test data number	5000
Sampling group stride	$\Delta L = 0.06$
Sampling factor	$\beta = 0.2$
Sampling parameter	$\alpha_1 = 0.3, \alpha_2 = 0.5$
Sampling times	4
Epoch number for each sampling	100
Total epoch number	$400 = 4 \times 100$
Batch size	20
Layer	(100)-(64 tanh)-(32 tanh)-(4 tanh)-(32 tanh)-(64 sigmoid)-(100)
Loss function	RMSE (Root Mean Squared Error)

Table 3. The changes of parameter settings for improved AE on C model simulation data

Sampling group stride	$\Delta L = 0.09$
Sampling factor	$\beta = 0.3$
Sampling parameter	$\alpha_1 = 0.5, \alpha_2 = 0.7$

Table 4. The changes of parameter settings for improved DAE on C model simulation data

Sampling group stride	$\Delta L = 0.08$
Sampling factor	$\beta = 0.3$
Sampling parameter	$\alpha_1 = 0.3, \alpha_2 = 0.8$

5-fold cross-validation

1	val	train	train	train	train	test
2	train	val	train	train	train	test
3	train	train	val	train	train	test
4	train	train	train	val	train	test
5	train	train	train	train	val	test
	18%	18%	18%	18%	18%	10%

Fig. 8. The splits of samples for A, B, C models simulation data through cross-validation

and do not change. The cross-validation is performed on the remaining 90%, therefore, for 5-fold cross-validation, 72% training data and 18% validation data are generated for the cross-validation of each time, at the same time, we always have the 10% test data to perform an additional validation.

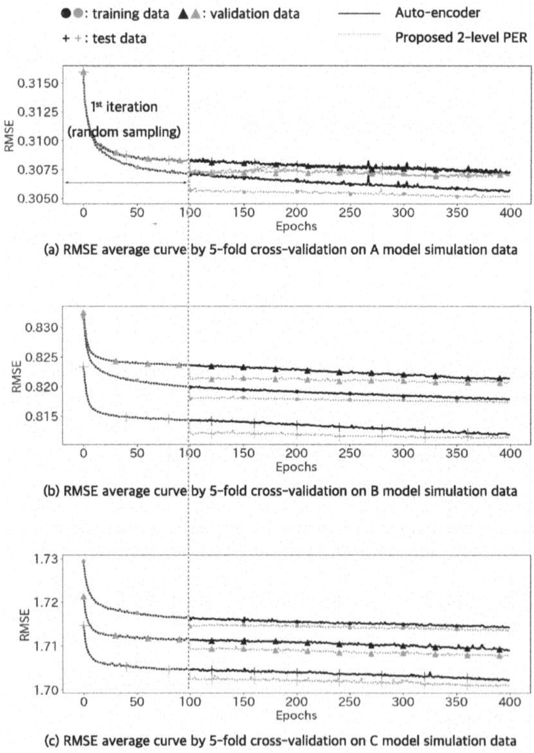

Fig. 9. RMSE average curve of AE and 2-level PER

4.3 Experiment Result

We performed AE, DAE, and proposed method on A, B, C models simulation data by 5-fold cross-validation. Figure 9 shows the comparison on RMSE average of reconstruction error between AE and proposed a method and Fig. 10 shows the comparison on RMSE average of reconstruction error between DAE and proposed method.

According to the experiment results, we can know that through using the proposed method, the RMSE average of reconstruction error by 5-fold cross-validation decreased for all the data of three models with different skewed degrees both on AE and DAE, and it also speeds up the convergence for learning.

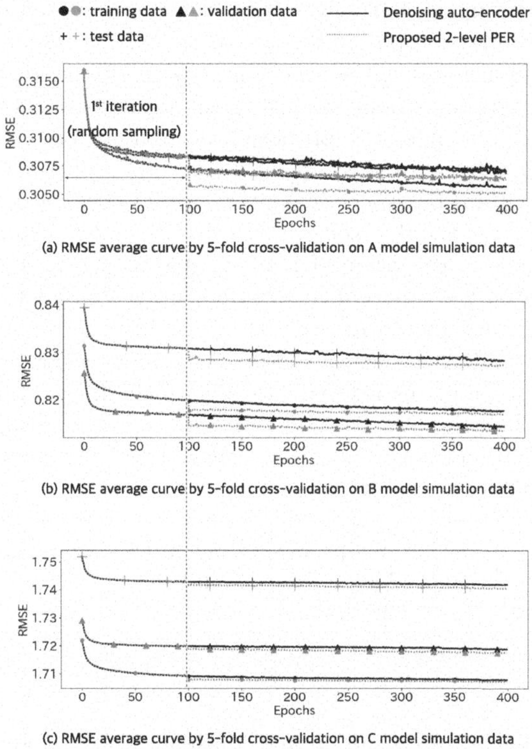

Fig. 10. RMSE average curve of DAE and 2-level PER

5 Conclusion

For the high dimension skewed data, a sampling method based on 2-level PER for AE and DAE is proposed. In order to evaluate the effectiveness, which without depending on a particular skewed degree, of this method, three models of high dimension skewed simulation data with a different skewed degree are generated.

Through appropriate parameter adjustments and evaluation experiments, the effectiveness of the proposed method, (1) reducing the reconstruction error caused by the skewed data effectively, (2) improving the generalization of the algorithm and the convergence speed of learning, have been confirmed.

It is expected to begin a practical problem in the future, and the proposed method will be applied to the actual high dimension skewed data, to improve the efficiency of feature extraction during learning, and further, verify the effectiveness of this method on a practical problem.

For example, there is a deep neural network (DNN) [20] used to a regression a value from some shape data, however, the data is skewed with some strange shape, in order to improve the learning rate and accuracy of DNN, the input layer always needs some compressed features with generalization. The features can be extracted through this study to improve the accuracy of the regression. Further, this is just an example, and more usages of the research are expected.

Of course, there are still many shortcomings in this study. For example, in the case of parameters, it is necessary to make adjustments through many preliminary experiments to make the proposed method get better performance, we treat this as a legacy topic in the current stage, and will consider a more intelligent optimization solution, such as parameters adjustment automation based on the skewed degree of the dataset in the future. Besides, when the dataset has a highly skewed degree, the original learner such as AE or DAE will get high reconstruction errors, we can reduce the high error caused because of skewed data, but can not fundamentally solve the high error problem of the learner. At the same time, for DAE, a learner that has some generalization performance itself, with the increase of the skewed degree and the high error problem of the learner, although the proposed method can achieve better results than the conventional methods, the performance is slightly reduced. These will be considered as the subject of future research.

Acknowledgment. Thanks to my professor Tomoki Hamagami, for helping me a lot in this research. Thanks to the China Scholarship Council (CSC) in 2016, pursue my study in Japan as a Ph.D. student under File no. 201608140085.

References

1. Bishop, C.M.: Pattern Recognition and Machine Learning. Springer, Heidelberg (2006)
2. LeCun, Y., Bengio, Y., Hinton, G.: Deep learning. Nature **521**(7553), 436 (2015)
3. Van Der Maaten, L., Postma, E., Van den Herik, J.: Dimensionality reduction: a comparative. J. Mach. Learn. Res. **10**(66–71), 13 (2009)
4. Guyon, I., Gunn, S., Nikravesh, M., Zadeh, L.A.: Feature Extraction: Foundations and Applications, vol. 207. Springer, Heidelberg (2008)
5. Jolliffe, I.: Principal Component Analysis. Springer, Heidelberg (2011)
6. Hinton, G.E., Salakhutdinov, R.R.: Reducing the dimensionality of data with neural networks. Science **313**(5786), 504–507 (2006)
7. Zurada, J.M.: Introduction to Artificial Neural Systems, vol. 8. West Publishing Company, St. Paul (1992)

8. Vincent, P., Larochelle, H., Bengio, Y., Manzagol, P.-A.: Extracting and composing robust features with denoising autoencoders. In: Proceedings of the 25th International Conference on Machine Learning, pp. 1096–1103. ACM (2008)
9. Zhang, B., Ma, K., Ji, Z.: The optimization of the hull form with the minimum wave making resistance based on rankine source method. J. Hydrodyn. **21**(2), 277–284 (2009)
10. Surendran, S., Venkata Ramana Reddy, J.: Numerical simulation of ship stability for dynamic environment. Ocean Eng. **30**(10), 1305–1317 (2003)
11. Machiraju, R., Yagel, R.: Reconstruction error characterization and control: a sampling theory approach. IEEE Trans. Visual Comput. Graph. **2**(4), 364–378 (1996)
12. Barlow, H.B.: Unsupervised learning. Neural Comput. **1**(3), 295–311 (1989)
13. Kingma, D.P., Welling, M.: Auto-encoding variational bayes. arXiv preprint arXiv:1312.6114 (2013)
14. Higgins, I., Matthey, L., Pal, A., Burgess, C., Glorot, X., Botvinick, M., Mohamed, S., Lerchner, A.: β-VAE: larning basic visual concepts with a constrained variational framework. ICLR **2**(5), 6 (2017)
15. Schaul, T., Quan, J., Antonoglou, I., Silver, D.: Prioritized experience replay. arXiv preprint arXiv:1511.05952 (2015)
16. Szegedy, M.: The DLT priority sampling is essentially optimal. In: Proceedings of the Thirty-Eighth Annual ACM Symposium on Theory of Computing, pp. 150–158. ACM (2006)
17. Chollet, F., et al.: Keras: the python deep learning library. Astrophysics Source Code Library (2018)
18. Rasmussen, C.E.: Gaussian processes in machine learning. In: Summer School on Machine Learning, pp. 63–71. Springer (2003)
19. Kohavi, R., et al.: A study of cross-validation and bootstrap for accuracy estimation and model selection. In: IJCAI, Montreal, Canada, vol. 14, pp. 1137–1145 (1995)
20. Schmidhuber, J.: Deep learning in neural networks: an overview. Neural Netw. **61**, 85–117 (2015)

Multi-task Learning Using Online Fine-Tuning Considering the Importance of Each Filter

Shota Ikawa[1(✉)] and Yuji Sato[2]

[1] Hosei University, 3-7-2 Chinocho, Koganei City, Tokyo 184-8584, Japan
shota.ikawa.6e@stu.hosei.ac.jp
[2] Hosei University, Tokyo, Japan
yuji@k.hosei.ac.jp

Abstract. Transfer Learning and Fine-Tuning are learning frameworks for dealing with the lack of labeled data in Deep Learning. These methods work effectively in a Convolutional Neural Network (CNN) that acquires common features in the convolutional layers. When transferring CNN layers, it is common to transfer multiple convolutional layers close to the input side excluding the identification layer after training the source task. On the other hand, there are few studies that focus on transfer during learning and higher-level transfer. In this paper, we propose Fine-Tuning by transferring convolutional filters during learning. Filters are ranked using pruning criteria, and only the low importance filters are overwritten as target filters. In the 10-class classification using a subset of CIFAR-100, we show that the proposed method can improve the test accuracy by up to 2% compared to training from scratch. We also show the tendency of the ratio of low importance filters for each layer changes with the progress of learning.

Keywords: Transfer Learning · Fine-Tuning · Convolutional Neural Network · Pruning

1 Introduction

Deep learning (DL) is currently one of the most remarkable machine learning methods for companies and researchers. It uses Deep Neural Network to extract common features contained in big data with high accuracy and has succeeded in various domains such as image recognition and natural language processing. DL requires a large amount of labeled data to adjust the trainable parameters. Therefore, it has been observed that performance decreased when it is difficult to collect data, or a few labeled data exist.

Transfer learning (TL) [1] is a promising technique for addressing said issues. TL is used to reuse a model pre-trained on a specific domain (source domain) to improve learning performance in the target domain related to the source domain. By transferring the parameters, i.e. knowledge of the source model obtained by

H. Sato et al. (Eds.): IES 2019, PALO 12, pp. 106–117, 2020.
https://doi.org/10.1007/978-3-030-37442-6_10

learning from the source model to the target model, it can learn efficiently by using common knowledge in both domains. Another technique similar to TL is Fine-Tuning. In TL, the transferred parameters are not updated by learning of the target task. In contrast, Fine-Tuning uses them as a substitute for the initial parameters and updates in the target task. It is known that CNN has a strong initial value dependency; therefore, these techniques are more efficient.

When applying TL or Fine-Tuning to CNN, it is common to transfer multiple convolutional layers close to the input side, excluding identification layers after learning of the source task. In [2] is concluded that CNN extracts low-level feature such as edges and highlights in the layer close to the input side, and extracts features specific to each data as it approaches the output side. Therefore, it works effectively. In addition, the pre-trained model has already acquired useful knowledge for the source task. Therefore, offline transfer (after learning) may have a better influence on performance than online transfer (during learning).

On the other hand, depending on the task, the combination of transfer layers showing the best performance may be different. Therefore, it is necessary to determine which layer to transfer by multiple experiments. In addition, online transfer and high-level transfer (e.g., in units of neurons) have only been considered relatively recently, so there are few studies. When applying TL or Fine-Tuning with the setting to learn multi-task with multi-model simultaneously, these studies may be important. If knowledge can be transferred online between related tasks, it can be reused without waiting for other tasks to finish, which may further improve learning efficiency. However, in the online transfer, it is necessary to prevent the useful knowledge obtained by learning from being overwritten and additionally, the knowledge of the pre-trained model may not be correct.

To address these issues, in this paper we propose online Fine-Tuning by transferring convolutional filters during learning. Each filter is calculated and ranked by the pruning criteria which is a popular method for compressing an NN and selects only the low-rank filters as the "not likely to be useful" filters as the target filter (overwritten by transfer). Filters are part of a convolutional layer, and each filter performs a convolution completely independently of the other filters, hence the extracted features differ from the filter to filter. Due to this, the parameters of each filter may only have partial knowledge. The purpose of this study was to prevent forgetting useful knowledge and improve transfer performance by overwriting only those filters that were considered to have little effect on the accuracy during learning as a target filter. We evaluate the proposed method in popular image recognition tasks: Two subsets of CIFAR-100 [3], Class: 1–10 and 11–20.

2 Related Research

2.1 Backpropagation Algorithms

Backpropagation (BP) algorithms are training methods in a multi-layer NN that updates the weights of neurons from the output layer to the input layer.

The error in each neuron is calculated based on the output error (in the case of supervised learning; the difference between the output value and the teacher signal) and the weights are adjusted so that the error is minimized. The error propagates to the previous neurons, the weight is updated from the output layer back to the input layer.

To update the weight, it is necessary to calculate the gradient of the loss function for the weight between each layer. (1) to (4) are formulas that update weight of neurons in a 3-layer network of input layer i, middle layer j, and output layer k. The output of the kth neuron is y_k, the total number of neurons in output layer is K, the activation function is φ, the loss function as $E = E(y_1, y_2, ..., y_K)$, the elements in the output layer and the middle layer are δ_k^{out} and δ_j^{hid}, the gradients $\frac{\partial E}{\partial w_{jk}}$ and $\frac{\partial E}{\partial w_{ij}}$ with weights w_{jk}, and w_{ij} as follows:

$$\delta_k^{out} = \varphi_k' \left(\sum_j w_{jk} y_j \right) \frac{\partial E}{\partial y_k} \tag{1}$$

$$\delta_j^{hid} = \varphi_j' \left(\sum_i w_{ij} y_i \right) \sum_k \delta_k^{out} w_{jk} \tag{2}$$

$$\frac{\partial E}{\partial w_{jk}} = \delta_k^{out} y_j \tag{3}$$

$$\frac{\partial E}{\partial w_{ij}} = \delta_j^{hid} y_i \tag{4}$$

By repeating the above calculation and updating the weight, the error is minimized efficiently.

2.2 Transfer Learning and Fine-Tuning

The problem with DL is that it may not be able to apply to a specific domain because a large amount of cost is required to preprocess training data such as data collection and labeling. Previous studies have used semi-supervised learning that combines a small amount of labeled and unlabeled data in a single task [4], but could not be applied to separate tasks with different domains. Transfer Learning (TL) and Fine-Tuning address this problem by reusing a model trained in a domain with abundant training data. The purpose of these methods is to improve the efficiency and accuracy of learning by using information obtained from related tasks for learning other tasks that contains few data. An example of transfer learning and fine-tuning in CNN is shown in Fig. 1. In TL, the transferred parameters are "frozen" and reused as a feature extractor. On the other hand, Fine-Tuning performs re-training in the target task. Empirically, the convolutional layers before the fully connected layers (or identification layers) are most often transferred. ImageNet [5] is known to be a suitable task for transfer learning using CNN in the field of image recognition.

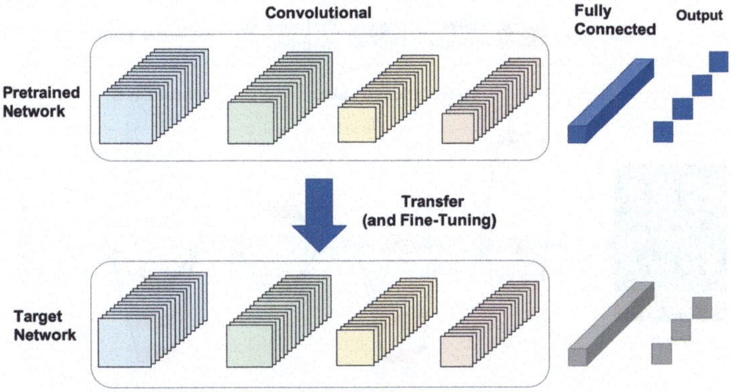

Fig. 1. Example of Transfer Learning (Fine-Tuning).

When these methods are assumed to be applied in a situation where different tasks are trained in parallel with multiple models, several problems must be considered.

The first problem is to choose what knowledge to transfer. When transferring after learning, knowledge is known to be useful in the source task. However, when transferring during learning, it is not known whether knowledge is useful in the source task. Therefore, transferring incomplete knowledge during learning can influence the learning negatively. In addition, the online transfer may overwrite useful knowledge during learning, and the accuracy may decrease. The second problem is the estimation of similarity between domains. These techniques work effectively when assuming that the low-level features are common to both the source and target domains. It is known that when a transfer is performed between domains that do not satisfy this or are weakly-related, the negative transfer occurs and decrease accuracy [6]. Previous studies have proposed a method for visualizing the efficiency of transfer learning between different domains in image recognition tasks [7]. In [7], transfer learning is performed using models pre-trained with two different source tasks, and a task similarity is calculated based on the difference in the accuracy of each model with respect to the target task. However, the transfer is offline and requires pre-training of the source task, thus it is difficult to apply it to the setting of learning the source task and target task in parallel.

2.3 PathNet

PathNet [8] is a learning algorithm related to transfer learning and multi-task learning (MTL) [9] using CNN. Figure 2 shows the network structure of PathNet. PathNet is a modular deep neural network. Each convolutional layer and linear layer are composed of M modules, and each layer is configured by activating up to $N(< M)$ modules. Each module translates to a layer (small neural network), only parameters in the active module are updated by the training.

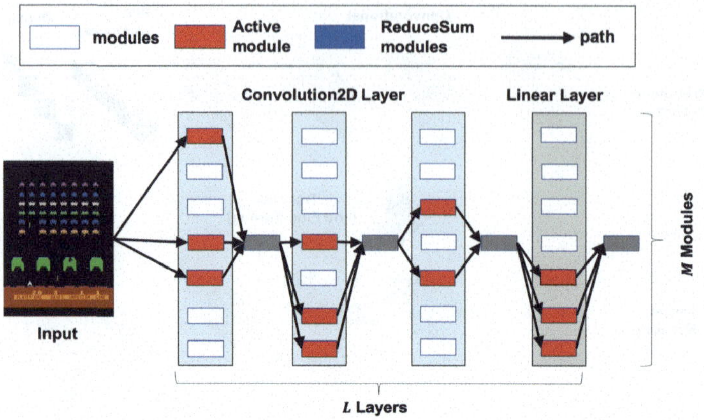

Fig. 2. Example of PathNet structure and pathway.

The outputs from the modules are summed for each layer by the ReduceSum module and then input into the next layer. In [9], the connections of active modules are called "pathway", and both parameters and pathways are updated simultaneously. In PathNet, pathways are randomly initialized before training a task, and paths are evolved using genetic algorithms. Each pathway is represented by a genotype using an N by L matrix of integers, and the fitness of the pathway is calculated from the results during the training period. PathNet supports image recognition and reinforcement learning (RL) tasks, fitness is accuracy or reward. During the training period, genotypes are selected multiple times by the tournament selection algorithm, and the losing-pathway genotype is overwritten by a copy of the winning-pathway genotype. In addition, each element is selected independently with a probability of $1/[N \times L]$, and a copy of the winning-pathway genotype is mutated. The RL algorithm is implemented Asynchronous advantage actor-critic (A3C) [10] which multiple agents (players) perform parallel learning asynchronously. The pathways are continuously evolved by training and tournament selection, and the best pathway and parameters are saved after converging on a single path. All pathways are reinitialized before training a new task, but the best pathway and parameters are inherited as one of the pathways. Therefore, PathNet can acquire new knowledge without forgetting the previous knowledge.

On the other hand, the training time increases as the number of tasks increases because tasks are trained sequentially. In addition, the result may depend on the order of tasks to be trained.

3 Online Fine-Tuning by the Filter in Multi-task Learning

Considering a setting like a parallel Multi-task Learning (MTL) that learns multiple tasks simultaneously with different models (agents), agents that learn similar tasks may have useful knowledge in common with each other, but they may be wrong and incomplete. To achieve the most effective transfer in parallel MTL, it is important to build a transfer method that can transfer only useful knowledge and avoid "Negative Transfer" by transferring the wrong knowledge. In addition, a method is needed to select only parameters that do not affect performance so that useful knowledge about the training task is not overwritten by the transfer.

To address these problems, this study considers Fine-Tuning that focuses on convolutional filters. The proposed method aims to improve learning efficiency by learning two image recognition tasks in parallel using different CNNs of the same structure and transferring by the filter. Before transferring, the importance of all filters is calculated based on pruning criteria. By selecting a filter with low importance as a target filter that has less influence on performance, it can be expected that efficient transfer without overwriting useful parameters.

3.1 Parameter Transfer Based on Filter Importance

Figure 3 shows the algorithm of the proposed method. In the proposed method, the task is divided into subsets (partial tasks) for a certain number of classes, and each subset is assigned as a CNN training data set. After we train e epoch, we calculates the importance for all convolutional filters of each CNN, and set the weight of the low importance $P\%$ filter to 0. The importance is calculated based on the Minimum weight (MW) method and Taylor expansion (TE) method proposed by Molchanov et al. [11] for pruning convolutional filters. Pruning is the method of compressing the model while minimizing the degradation of accuracy, and these criteria are used for selecting the filter which is to be pruned. In this paper, we think that efficient transfer learning is possible while maintaining the filter (partial structure) necessary for maintaining accuracy by calculating the importance of the filter based on these criteria. Therefore, we compare and evaluate the MW and TE method. The source filter and target filter are determined as follows.

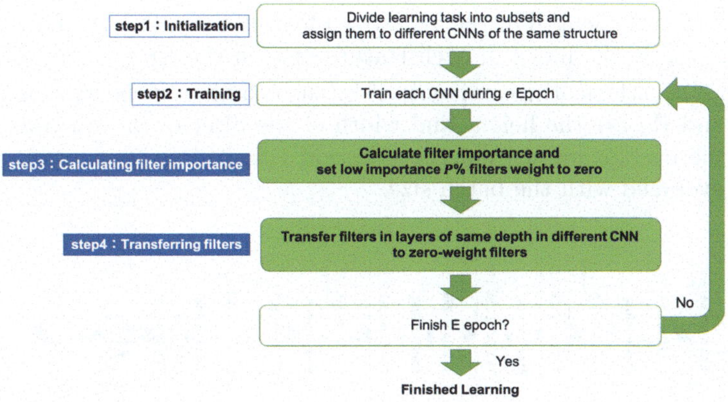

Fig. 3. Online Fine-tuning algorithm by the filter.

Source filter
A filter that exists in a different CNN from target model, has the same depth as the target filter, and a randomly selected that weight is not zero.
Target filter
A filter that weight is zero.

Learning and transfer are repeated every e epoch, and training ends when $E(> e)$ epoch has passed.

3.2 Calculation Criteria for Filter Importance

In this section, we explain the Minimum weight (MW) and Taylor expansion (TE) methods are used as the calculation criteria for the importance of the filter. [11] defines the importance of the filter Θ_{MW} in the MW method as follows:

$$\Theta_{MW} : \mathbb{R}^{C_{l-1} \times p \times p} \rightarrow \mathbb{R}, \Theta_{MW}(\mathbf{w}) = \frac{1}{|\mathbf{w}|} \sum_i w_i^2 \tag{5}$$

where C_l is the number of filters in the lth convolutional layer, $l \in [1, 2, ..., L]$ is an index of the convolutional layer, p is the size of the filter. w_i is the weight of ith filters in each convolutional layer, and $|\mathbf{w}|$ is dimensionality of the set of weights after vectorization. According to this criterion, a filter with a smaller absolute value of weight is evaluated as less important. In [11], the authors describe that the motive for proposing the MW method is that filters with smaller l_2 norms tend to detect less important features.

The TE method expresses pruning as an optimization problem and approximates changes in the loss function due to removal specific parameters by Taylor expansion. In [11], the filter importance Θ_{TE} in the TE method is calculated as follows:

$$\Theta_{TE}(z_l^{(k)}) = \left| \frac{1}{M} \sum_m \frac{\delta C}{\delta z_{l,m}^{(k)}} z_{l,m}^{(k)} \right| \tag{6}$$

Where M is the length of the vectorized feature map, C is the cost function, $k \in [1, 2, ..., C_l]$ is the index of each feature map in the lth convolutional layer, $z_l^{(k)} \in \mathbb{R}^{H_l \times W_l}$ is the feature map output by the kth filter of the lth convolutional layer, H_l and W_l are the height and width of the filter in the lth convolutional layer. In the case of minibatch, the criterion is calculated separately for each data and averaged with the batch size.

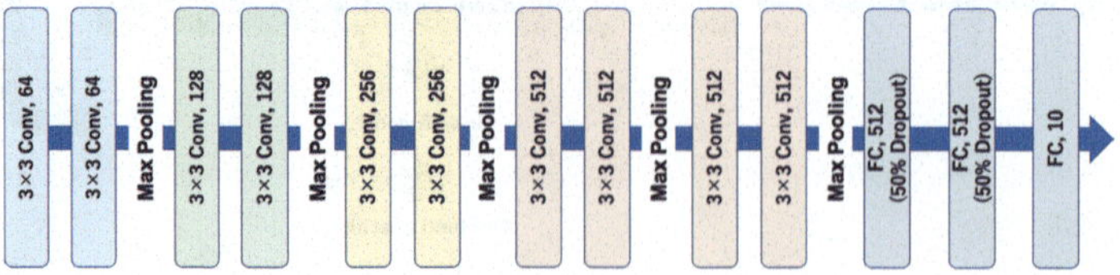

Fig. 4. VGG-13 network structure used in this experiment.

4 Evaluation

In this section, we examined whether the improvement in accuracy and learning efficiency can be achieved by online transfer using the proposed method by comparison experiments with training from scratch. Scratch means training the CNN model used in this experiment from the initial value using Backpropagation algorithms described in Sect. 2.1. Figure 4 shows the structure of the model used in this experiment. The model is based on VGG-13 [12] and consists of 10 convolutional layers and 3 fully connected layers. This study makes the following changes to the original model:

1. Reduce the parameters of the fully connected layers.
2. Apply ReLU [13] to each layer except the output layer.
3. Add Batch Normalization [14] to each convolutional layer.
4. Apply 50% Dropout [15] to each fully connected layer.

4.1 Methods

We compared with the proposed method and training from scratch using CIFAR-100. CIFAR-100 is a typical image classification task, and the datasets are classified into 100 classes. Each class also belongs to 20 more abstract superclasses. In this paper, we divided CIFAR-100 datasets into 10 class subsets and assigned the 1–10 and 11–20 class subsets into two models for training. The reason why only the first 20 classes were used in this experiment is that there is no large bias in the number of superclasses belonging to the images of each subset when the datasets divided into every 10 classes. We apply The ME and TE method as the criteria of filter importance, and the average of the results of 10 experiments with $E = 600$ epoch for each criterion is compared with scratch. The model is implemented using Pytorch, an open-source machine learning framework in Python. Hyperparameter settings are shown in Table 1.

Table 1. Hyperparameter settings in the experiment.

Item	Settings
Model	VGG13 (Batch Normalized)
Train batch size	128
Optimizer	SGD
Learning rate	0.05
Momentum	0.9
Weight decay	0.0005
e (interval between training and transferring)	25/50/100
P (% of filter weights to be zero)	10.0

Table 2. Accuracy, variance and standard deviation for test data. Bold indicates better results than training from scratch.

		Class: 1–10			Class: 11–20			
Method	e	Avg_acc (%)	Variance	Standard deviation	Avg_acc (%)	Variance	Standard deviation	Execution time (m)
Scratch		85.03	8.609	2.934	79.54	12.98	3.603	57.6
Minimum weight	25	**85.92**	**5.317**	**2.306**	**81.83**	**6.320**	**2.514**	64.5
	50	83.67	10.97	3.312	**81.40**	15.63	3.953	61.5
	100	83.64	9.276	3.046	**80.59**	29.45	5.426	59.6
Taylor expansion	25	**86.13**	**4.773**	**2.185**	**81.82**	**7.395**	**2.719**	61.5
	50	**86.21**	**3.879**	**1.970**	**82.41**	**8.214**	**2.866**	60.1
	100	**85.13**	**2.160**	**1.470**	**81.27**	13.35	3.654	58.1

4.2 Results

Tables 2 and Fig. 5 shows the average accuracy, variance, and standard deviation for the test data at the end of the training. Execution time is the average execution time required by the program in 10 experiments in each case. For proposed method, it includes the filter importance calculation using each pruning criterion and Fine-Tuning.

The accuracy for the test data when the filter is transferred every 25 epochs in the proposed method was about 1–2% higher than scratch in both MW method and TE method, see Table 2. In addition, the variance was also small. On the other hand, the accuracy was worst when both MW and TE were transferred every 100 epochs, and the accuracy in ME was about 1.5% lower than scratch.

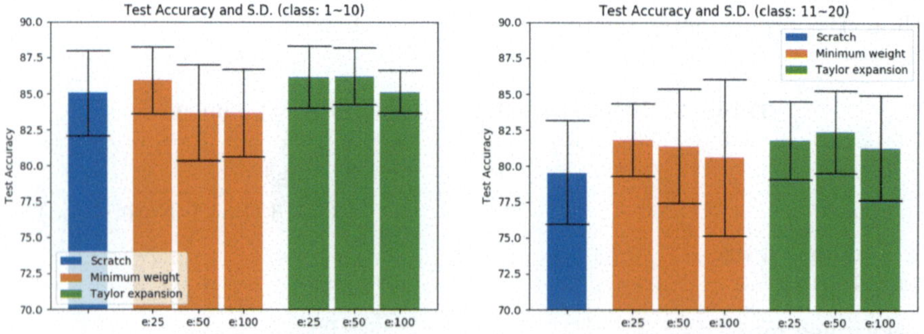

Fig. 5. Accuracy and standard deviation for test data. The vertical axis represents the correct answer rate, the horizontal axis represents each condition, and the error bar represents the standard deviation.

Comparing the results of each condition. In the TE method, all evaluation indices (accuracy, variance, and standard deviation) were better than scratch except in the case of transitions every 100 epochs in a subset of class 10–20. In the ME method, variance and standard deviation were worse than scratch when transferring every 50 epoch and 100 epochs, and all indices were superior to scratch only when every 25 epochs, see Table 2 and Fig. 5.

4.3 Discussion

Table 2 shows the results of proposed method and training from scratch when using a subset of the CIFAR-100 dataset. The proposed method achieved a result that is similar to or better than from scratch in average accuracy in most cases. Both methods gave better results than from scratch in all indicators when the epoch of transition interval "e" is 25. This suggests that frequent transfers in the early stages of training can obtain common features and prevent convergence to local solutions and deterioration of accuracy. By contrast, both showed worse results than scratch in some or all indicators when the epoch of transition interval is 100. This suggests that the number of filters to be transferred and the number of times of transfer were less, and parameters were not transferred enough to obtain common features.

Figures 6 and 7 shows the transition of the target filter (which is less important) in each layer during the learning process in the MW and TE method. Figure 6 shows that the target filters moved from the input side to the output side as the learning progressed, and it showed that this tendency was strong especially in the third and final layer filters as seen from the input side. On the other hand, the layer closest to the input side shows little change in the ratio of the transfer filter regardless of the number of epochs, indicating that some layers have different tendencies from the overall tendency. By contrast, Fig. 7 shows that the target filters moved from the output side to the input side, which is opposite tendency from MW method. In addition, this tendency was strong in the last layer and the 5th to 7th layer filters as seen from the input side, and the ratio of the target filter was hardly changed from the first layer to the 4th layer.

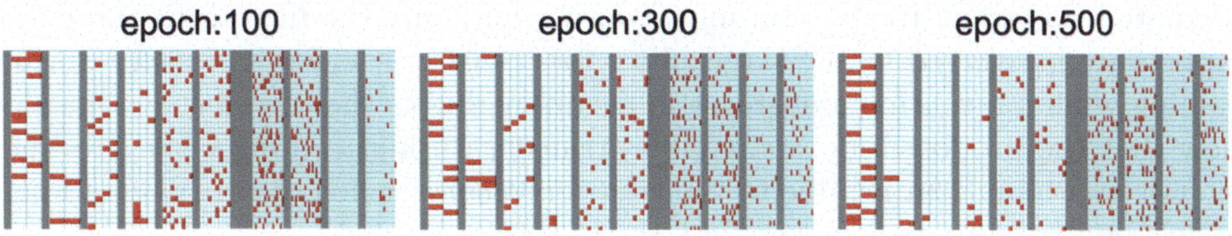

Fig. 6. Transition of the destination filter during learning in the MW method. Vertical white frames represent each convolutional layer, and the leftmost layer represents the input layer. The white frames in the layer represents filters, and the source filters (weight is zero) is painted red.

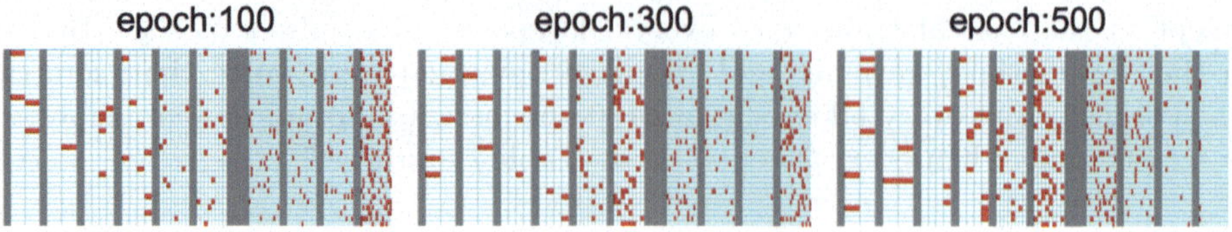

Fig. 7. Transition of the target filter during learning in the TE method.

The problem with our method is that the accuracy may temporarily decrease for a few epochs after pruning. Furthermore, the execution time will increase if the pruning interval is shortened because it is necessary to calculate the importance every time pruning.

Further studies are needed to clarify the causes of differences in the filter importance trends for each layer during the learning process. In this experiment, we used two subsets of CIFAR-100 for the dataset, so the strong relationship between tasks may have affected performance. Therefore, we will examine experiments in the task of reinforcement learning and natural language processing that are not covered in this study and comparisons with the proposed method applied to other models.

5 Conclusion

When Transfer Learning or Fine-Tuning is used for CNN, it is established that good performance is likely to be achieved if the convolutional layers before the identification layers are transferred after training the source task. On the other hand, considering the case of learning multiple tasks in parallel with different models, there is a possibility that the performance may be improved by transferring the knowledge of another task without overwriting the useful knowledge. This paper proposed an online Fine-Tuning method transfers convolutional filters during training in order to verify the effectiveness of transfer during training and high-level transfer. In the proposed method, the importance of the filter is calculated according to the pruning criterion, and only the filter with the lower importance is selected as the target filter for transfer, it enables transfer during training while preventing a decrease in accuracy due to the transfer. The results show that the proposed method enables transfer during training while preventing a decrease in accuracy due to the transfer When applied to two VGG-13-like models and training two subsets of CIFAR-100 in parallel.

Further investigation is required to clarify why the ratio of less important filters in each layer changes during the training process. We also wish to apply the proposed method in multi-task learning with 3 or more tasks and verify the effectiveness of the method in other domains such as reinforcement learning. In addition, it is needed to clarify the examples of classes correctly classified by our method by comparing the correct answer rate for each class with the conventional method.

Acknowledgements. The authors would like to thank Dr. Ryuji Mine, Mr. Tadayuki Matsumura and Dr. Atsushi Miyamoto from Hitachi Ltd., for their feedback. This research is partly supported by the collaborative research program 2018, Hitachi Kyoto University Laboratory, Center for Exploratory Research, Hitachi Ltd. Advices given by M. Sato from Tokai University has been also a great help in writing the paper.

References

1. Pan, S.J., Yang, Q.: A survey on transfer learning. IEEE Trans. Knowl. Data Eng. **22**(10), 1345–1359 (2010)
2. Zeiler, M.D., Fergus, R.: Visualizing and understanding convolutional networks. In: ECCV (2014)
3. Krizhevsky, A., Nair, V., Hinto, G.: The CIFAR-10 dataset. https://www.cs.toronto.edu/kriz/cifar.html. Accessed 12 July 2019
4. Zhu, X.: Semi-supervised learning literature survey. Technical report 1530, Computer Sciences, University of Wisconsin-Madison (2005)
5. Deng, J., Dong, W., Socher, R., Li, L.-J., Li, K., Fei-Fei, L.: ImageNet: a large-scale hierarchical image database. In: IEEE CVPR (2009)
6. Rosenstein, M.T., Marx, Z., Kaelbling, L.P., Dietterich, T.G.: To transfer or not to transfer. In: NIPS 2005 Workshop on Transfer Learning (2005)
7. Zamir, A., Sax, A., Shen, W., Guibas, L., Malik, J., Savarese, S.: Taskonomy: disentangling task transfer learning. In: IEEE CVPR (2018)
8. Fernando, C., Banarse, D., Blundell, C., Zwols, Y., Ha, D., Rusu, A.A., Pritzel, A., Wierstra, D.: Pathnet: evolution channels gradient descent in super neural networks. arXiv preprint arXiv:1701.08734 (2017)
9. Caruana, R.: Multitask learning: a knowledge-based source of inductive bias. In: Proceedings of the Tenth International Conference on machine Learning, University of Massachusetts (1993)
10. Mnih, V., Badia, A.P., Mirza, M., Graves, A., Lillicrap, T.P., Harley, T., Silver, D., Kavukcuoglu, K.: Asynchronous methods for deep reinforcement learning. arXiv preprint arXiv:1602.01783 (2016)
11. Molchanov, P., Tyree, S., Karras, T., Aila, T., Kautz, J.: Pruning convolutional neural networks for resource efficient inference. In: ICLR (2017)
12. Simonyan, K., Zisserman, A.: Very deep convolutional networks for large-scale image recognition. In: ICLR (2015)
13. Xavier, G., Bordes, A., Bengio, Y.: Deep sparse rectifier neural networks. In: AISTATS (2011)
14. Ioffe, S., Szegedy, C.: Batch normalization: accelerating deep network training by reducing internal covariate shift. In: ICML (2015)
15. Srivastava, N., Hinton, G., Krizhevsky, A., Sutskever, I., Salakhutdinov, R.: Dropout: a simple way to prevent neural networks from overfitting. In: JMLR (2014)

Visualizing Deep Q-Learning to Understanding Behavior of Swarm Robotic System

Xiaotong Nie(✉), Motoaki Hiraga(✉), and Kazuhiro Ohkura(✉)

Graduate Shool Engineering, Hiroshima University, Hiroshima, Japan
{nie,hiraga}@ohk.hiroshima-u.ac.jp, kohkura@hiroshima-u.ac.jp

Abstract. Swarm robotic systems (SRS) are a type of multi-robot systems that consist of many homogeneous autonomous robots inspired by social insects. In our pervious study, we succeeded in developing end-to-end control policies for SRS using Deep Q-Network (DQN) algorithm. However, since DQN is totally a black box, it is difficult to understand what were learnt through the learning process. Therefore, in this paper, a novel method of visualizing the decision making process in the DQN is proposed by combining Deconvolutional Network (Deconvnet) and Gradient-weighted Class Activation Mapping (Grad-CAM). Then we show what are being preserved as the deep features and which part of input image is concerned to make an action decision. The proposed method is demonstrated by conducting the computer simulations of a round trip task, in which the swarm robots need to visit two different locations alternatively as many times as possible. The computer simulations might also be explained that the proposed method visualizes the policies learned by DQN.

Keywords: Deep Q-Network · Deconvnet · Grad-CAM · Swarm Robotic System

1 Introduction

Swarm robotic systems (SRS) are a type of multi-robot systems that consist of many homogeneous autonomous robots can be used to accomplish collective tasks that beyond the capability of a single robot, such as aggregation, chain formation or collective decision making [2,3]. There exists no centralized coordination mechanisms behind the synchronized operation. SRS has three desired functional properties: robustness, flexibility and scalability [1].

The design methods in SRS can be mainly divided into two approaches, i.e., behavior-based design method and automatic design method. It has been recognized that automatic design [3] is a promising approach for SRS.

In our previous study [4], we succeeded in proving that Deep Q-Network is also a good approach to SRS problems by developing end-to-end control strategies. However, due to the fact that DQN is totally a black box, it is not easy

H. Sato et al. (Eds.): IES 2019, PALO 12, pp. 118–129, 2020.
https://doi.org/10.1007/978-3-030-37442-6_11

to explain why DQN gives us successful results. Therefore, we propose a novel method of visualizing and interpreting the decision making process of DQN.

In this paper, we present a method of analyzing the latent representations learned by DQN on SRS. The proposed method records the neural activations of the last hidden layer, and then apply Grad-CAM and Deconvolutional Network (Deconvnet) for visualization. Gradient-weighted Class Activation Mapping (Grad-CAM) is an Explainable Artificial Intelligence (XAI) technique [5]. It creates a heat map that highlights important region in the input image, so that we can describe which parts of an input image led the model to its final decision. Deconvnet [6] is adopted to generate the reconstructed image and display what information is being preserved in the deep features. We discuss the interpretability by changing the weight of hyperparameter. We also test our method in a round trip task, where the robotic swarm need to visit two locations alternatively as many times as possible.

The following sections are composed of five sections. Related work is discussed in Sect. 2. Section 3 describes the Deep Q-learning algorithm. Section 4 explains the experimental settings. Simulation results are discussed in Sect. 5. Finally, the conclusion of this paper and the future work are presented in Sect. 6.

2 Related Work

Explainable Artificial Intelligence (XAI) [5] is an emerging field that is dedicated to developing methods to make deep learning more understandable. It contrasts with the concept of the "black box" in machine learning. In the filed of XAI, there are growing attempts have been made on the problem of interpretability with CNNs, largely in classification tasks. Visualizing the internal representation is the common choice, such as Zeiler et al. [6] use Deconvnet to visualize what patterns activate each unit. Another popular choice is saliency map, which can be divided into two main categories: gradient-based methods and perturbation-based methods.

In gradient-based methods, the gradient of the output with respect to the input is used for constructing the saliency maps. It can generate local explanation i.e. explaining network predictions for specific input examples. A simple approach to building understanding in CNNs is Class Activation Map (CAM) [7]. The CAM method has strict requirements on the network architecture, it requires a Global Average Pooling (GAP) [8] layer, and uses this layer's activations, creates a class discriminative heat map that highlights class-specific regions. Unlike CAM, Grad-CAM [9] is applicable to a wide variety of CNN model-families, without architectural changes or re-training. It is also adapted to work with DRL methods.

Perturbation methods modify a given input to understand the importance of individual image regions. Zeiler et al. [6] slide an occluding rectangle across the image and measure the change in the prediction, which results in a heatmap of importance for each occluded region. Fong and Vedaldi [10] revised this technique by introducing blurring/noise into the image. Basically, Perturbation methods quantify how input modification affects the output of the model.

Although explainable technology is relatively undeveloped in DRL, many studies have made outstanding attempts in the past. Mnih et al. apply t-SNE [11] on Atari 2600 game to understand how an agent ranks environments based on the value function in Q-Learning. Recent work by Zahavy et al. [12] has developed tools for explaining DRL policies in visual domains. However, most current works around explainable methods in DRL are based on single agent scenarios. Our approach is focused on explaining how multi-agent learn and execute the learned policy. In multi-agent scenarios, the problem can be described either globally, where the agents are treated as a whole, or locally, that is, describe the problem from the perspective of single agent. In this paper, our method is based on a local view, where the problem is dealt with at single-robot level and does not access global information. In this way, other robots are considered part of the environment. Dynamic environment makes the task more difficult.

3 Method

3.1 Deep Q-Learning

The reinforcement learning problem can be modeled as a Markov Decision Process (MDP) represented by the 5-tuple (S, A, R, P, γ), include a set of states S, a set of actions A, a reward function R, state transition probability function P and discount factor γ. At each timestep, agent observes a state s $\in$ S, take an action a $\in$ A, and gets a reward r $\in$ R then moves to the state s' $\in$ S. The value function $V^{\pi}(s_t)$ is the expected discounted sum of rewards by following policy π from state s at time t, The final objective is to find an optimal policy π*, which maximize the sum of rewards. Similarly, in Q-learning, the action value function, which is also referred to as the Q-function, is used to predict future reward. The update formula is expressed by the equation:

$$Q(s,a) \rightarrow Q(s,a) + \alpha[r + \gamma max_a' Q(s',a') - Q(s,a)] \quad (1)$$

The Deep-Q network was initially proposed in [13], which is able to extract features directly from high-dimensional raw inputs. DQN approximate the optimal Q-function using a Convolutional Neural Network. In the case of performing the optimal strategy, the deep neural network is updated by minimizing the following loss function:

$$L(\theta) = (r + \gamma max_{a'} Q(s',a';\theta) - Q(s,a;\theta))^2 \quad (2)$$

DQN has been extended by the authors in [14] to gain human level results on Atari 2600 games.

4 Experiments

4.1 Implementation Details

The experimental environment is shown in Fig. 1. The field is a rectangular area of 40 m × 30 m surrounded by a 6 m high wall. Two landmarks with different colors are placed on the left and right sides of the field, which show the direction of the destination of the robot. Two orange obstacles are located in the outer area, which limits the robotic swarm to observe and select their paths. In the beginning, 36 robots are located in the center of the field, and the initial destination of each robot is set randomly. Once a robot arrives at its current destination, i.e., get into the corresponding semicircle area, the destination of the robot is automatically set to another one. The goal of the robotic swarm is to travel between two destinations as much as possible.

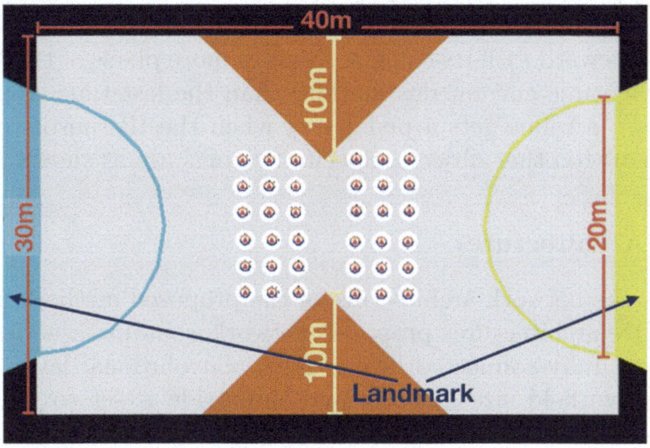

Fig. 1. Experimental environment.

The robot in this experiment is illustrated in Fig. 2. The diameter of robot is 1.0 m. Each robot is equipped with eight IR sensors, one LED, one camera and two motors. Eight IR sensors are placed in different directions to detect the distance between the robot and other objects within 1.0 m. The LED is used to show the current destination of the robot. The camera produces an RGB image of 128 × 128 pixels whose angle of view is horizontal 90 and vertical 90. Figure 2(b) displays an example of an image obtained by the camera. The output is selected from the six target velocities shown in equation (3) and controls motors to achieve the target velocity, v_{right} is the target velocity of the right wheels, v_{left} is the target velocity of the right wheels.

$$
\begin{aligned}
(v_{left}, v_{right}) &= ((0,0),(1,1),(0,1),(1,0) \\
&= (0,0.5),(0.5,0))
\end{aligned}
\tag{3}
$$

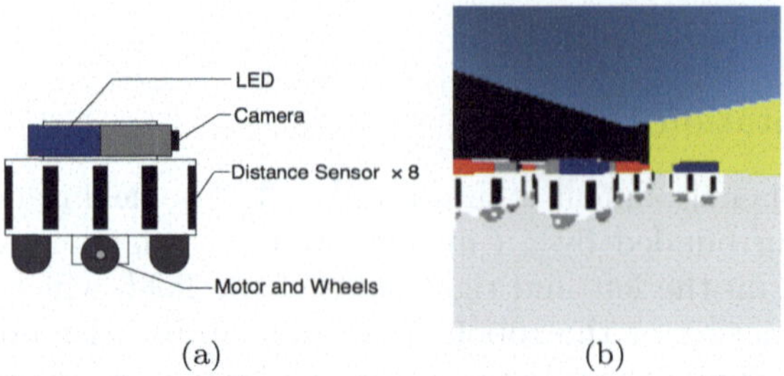

Fig. 2. (a) The robot specification. (b) An example of the images obtained by the camera.

The reward given to the robots at each time step is composed of four parts: (1) destination (r_d); a robot receives the reward r_d when arrives at its current destination. (2) landmark (r_l); a robot gets the reward r_l if its camera observes more pixels of its current landmark than the last time step. (3) fellow robot (r_f); a robot gets the reward r_f if its camera observes more pixels of the LED of fellow robots (with the same current destination) than the last time step. (4) penalty for collision (r_p); a robot gets a penalty r_p when the IR sensors detect a close distance (<0.5 m) to other objects. Specific settings are as shown in Table 1.

4.2 Model Architecture

Figure 3 shows the network architecture of the proposed method. Similar to the traditional DQN architecture, proposed network contains a series of convolutional layers and fully-connected layers. Each convolutional layer has 32 filters of which receptive field size is 4×4 and the stride is set to 2. The input to the convolutional layer is raw RGB pixels from the camera of the past two time steps ($128 \times 128 \times 3 \times 2$). There are three hidden layers, with 63, 30, 14 units respectively. After the last convolutional layer, we add a series of deconvolutional layers to synthesize a reconstructed image with the same size as the input. The output of the last convolutional layer is taken along with the distance information from IR sensors and destination information as inputs by the first fully-connected layer (128 neurons). The second full-connected layer (6 neurons) holds the estimated Q-values corresponding to the six motor patterns.

Table 1. Reward settings

r_d	r_l	r_f	r_p
5.0	1.0	5.0	−1.0

4.3 Loss Function

We define different loss components that can be used to train the network.
Bellman Error ($L_{bellman}$)**:** The usual value function estimation error.

$$L_{bellman}(\theta) = (r + \gamma max_{a'} Q(s', a'; \theta) - Q(s, a; \theta)^2 \tag{4}$$

Reconstruct Error ($L_{reconstructed}$)**:** Reconstruct Error is the mean squared error between reconstructed image s_t^* and original input s_t.

$$L_{reconstructed}(\theta) = \frac{1}{2}||s_t^* - s_t||_2^2 \tag{5}$$

Finally, the model is trained to minimize a weighted linear combination of two losses. Hyperparameter λ is introduced for weighting different loss components.

$$L_{final}(\theta) = \lambda L_{bellman}(\theta) + (1 - \lambda) L_{reconstruct}(\theta) \tag{6}$$

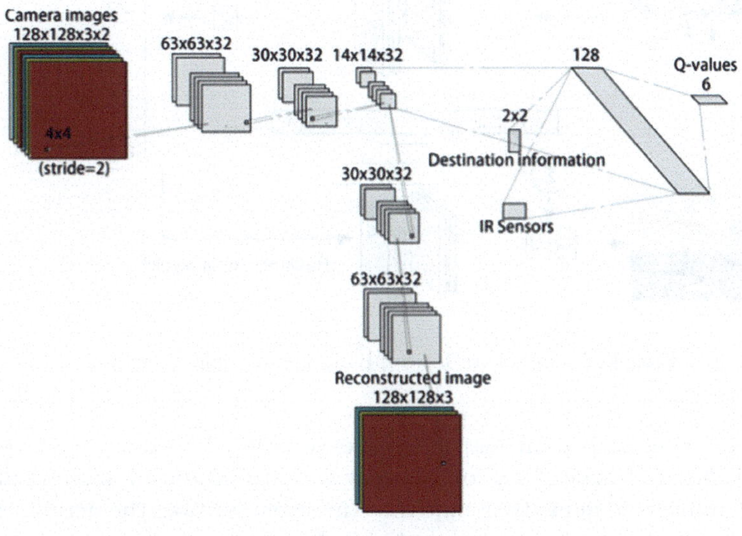

Fig. 3. Architecture of the proposed method.

4.4 Grad-CAM Procedure for DRL

We adopt Grad-CAM as our visualization technique, because it is applicable to any CNN-based architectures and needs no re-training. Grad-CAM is a generic interpretation method that combines gradient information with Class Activation Maps to visualize the importance of each input. In order to apply the Grad-CAM to DRL, it goes through several steps. We first prepared well-trained model in the round-trip task and record the neural activations of the last hidden layer. The Grad-CAM technique computes the gradient of the output Y_c with respect to feature map A^k of the last hidden layer. And then we calculated a global average pooling to obtain weights α_c^k.

$$\alpha_c^k = \frac{1}{Z}\sum_i \sum_j \frac{\partial y^c}{\partial A_{ij}^k} \tag{7}$$

where Z is the size of feature map and k is the number of feature map. α_c^k captures the importance of the activation layer k for predicted action c.

The heatmap $L_{Grad-CAM}^c$ is the weighted combination of feature maps with ReLU.

$$L_{Grad-CAM}^c = ReLU(\sum_k \alpha_c^k A^k) \tag{8}$$

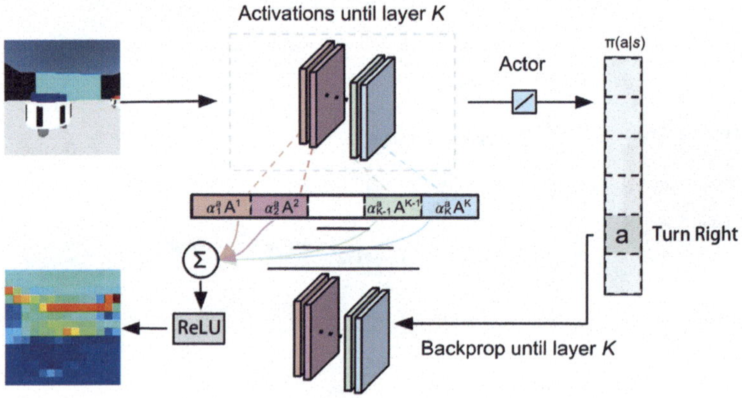

Fig. 4. Architecture for applying Grad-CAM to DQN.

The resulting activation map can be extrapolated to the size of the input state and then overprinted on top of input state to produce a high quality heat map that indicate regions that motivate the agent to take the action. A visual representation of this process is depicted in Fig. 4.

5 Results

In the round-trip task, the robotic swarm forms a chain-like pattern, in which most robots follow fellow robots in front of them and move in a clockwise circle toward their destinations. Figure 5 show the simulation snapshots of behaviors in experiments.

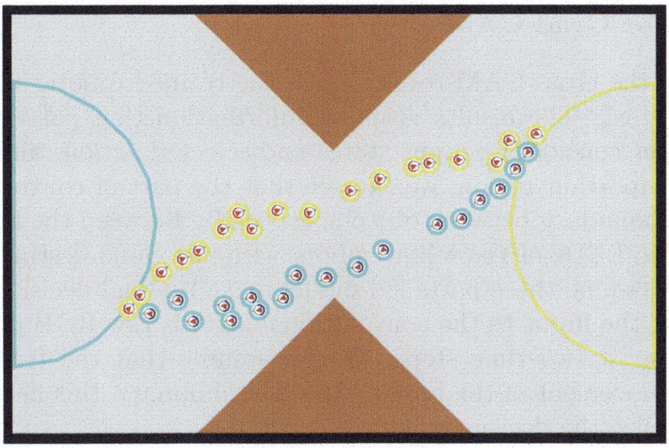

Fig. 5. Simulation snapshots of the best controller developed by DQN.

5.1 Result for Reconstructed Image

In order to examine the interpretability under different loss components weights, three settings are performed in the loss function we mentioned last section.

$$L_{final}(\theta) = \lambda L_{bellman}(\theta) + (1 - \lambda)L_{reconstruct}(\theta) \tag{9}$$

Five trials are performed for each setting. To evaluate the performance of different λ setting, which $\lambda = 1.0, 0.5, 0.1$ respectively, we recorded how many times the robots have arrived at their destinations. Figure 6 show the trajectories of arrivals during the learning process in experiments. In general, we found $\lambda = 1.0$ (which has no interpretability) can work well on round-trip task, where the robots arrived at their destinations 74 times at maximum (averaged by 5 trials). Decreasing λ to 0.5 and 0.1 can get better reconstruction image in visually. The best performance was obtained in the experiment $\lambda = 0.5$, in which robots reached maximum of 81 arrivals. However, decreasing λ to 0.1 results in slower convergence and slight loss in performance, where the maximum arrivals were 69 (drop by 15%). By considering the converge rate, swarm performance and reconstructed image quality, the setting $\lambda = 0.5$ seems can keep the balance between the bellman loss and reconstruct loss. Examples for $\lambda = 1.0, 0.5$ are shown in Figs. 7 and 8. For $\lambda = 1.0$, the reconstructed image contains only some textures that hard to interpret. For $\lambda = 0.5$, with the training process, the interpretability of reconstructed images have improved. As the reconstructed images change, we can observe the learning process of CNN. Figure 8(a) shows the input and reconstructed image in episode 14. CNN can simply segment different regions in the input image, reconstructed image contains color information, which is an improvement over the result of $\lambda = 1.0$. It can be clearly observed from Fig. 8(b), the edges of the landmarks and fellow robot LEDs are preserved by the trained model in deep features.

5.2 Result for Grad-CAM

Figure 9 shows the Grad-CAM results when the trained robotic swarm perform the round-trip task. The results visualize information that passed through the last hidden layer considering input states and selected action. Since the results highlight the important region, we can see that the part of convolutional layers concern is the boundary between objects, especially between the landmarks and the sky, and the LEDs of the fellow robots with the same destination. So that we are able to assume these parts led model make its final decision.

Considering the input to the convolutional layer is raw RGB pixels from the camera of the past two time steps, we can assume that the trained model is observing the movement of the found object. In summary, this heat map can be used to rationalize the decisions made by the DQN.

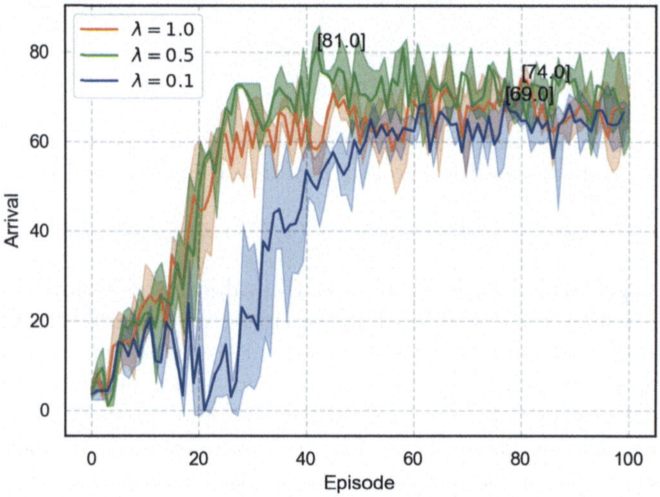

Fig. 6. Trajectories of arrivals, where each data point is the average of 5 trials (with standard deviation).

Comparing Figs. 8(b) and 9, although Deconvnet and Grad-CAM have different aspect of calculation, (i.e. Deconvnet focus on what information preserved by model in deep features, Grad-CAM focuses on which regions that motivate agent to take the selected action) they all reflect Landmark and LEDs of the fellow robots (robots with the same destination) as important information. In addition, unlike Deconvnet, which takes a lot of time to observe the learning process. Grad-CAM can be used in trained models to improve efficiency when testing and optimizing models.

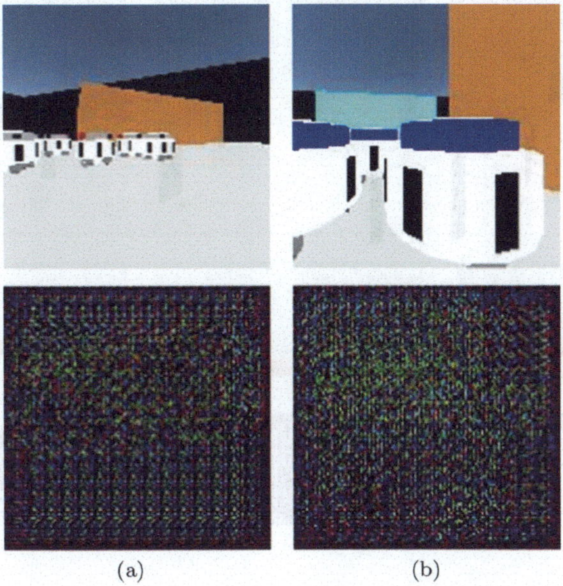

Fig. 7. Experiments using $\lambda = 1.0$. (a) In episode 9. (b) In episode 88. The experiment consists of 100 episodes.

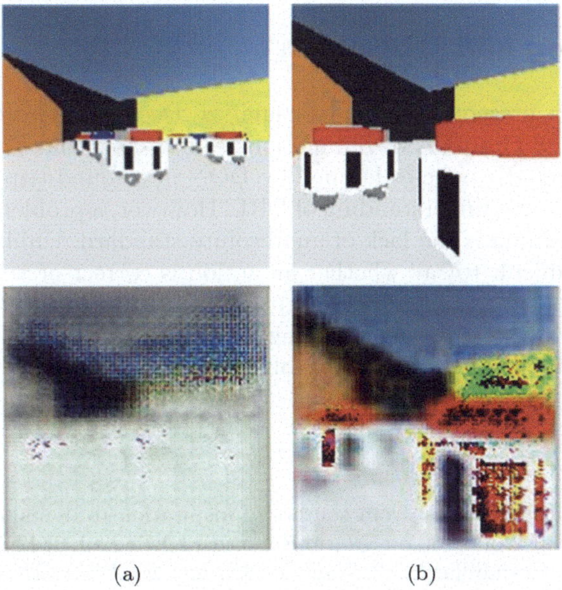

Fig. 8. Experiments using $\lambda = 0.5$. (a) In episode 14. (b) In episode 66. The experiment consists of 100 episodes.

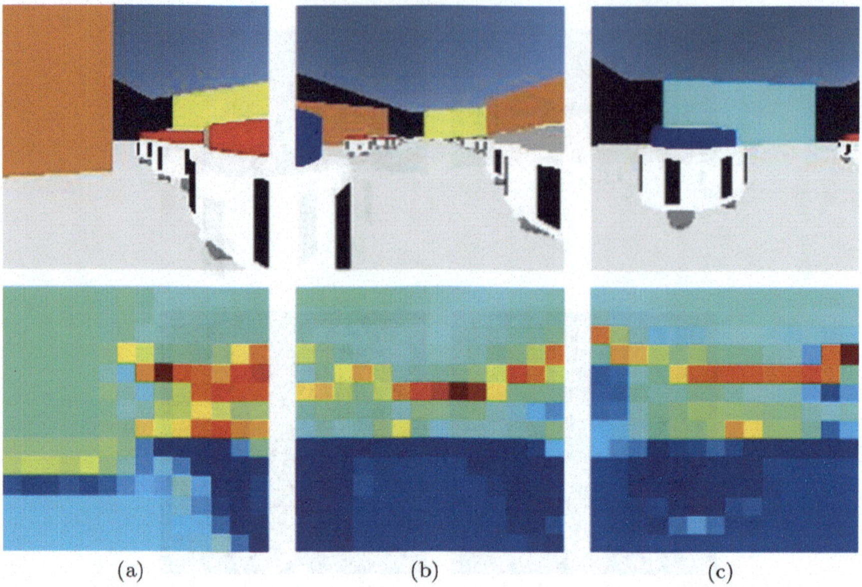

Fig. 9. Grad-CAM results for (a) turn right with half speed, (b) turn left with half speed and (c) turn left with full speed. The Grad-CAM model outputs a class activation map with colors ranging form red to blue. Red/orange indicate a high activation, yellow/green indicate a medium activation and blue indicates a low activation.

6 Conclusion and Discussion

In this paper, we present a method to analyze Deep Q-networks by introducing Deconvnet and Grad-CAM. Simulation results show that proposed method is able to interpret the policies learned by DQN in a round-trip task. It is very valuable for increasing understanding of DRL. However, a problem in explainable reinforcement learning is the lack of an absolute standard. Unlike the classification task, it is difficult to say whether an action is correct or wrong in DRL. It might be an important direction for explainable reinforcement learning in the future. For the next step, we are planning to use the proposed method to debug problems with different DRL model to obtain better results.

References

1. Sahin, E.: Swarm robotics: from sources of inspiration to domains of application. In: International Workshop on Swarm Robotics. LNCS, vol. 3342, pp. 10–20 (2004)
2. Francesca, G., Brambilla, M., Trianni, V., Dorigo, M., Birattari, M.: Analysing an evolved robotic behaviour using a biological model of collegial decision making. In: International Conference on Simulation of Adaptive Behavior, pp. 381–390 (2012)
3. Brambilla, M., Ferrante, E., Birattari, M., Dorigo, M.: Swarmrobotics: a review from the swarm engineering perspective. Swarm Intell. **7**(1), 1–41 (2013)

4. Wei, Y., Nie, X., Hiraga, M., Ohkura, K., Car, Z.: Developing end-to-end control policies for robotic swarms using deep Q-learning. J. Adv. Comput. Intell. Intell. Inf. **23**, 920–927 (2019)
5. Gunning, D.: Explainable artificial intelligence (XAI). Defense Advanced Research Projects Agency (DARPA) (2017)
6. Zeiler, M.D., Fergus, R.: Visualizing and understanding convolutional networks. In: European Conference on Computer Vision (ECCV), pp. 818–833 (2014)
7. Zhou, B., Khosla, A., Lapedriza, A., Oliva, A., Torralba, A.: Learning deep features for discriminative localization. In: IEEE Conference on Computer Vision and Pattern Recognition (CVPR), pp. 2921–2929 (2016)
8. Lin, M., Chen, Q., Yan, S.: Network in network. In: International Conference on Learning Representations (ICLR) (2014)
9. Selvaraju, R.R., Das, A., Vedantam, R., Cogswell, M., Parikh, D., Batra, D.: Grad-CAM: why did you say that? Visual explanations from deep networks via gradient-based localization. arXiv preprint arXiv:1610.02391 (2016)
10. Fong, R.C., Vedaldi, A.: Interpretable explanations of black boxes by meaningful perturbation. arXiv preprint arXiv:1704.03296 (2017)
11. Zahavy, T., Zrihem, N.B., Mannor, S.: Graying the black box: understanding DQNs. In: International Conference on Machine Learning (ICML), pp. 1899–1908 (2016)
12. Van der Maaten, L., Hinton, G.E.: Visualizing data using t-SNE. J. Mach. Learn. Res. **9**, 2579–2605 (2008)
13. Mnih, V., Kavukcuoglu, K., Silver, D., Graves, A., Antonoglou, I., Wierstra, D., Riedmiller, M.: Playing Atari with deep reinforcement learning. arXiv preprint arXiv:1312.5602 (2013)
14. Mnih, V., Kavukcuoglu, K., Silver, D., Rusu, A.A., Veness, J., Bellemare, M.G., Graves, A., Riedmiller, M., Fidjeland, A.K., Ostrovski, G., Petersen, S.: Human-level control through deep reinforcement learning. Nature **518**(7540), 529–533 (2015)

A Study of Upper Tropospheric Circulations over the Northern Hemisphere Prediction Using Multivariate Features by ConvLSTM

Ekasit Phermphoonphiphat[1(✉)], Tomohiko Tomita[2], Masayuki Numao[3], and Ken-ichi Fukui[3]

[1] Graduate School of Information Science and Technology, Osaka University, Suita, Japan
ekasit@ai.sanken.osaka-u.ac.jp
[2] Faculty of Advanced Science and Technology, Kumamoto University, Kumamoto, Japan
t-tomita@kumamoto-u.ac.jp
[3] The Institute of Scientific and Industrial Research, Osaka University, Suita, Japan
numao@sanken.osaka-u.ac.jp, fukui@ai.sanken.osaka-u.ac.jp

Abstract. Spatiotemporal prediction on climate data is aiming to predict future spatial data by learning from prior spatial sequence data. In this paper, we are interested in a prediction of upper tropospheric circulations over the Northern Hemisphere by predicting a geopotential height at 300 hPa (Z300) variable. We proposed a predictive model by constructing an architecture with convolutional layers and deconvolutional layers and applied to convolutional long short-term memory (ConvLSTM) network. The results show that our model obtained root mean square error (RMSE) of 77.36 m (0.84% comparing to average Z300 value) in short-term prediction. While, a convolutional neural network (CNN) and a linear regression (LR) model obtained RMSE of 109.35 (1.19%) and 153.61 (1.67%), respectively. The ConvLSTM maintains RMSE even in long-term prediction. Furthermore, the prediction features' investigation result shows that temperature at 300 hPa (T300) and self prior Z300 features are important for Z300 prediction.

Keywords: Upper tropospheric circulations · Spatial correlation · ConvLSTM

1 Introduction

Due to increase of recent vast amount climate observations, such as satellite image, radar echo, Automated Meteorological Data Acquisition System, radiosonde, etc., a data-centric method including machine learning is currently quite attractive in the field of climate forecasting.

H. Sato et al. (Eds.): IES 2019, PALO 12, pp. 130–141, 2020.
https://doi.org/10.1007/978-3-030-37442-6_12

Many research papers are trying to apply machine learning to predict climate data. Support vector machine controlled autoregressive (SVM-CAR) [12] is used to predict precipitation in Chifeng region, this time series prediction model can predict only for one-step ahead and shows the average value of an entire area; therefore, we do not know exactly where inside the area has high or low precipitation. There are works on time series prediction on temperature [1] and precipitation [11]; they all do not include spatial correlation into the model. The information on space and time should be considered together to predict the value of future spatial data.

Spatiotemporal prediction is a predictive model on space and time, trying to get a spatial output by learning from prior spatial data. We can find examples of spatiotemporal prediction model on climate data such as predicting precipitation [6,14], tracking typhoon [5] and hurricane [2,7]. The important matter on spatiotemporal prediction is a spatial correlation, which is a relationship between datapoint nearby that appears in climate data, the model that can maintain the spatial correlation through the networks will be able to capture the patterns in prediction output.

Convolutional long short-term memory (ConvLSTM) [14] is a predictive model that can capture spatial correlation through the network; moreover, the LSTM [4] structure can remember an important sequence information of prior data. In their work, this model is implemented for precipitation nowcasting in Hongkong from radar echo dataset, and the results show that ConvLSTM outperforms fully connected LSTM (FC-LSTM) [3] and Real-time Optical flow by Variational methods for Echoes of Radar (ROVER) algorithm.

ConvLSTM is widely used as a baseline method for climate prediction and becomes an idea to implement new spatiotemporal predictive models. The predictive recurrent neural network (PredRNN) [19] has new spatiotemporal LSTM (ST-LSTM) unit inside ConvLSTM memory cell and the spatiotemporal memory flow with a zigzag direction which makes the model better to capture temporal information. PredRNN++ [18] was using spatiotemporal memory flow of PredRNN and added a new recurrent structure named Causal LSTM with cascaded dual memories. Moreover, PredRNN++ proposed a gradient highway unit (GHU) to be a quick alternative route for a gradient flow that alleviates gradient propagation difficulties in deep predictive models.

Motivated by ConvLSTM, we need to study the availability of having the LSTM structure inside the CNN network can improve the prediction result. We investigated a new application to predict upper tropospheric circulations over the Northern Hemisphere by constructing an architecture with convolutional layers and deconvolutional layers and applied to ConvLSTM model to improve the capability of capturing spatiotemporal climate data. We are interested in the prediction of geopotential height at 300 hPa (Z300) on the Northern Hemisphere which is hight of polar jet stream (9–12 km above the sea level) because Z300 relates to position of a polar jet stream that controls the strength and the path of mid-latitude cyclones.

The atmospheric variation in the upper troposphere is used for the weather prediction at the surface because the upper-tropospheric circulations can be precursors of the weather prediction [17]. Moreover, the change of upper troposphere can cause the extreme winter cold spells [16] over northern Europe and the cold air outbreaks in the Northern Hemisphere [8].

In this paper, we have experimented short- and long-term prediction to predict Z300 comparing with 3 models; ConvLSTM, convolutional neural networks (CNN), and linear regression (LR) model. We also investigated the importance of prediction features that affect to the prediction accuracy.

2 Convolutional LSTM

ConvLSTM [14] has been developed and extended from fully connected LSTM (FC-LSTM) by making every gate in LSTM memory cell to be a 3D tensor that is able to maintain spatial correlation through the network. The advantage of LSTM structure inside ConvLSTM is an ability to remember important information while learning from long-sequence data.

Unlike recurrent neural network (RNN) that the error term in the backpropagation process of the long-term recurrent network is multiplied by less than one for many times along the recurrent network, causing long-term of the error term will not be back-propagated, this problem is called "vanishing gradient problem."

The spatial input data will be fed to ConvLSTM in each time step and can be seen as 3D tensor, where the first dimension is temporal information, and the last two dimensions are spatial information. Convolution operator has been used in this network to capture spatial correlation in every gate of the networks. Equation (1) are equations of ConvLSTM, where W_x and W_h denote set of weights (kernel) of current input and hidden layer, respectively. X_t, C_t, H_t, and gates i_t, f_t, o_t are input, cell output, hidden state, input gate, forgot gate and output gate, respectively, which are all 3D tensors, 'σ' denotes an activation function, '$*$' denotes a convolution operator and '$\circ$' denotes Hadamard product:

$$\begin{aligned} i_t &= \sigma(W_{xi} * X_t + W_{hi} * H_{t-1} + W_{ci} \circ C_{t-1} + b_i) \\ f_t &= \sigma(W_{xf} * X_t + W_{hf} * H_{t-1} + W_{cf} \circ C_{t-1} + b_f) \\ C_t &= f_t \circ C_{t-1} + i_t \circ \tanh(W_{xc} * X_t + W_{hc} * H_{t-1} + b_c) \\ o_t &= \sigma(W_{xo} * X_t + W_{ho} * H_{t-1} + W_{co} \circ C_t + b_o) \\ H_t &= o_t \circ \tanh(C_t) \end{aligned} \tag{1}$$

We applied zero-padding to remain an output to be the same size as the input after convolution operation inside the network. All hidden states were initialized as zero to ignore the state before training.

3 Dataset

We use ERA-Interim[1] dataset from the European Centre for Medium-Range Weather Forecasts (ECMWF) which is a global atmospheric reanalysis from 1979 as the main dataset. This dataset contains a monthly average of climate variables in spatial grid data. We also use NINO 3 SST Index which is anomaly area-averaged sea surface temperature (SST) from 5°S–5°N and 150°–90°W, that indicates El Niño event. We selected four relevant features for Z300 prediction: temperature at 300 hPa (T300), T850, Z300, and NINO 3 index. All features are extracted to 360×90 image resolution, and 480 months started from 1979 until 2018.

4 Experiment

We study the prediction availabilities of Z300 on the Northern Hemisphere, which has a spatial correlation for an entire area. It will be difficult for a model that is not able to capture spatial correlation to turn out the accurate pattern on prediction output. We experiment on three models as follows:

1. Linear regression model (LR): This model is the most straightforward linear prediction algorithm that works well on linear relationship data. This model is considered as a baseline method for linear prediction algorithm. In this study, we trained separated 360×90 LR models so that the model cannot capture the spatial correlation of the input data. We train the model by learning from 300 months prior to predict the next month. In conclusion, this model uses 4 features of 300 months, i.e., 1,200 prediction features.
2. Convolutional neural networks (CNN) [9]: We use four convolutional layers to encode four channels of input data and decode it with four deconvolutional layers to get one output channel. 3×3 kernel size is applied to this model. The entire structure is shown in Fig. 1 but without ConvLSTM structure.
3. ConvLSTM: We use the same structure as CNN, but in the middle between convolutional and deconvolutional layers is ConvLSTM structure as shown in Fig. 1.

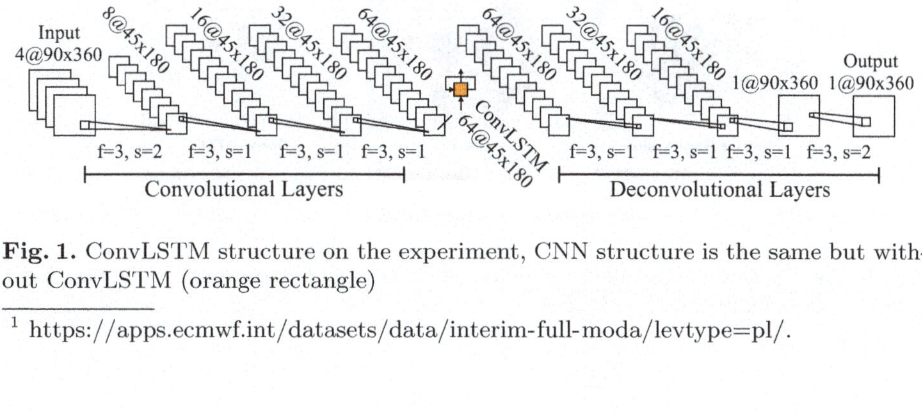

Fig. 1. ConvLSTM structure on the experiment, CNN structure is the same but without ConvLSTM (orange rectangle)

[1] https://apps.ecmwf.int/datasets/data/interim-full-moda/levtype=pl/.

The input data that we feed into both CNN and ConvLSTM are spatial grid data. NINO 3 index is a scalar variable in each month, in order to apply the NINO 3 index to be a spatial grid data, we use NINO 3 variable as the same value for all grid points. In this experiment setting, CNN learns one-month data to predict next month, but ConvLSTM learns six-months to predict next month. The six-months input has been considered because it is a long period enough to the change of seasons. We trained the models with 462 months as a training set and 18 months as a test set aiming to predict 12 months in 2018. We trained the models until converged or reached the limit of 200,000 steps. We applied Dropout to the models to reduce overfitting and used adaptive moment estimation optimizer (Adam optimizer) as an optimizer.

5 Result

5.1 Short-Term Prediction

Fig. 2a shows the output result of April 2018, which is the best month of ConvLSTM method. LR does not capture Z300 pattern properly, while CNN and ConvLSTM perform better in capturing the spatial pattern.

The spatial error distribution of CNN (Fig. 2b) clearly indicates that the error distribution of CNN is large at the paths of the subtropical and polar jets where the meridional Z300 gradient is large, and the variability of Z300 is also large.

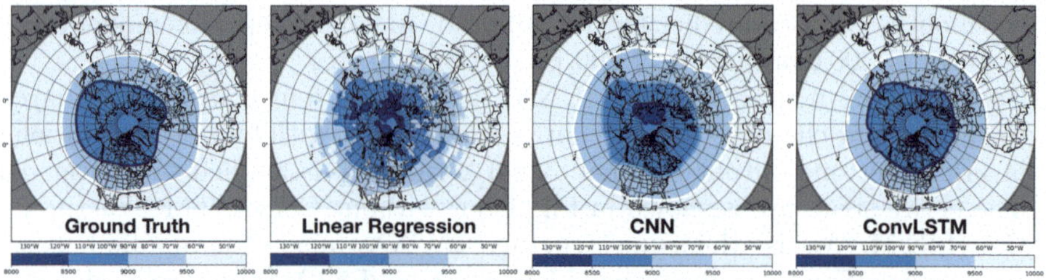

(a) Ground truth and prediction result

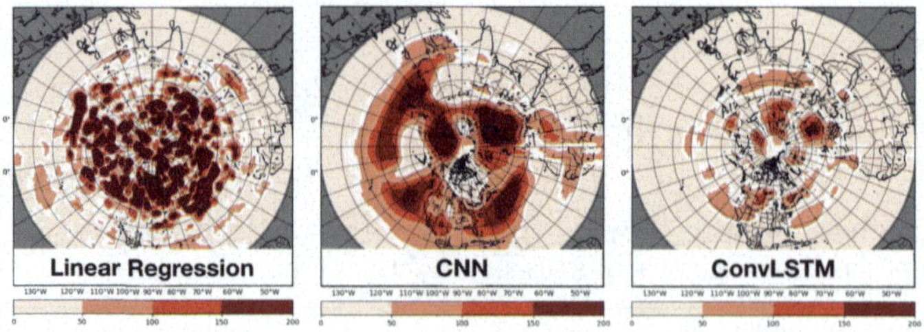

(b) Error distribution

Fig. 2. The Northern Hemisphere Z300 on prediction and error distribution on April, 2018 (the best month of ConvLSTM)

The large variability of Z300 is coherent with the strong activity of extratropical cyclones there.

Thus, the CNN prediction with the one-month earlier data can not predict the variability of the subtropical and polar jet streams correctly, which is critical for climate prediction. The spatial error distribution of ConvLSTM, on the other hand, shows much smaller errors even on the paths of the subtropical and polar jets. It is conceivable that this improvement was achieved by considering the earlier six-months' data and these temporal variations in the ConvLSTM layer.

We also tried six-months input to CNN, which is equivalent to ConvLSTM input. The four features of six-months spatial data are arranged as the sequential input of CNN. The result was worse than learning with one-month in short- and long-term prediction. This might because of the 24-channels of input data is too much for the network, and the network can not capture a transition of the same feature in different month input.

The spatial error distribution of LR seems to be like the distribution of white noise in mid- and high-latitudes, and does not show a systematic distribution pattern. This distribution is probably because the LR prediction of this study did not include the spatial correlation.

Table 1. RMSE (meters) of 12 months in 2018.

Month	LR	CNN	ConvLSTM
1	156.40	57.11	78.73
2	164.15	112.60	117.36
3	143.30	113.76	98.60
4	145.93	117.70	58.12
5	155.28	142.50	86.86
6	143.54	127.54	61.41
7	153.44	130.47	76.97
8	140.94	68.35	59.97
9	162.33	97.65	84.04
10	148.86	111.71	51.53
11	158.88	120.64	76.12
12	170.32	112.13	78.60
Average	**153.61**	**109.35**	**77.36**

Figure 3 shows the result of February 2018, which is the worst month of ConvLSTM. The error distribution on Fig. 3b shows that high error of CNN and ConvLSTM are in the same area. It might because of a Sudden Stratospheric Warming (SSW) that occurred on February, 2018 [13] that exceed the limit of CNN and ConvLSTM model to capture such kind of change.

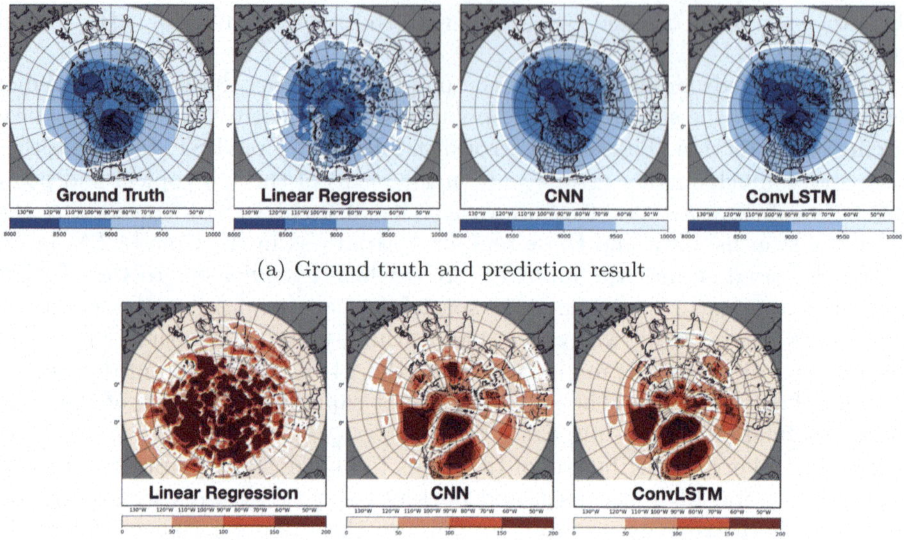

(a) Ground truth and prediction result

(b) Error distribution

Fig. 3. The Northern Hemisphere Z300 on prediction and error distribution on February, 2018 (the worst month of ConvLSTM)

Table 1 is RMSE of 12 months (compare to the ground truth) in 2018. The result shows that ConvLSTM outperforms with RMSE of 77.36 m (0.84% comparing to average Z300 value of 9,225.19 m) over CNN and LR that have RMSE of 109.35 (1.19%) and 153.61 (1.67%) m, respectively. LSTM network inside ConvLSTM helps the networks remember important information for a long period of time. However, the RMSE in each month of CNN and ConvLSTM do not indicate high accuracy on certain months.

We plotted the predicted value minus the ground truth for all months in 2018 to investigate over and underestimation of CNN and ConvLSTM models, as shown in Fig. 4. The overestimated months mostly appear in winter and underestimated months appear in summer for CNN. On the other hand, the ConvLSTM showed the over and underestimated at about the same amount in each month (the amount of red and blue area are similar for every month). Simple CNN would not even have an ability to predict the regular (or climatological) seasonal change in Z300 which increases to summer and decreases to winter over the Northern Hemisphere polar region.

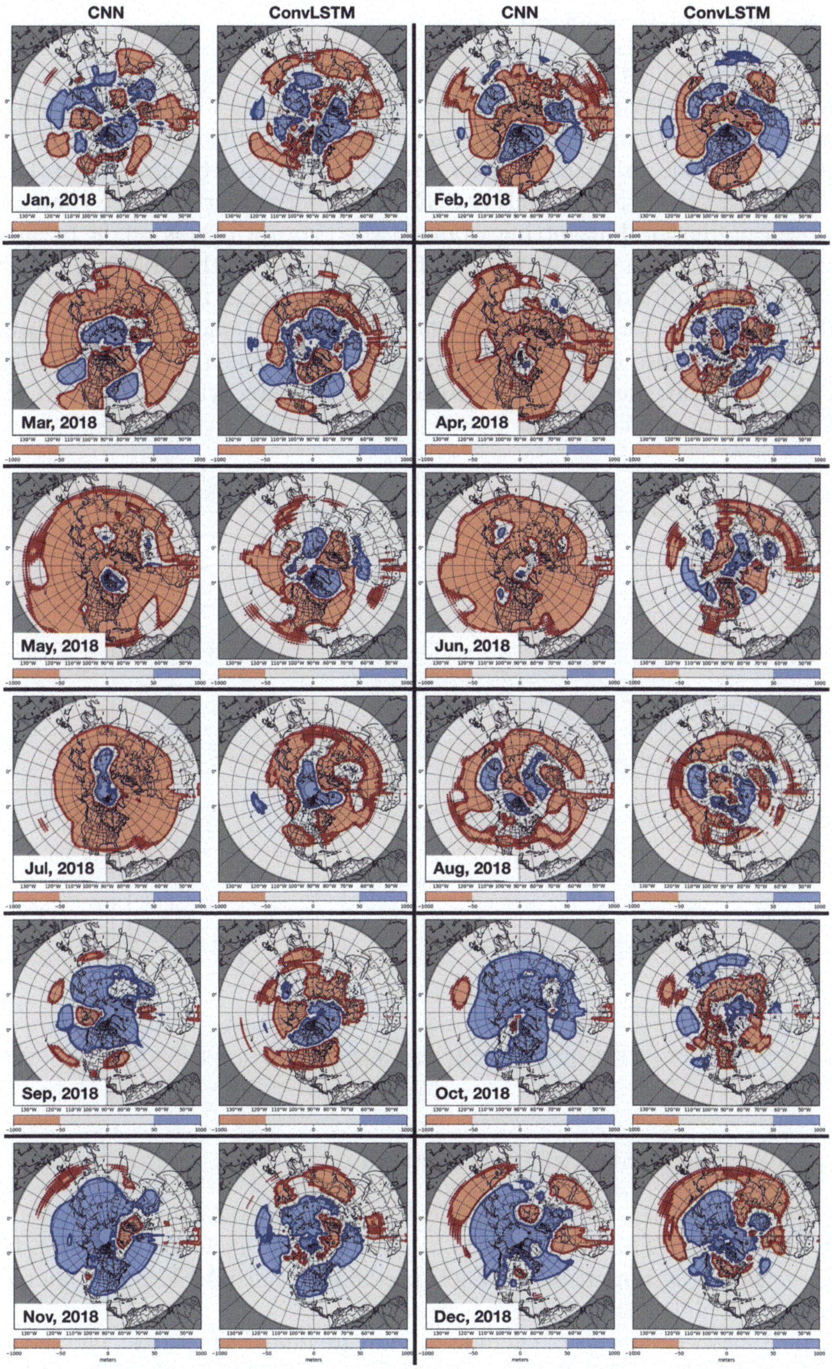

Fig. 4. The predicted value minus ground truth in every month of 2018 (blue color is overestimated and red color is underestimated). In every month, left figure is CNN and right figure is ConvLSTM.

5.2 Long-Term Prediction

Next, we have applied the models to predict the next 1–5 months. The result is shown in Fig. 5, RMSE of ConvLSTM is only slightly increased, showing the robustness over the long-term prediction. The RMSE of ConvLSTM is less than 1% to average Z300 value even for long-term prediction (the next 5 months prediction has RMSE of 1.02%, only). CNN has a better performance than LR in short-term prediction but getting worse in the long-term, this might because of the number of features for LR is 1,200 features (300 months prior × 4 features), while CNN uses only one month of 4 features. Although the LR maintains the RMSE well in long-term prediction, the output result shows poorly capturing Z300 pattern. We also tried 6 months input to CNN, which is equivalent to ConvLSTM input, but the result was worse than learning with 1 month in short- and long-term prediction.

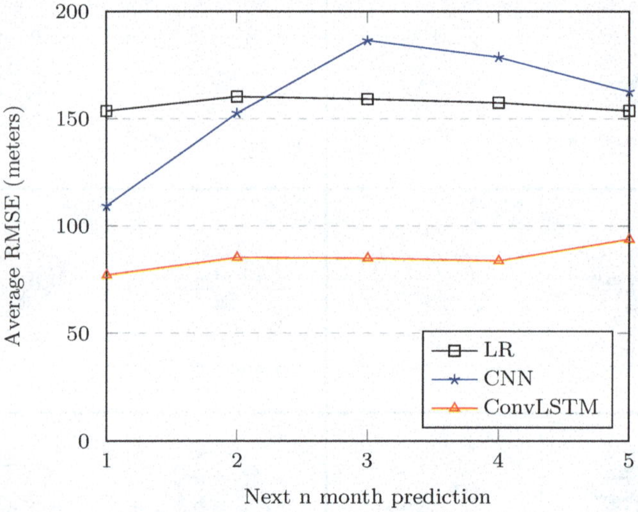

Fig. 5. RMSE of long-term prediction

5.3 Prediction Features' Investigation

We also investigate the importance of features to predict Z300. We run an experiment with ConvLSTM on 4 datasets; each of them is excluded 1 feature. If the excluded feature is important to Z300 prediction, the RMSE result should be worse (RMSE higher than using 4 features).

Table 2 shows the result of excluding 1 feature out of 4 features. The standard deviation (SD) shows the dispersion to the average RMSE value. By excluding T300 or Z300, which are temperature at the same hight as Z300 and prior Z300, have higher RMSE of 79.42 and 81.30 m than that of using all the 4 features, respectively. Therefore, these 2 features are relevant to Z300 prediction. On the other hand, excluding T850 shows the lower RMSE of 74.59 m, which means that T850 interferes with Z300 prediction and should be excluded from the dataset.

Table 2. Excluded features result of average RMSE of 12 months in 2018.

Dataset (Exclude)	Avg. RMSE (m)	SD of RMSE (m)
Z300	81.30	17.21
T850	74.59	15.11
T300	79.42	17.32
NINO 3	77.08	18.13
Includes all 4 features	**77.36**	**18.55**

Moreover, SD of excluding T850 is lowest that shows the error in every month is close to the average RMSE, which is good for prediction (low RMSE and low SD). The result of excluding NINO 3 has about the same RMSE of 77.08 comparing to included 4 features, which seems less relevance to Z300 prediction. However, it might be the cause of how we implemented the scalar value of NINO 3 to be a spatial grid data that is the same value on every data point; different implementation of NINO 3 index may affect to the result.

Lastly, we run another experiment using ConvLSTM to predict Z300 in short- and long-term prediction (next 1–5 months prediction) by using the most 2 relevant prediction features (T300 and Z300). The result shows that using 2 relevant prediction features obtained the lower average RMSE of 82.92 (0.9%) comparing to average RMSE of 85.194 (0.92%) using 4 features.

6 Conclusion

In this research, we studied the prediction of upper tropospheric circulations over the Northern Hemisphere by predicting Z300, which relates to the polar jet stream affecting the strength and the path of mid-latitude cyclones. Our constructed predictive model ConvLSTM outperforms CNN and LR models in short- and long-term predictions. This result shows that ConvLSTM can maintain spatial correlation and remember important information through the network.

The spatial error distribution of prediction output shows that ConvLSTM has much smaller errors at the paths of subtropical and polar jets. It is conceivable that by considering the earlier six-months' data as the input of ConvLSTM and temporal variations in the ConvLSTM layer can achieve the accuracy improvement comparing to CNN and LR models.

We investigated the importance of features to predict Z300. The result shows that T300 and self prior of Z300 are important features. Moreover, using only the important features can increase accuracy, which achieved less than 1% RMSE against Z300 original value even for short- and long-term prediction.

As for future work, not only the prediction accuracy but also spatial pattern would be important to interpret the output prediction results. The prior studies have already revealed the spatial patterns predominant in the upper tropospheric circulations, such as North Atlantic Oscillation (NAO) [15], West Pacific (WP) [10] patterns and so on [17].

Acknowledgment. This work was supported in part by the Network Joint Research Center for Materials and Devices and by JSPS KAKENHI Grant Number 19K22876.

References

1. Chen, P., Niu, A., Liu, D., Jiang, W., Ma, B.: Time series forecasting of temperatures using SARIMA: an example from Nanjing. In: IOP Conference Series: Materials Science and Engineering (2018)
2. Giffard-Roisin, S., Yang, M., Charpiat, G., Ke′gl, B., Monteleoni, C.: Fused deep learning for hurricane track forecast from reanalysis data. In: Proceedings of International Workshop on Climate Informatics, pp. 69–72 (2018)
3. Graves, A.: Generating sequences with recurrent neural networks. arXiv:1308.0850 (2013)
4. Hochreiter, S., Schmidhuber, J.: Long short-term memory. Neural Comput. **9**(8), 1735–1780 (1997)
5. Hong, S., Kim, S., Joh, M., Kwang Song, S.: GlobeNet: convolutional neural networks for typhoon eye tracking from remote sensing imagery. In: Proceedings of International Workshop on Climate Informatics, pp. 69–72 (2017)
6. Kim, S., Hong, S., Joh, M., Kwang Song, S.: DeepRain: ConvLSTM network for precipitation prediction using multichannel radar data. In: Proceedings of International Workshop on Climate Informatics, pp. 89–92 (2017)
7. Kim, S., Yoon, S., Lee, J., Kahou, S., Kim, H., Kashinath, K., Prabhat, M.: Deep-Hurricane-Tracker: tracking extreme climate events. In: Proceedings of International Workshop on Climate Informatics, pp. 16–18 (2018)
8. Kolstad, E.W., Breiteiga, T., Scaife, A.A.: The association between stratospheric weak polar vortex events and cold air outbreaks in the northern hemisphere. Q. J. R. Meteorol. Soc. **136**, 886–893 (2010)
9. Krizhevsky, A., Sutskever, I., Hinton, G.E.: Imagenet classification with deep convolutional neural networks. In: Neural Information Processing Systems, pp. 1097–1105 (2012)
10. Linkin, M.E., Nigam, S.: The North Pacific oscillation-West Pacific teleconnection pattern: mature-phase structure and winter impacts. J. Clim. **21**, 1979–1997 (2008)
11. Papacharalampous, G.A., Tyralis, H., Koutsoyiannis, D.: Predictability of monthly temperature and precipitation using automatic time series forecasting methods. Acta Geophysica **66**(4), 807–831 (2018)
12. Qing, C., Xiaoli, Z., Kun, Z.: Research on precipitation prediction based on time series model. In: Computer Distributed Control and Intelligent Environmental Monitoring, pp. 568–571 (2012)
13. Rao, J., Ren, R., Chen, H., Yu, Y., Zhou, Y.: The stratospheric sudden warming event in February 2018 and its prediction by a climate system model. J. Geophys. Res. Atmos. **123**, 332–345 (2018)

14. Shi, X., Chen, Z., Wang, H., Yeung, D.Y.: Convolutional LSTM network: a machine learning approach for precipitation nowcasting. Neural Inf. Process. Syst. **1**, 802–810 (2015)
15. Thompson, D.W.J., Wallace, J.M.: The arctic oscillation signature in the wintertime geopotential height and temperature fields. Geophys. Res. Lett. **25**, 1297–1300 (1998)
16. Tomassini, L., Gerber, E.P., Baldwin, M.P., Bunzel, F., Giorgetta, M.: The role of stratosphere-troposphere coupling in the occurrence of extreme winter cold spells over northern europe. J. Adv. Model. Earth Syst. **4**, 1–14 (2012)
17. Tomita, T., Yamaura, T.: A precursor of the monthly-mean large-scale atmospheric circulation anomalies over the North Pacific. SOLA **13**, 85–89 (2017)
18. Wang, Y., Gao, Z., Long, M., Wang, J., Yu, P.S.: PredRNN++: towards a resolution of the deep-in-time dilemma in spatiotemporal predictive learning. In: Proceedings of International Conference on Machine Learning, pp. 5123–5132 (2018)
19. Wang, Y., Long, M., Wang, J., Gao, Z., Yu, P.S.: PredRNN: recurrent neural networks for predictive learning using spatiotemporal LSTMs. In: Neural Information Processing Systems, pp. 879–888 (2017)

Learning Localization Skills of Soccer Robot Using by Simulated Omni-Vision Camera

Naoya Horio(✉), Masao Kubo, and Hiroshi Sato

Department of Computer Science, National Defense Academy of Japan, 1-10-20 Hashirimizu, Yokosuka, Kanagawa 239-8686, Japan
em58032@nda.ac.jp

Abstract. In the RoboCup, each robot must perform localization, strategy selection, etc. within its limited resources. For this reason, the problem is that the programming cost is high for performing optimization according to the characteristics of the opponent, role, and environment so that resources can be utilized to the limit. Therefore, in order to reduce the cost of this tuning, we will consider a transfer learning method using a simulator. Since a soccer field can be regarded as a semi-structured environment, we study how to learn these features using CNN. Also, we think a method for generating test images used for CNN learning, in particular, a method for generating omnidirectional images using the simulator. We performed a simple computer experiment for localization using omnidirectional images and CNN, and then examined learning data suitable for tuning.

Keywords: RoboCup · Semi-structured environment · CNN · Transfer learning · V-REP

1 Introduction

In the RoboCup, each robot must perform localization, strategy selection, catch a ball, pass, shoot, and so on. within its limited resources. For this reason, the problem is that the programming cost is high for performing optimization according to the characteristics of the opponent, role, and local condition so that resources can be utilized to the limit. Therefore, we expect to reduce the tuning cost by conducting pre-learning in a simulated environment, considering the localization of the robot as an image identification problem using CNN. The reason for using CNN is that there is a possibility that the tuning cost can be quantified from the time required when learning is performed in the real environment using learning results in pre-learning environments [1]. In this study, we consider the method for generating learning data and an environment suitable for pre-learning.

In this paper, as a first step, we performed a simple computer experiment for localization using omnidirectional images such as [2–5]. A soccer field can be regarded as a semi-structured environment with highly consistent structures in the environment while lights, audience, etc. are changing dynamically. So, we examined requirements suitable for tuning by changing the color of the wall at the boundary with the field.

In the following, we will first explain the RoboCup middle size league and the V-REP, which is the simulator to be used, and then explain a method for generating

H. Sato et al. (Eds.): IES 2019, PALO 12, pp. 142–155, 2020.
https://doi.org/10.1007/978-3-030-37442-6_13

test images used for CNN learning. In particular, we propose the method for simulated omnidirectional images often used in RoboCup using V-REP. Next, a simple computer experiment is performed and it is shown that simple localization is possible, and then learning data suitable for tuning is examined.

1.1 RoboCup Middle Size League

This paper treat RoboCup Middle Size League[msl.robocup.org], this is a league of autonomous mobile robots. In a 22 m × 14 m field, teams of five fully autonomous robots less than 50 cm in length and width play soccer with a regular size FIFA soccer ball. Figure 1 shows illustrations the feature point of the RoboCup soccer field. Although there are differences in the soccer field for each stadium, such as the venue, the audience, the color of the turf, and the size, the soccer field is a semi-structured environment with common features as shown in Fig. 1. These *features* can be used to estimate the position of the robot. In the following, we will conduct research to estimate the current pose of the robot by combining the relationship with each feature and the relative distance obtained from the omnidirectional image.

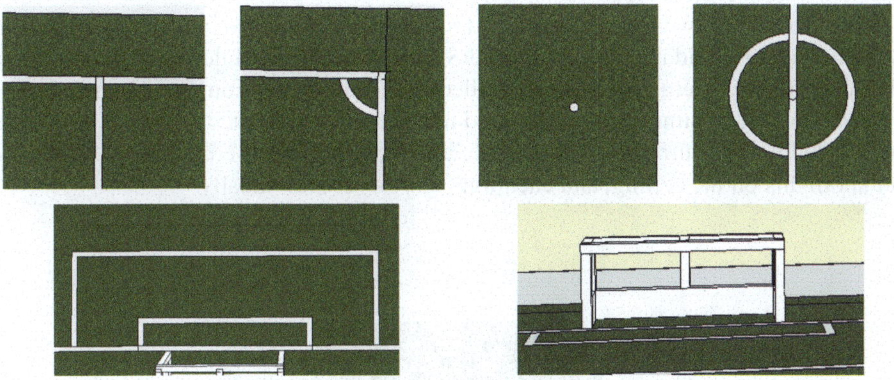

Fig. 1. Feature points of soccer field

Figure 2 shows the robot used in this league. The size of the robot is about 43 cm × 43 cm × 72 cm. Most soccer robots currently in use are equipped with a camera system that can look at all directions at once (camera mounted upward hyperboloidal mirror) at the top, and move by judging the relationship between itself and the opponents and the ball based on the information obtained by this camera system. Figure 3 shows an example of a real image using the omni-vision camera with hyperboloidal mirror. The camera can shoot all directions at once, centering on itself. However, things that are far away are difficult to be seen because they are reduced. In the following, this image centering on camera is called the *original* image.

Fig. 2. A soccer Robot

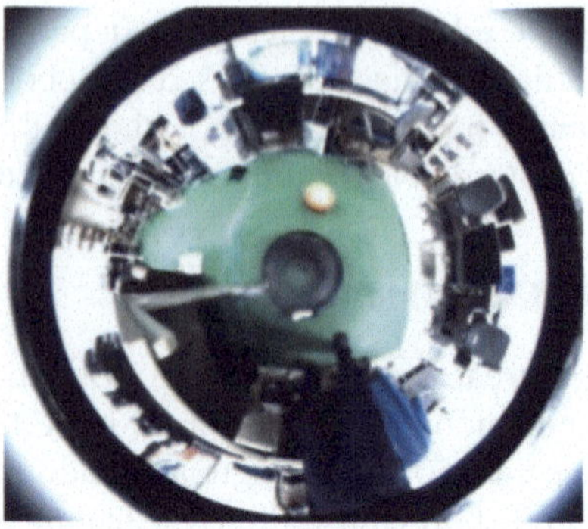

Fig. 3. The image by omni-vision camera system

1.2 V-REP

In this study, we use V-REP[www.coppeliarobotics.com] for simulation. V-REP is a robot simulator with an integrated development environment provided by Coppelia Robotics. Each object in this simulator can be controlled via embedded scripts, plug-ins, ROS, BlueZero nodes, remote API clients, etc.

In the following, we propose a method to reproduce the above an omni-vision camera with a hyperboloidal mirror using the SphericalVision module of this simulator, a spherical virtual camera function is installed in the simulator from the beginning. The omni-vision camera simulated is evaluated in a soccer field created on V-REP shown in Fig. 4. This field is surrounded four walls and a ceiling of different colors. There are four spotlights on the ceiling, and each light can be set individually.

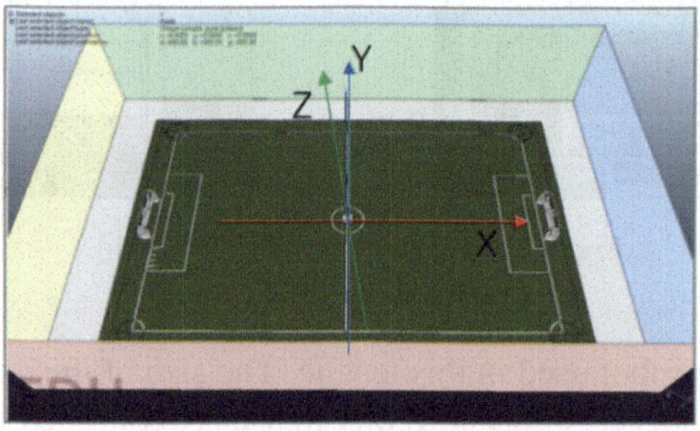

Fig. 4. The soccer field created in the simulation

2 The Proposed Method

The purpose of this study is to examine the method for generating learning data used for transfer learning and the environment suitable for pre-learning in order to reduce the tuning cost. So first, we propose the method for simulating an omnidirectional image using V-REP, which is CNN training data. The following describes the configuration of the CNN used.

2.1 Optical Properties of the Omni-Vision Camera

The appearance of the omni-vision camera that is mainly used in this study is shown in Fig. 5. Figure 6 shows the relationship between an arbitrary point P (X, Y, Z) in space and the mapping p (u, v) by the omni-vision camera. In the figure, f represents the focus of the camera.

Fig. 5. The omni-vision camera to be simulated

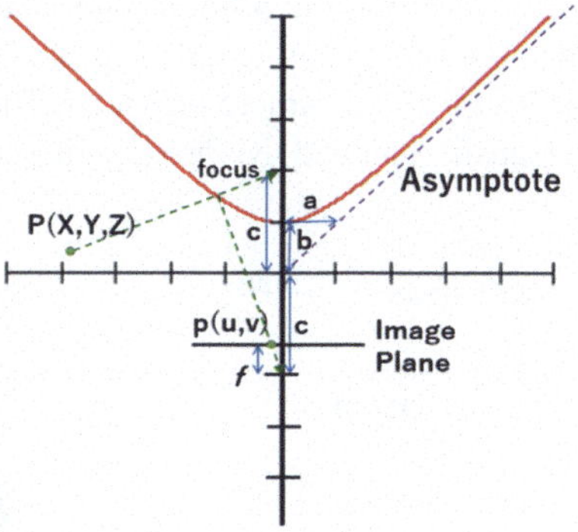

Fig. 6. Mapping by the omni-vision camera.

The hyperboloid in Fig. 6 is expressed by the following equation.

$$\frac{x^2 + y^2}{a^2} - \frac{z^2}{b^2} = -1 \tag{1}$$

$$c^2 = a^2 + b^2 \tag{2}$$

The mapping p (u, v) can be obtained by the following equation from an arbitrary point P (X, Y, Z) in the space and parameters of the omni-vision camera.

$$u = \frac{a^2 f X}{-(b^2 + c)Z + 2bc\sqrt{X^2 + Y^2 + Z^2}} \tag{3}$$

$$v = \frac{a^2 f Y}{-(b^2 + c)Z + 2bc\sqrt{X^2 + Y^2 + Z^2}} \tag{4}$$

There is the following relationship between an arbitrary point P (X, Y, Z) in the space and the map p (u, v).

$$\tan\theta = \frac{Y}{X} = \frac{v}{u} \tag{5}$$

2.2 Overview of SphericalVision Camera

V-REP has a spherical camera module SphericalVision. This virtual camera is composed of a total of 7 cameras, and the images of the front, back, left, right, top and bottom seen from the 6 cameras are integrated into one camera so that all directions can be seen.

2.3 Generation Process of SphericalVision Camera

The images from each camera of SphericalVision are designed to be projected onto each surface of a 2 m cube that surrounds a sphere with a radius of 1 m centered on the origin. Figure 7 shows an image of transcription. Using this relationship, the intersection point mapP (xm, ym) between the extension line of the point (X, Y, Z) on the spherical surface and each surface mapV of the cube can be easily obtained. The camera image corresponding to the spherical surface is re-projected on the 2D plane of distance-relative angle and provided to the robot as one image.

In the following, this is called a *panoramic* image. Figure 8 shows a panoramic image of the spherical camera provided. These processes are implemented by a program CVisionTransf::_calculateMapping() in V-REP.

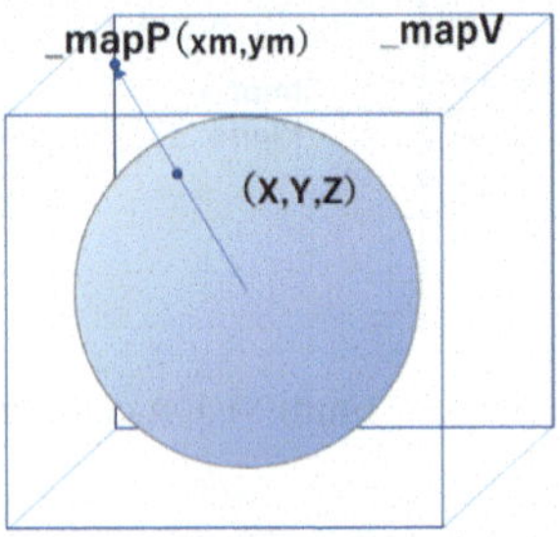

Fig. 7. Image of transcription by SphericalVision camera

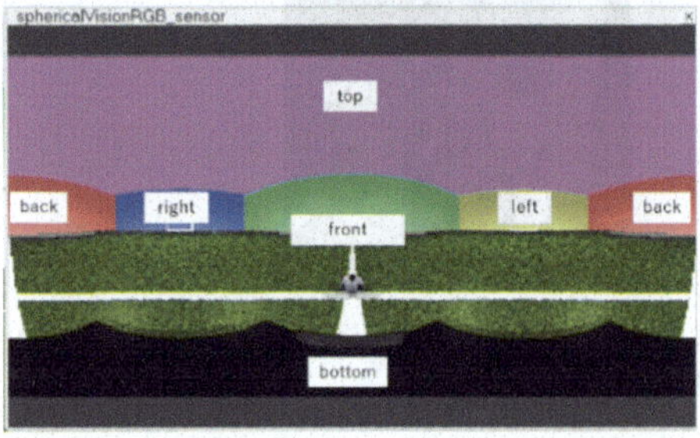

Fig. 8. A panoramic image of the spherical camera provided

2.4 The Proposed Method to Create the Omni-Vision Camera

Figure 9(a) shows the proposed the omnidirectional image production method, and Fig. 9(b) shows the relationship between mapP and mapP's coordinates P′ (x, y, z).

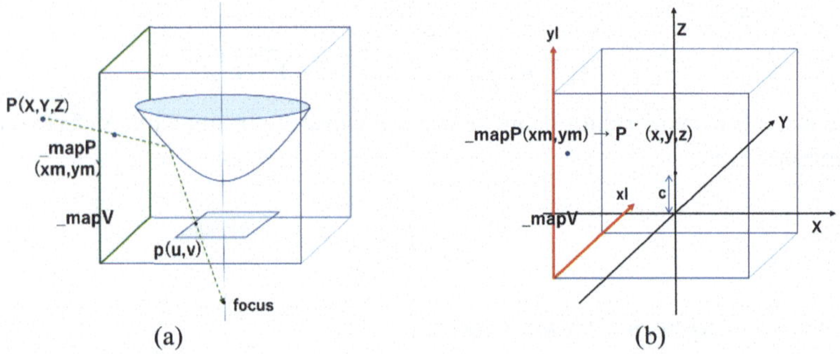

Fig. 9. The proposed the omnidirectional image production method (a), and relationship between mapP and mapP's coordinates P′ (x, y, z) (b)

In the following, we propose a method to create the omni-vision camera using SphericalVision module.

First, each image of the SphericalVision camera (front, left, right, back, top, bottom) is placed around the focal point of the hyperboloidal mirror as the side of a 2 m cube.

Next, an arbitrary point P(X, Y, Z) in the space is considered to be mapped as an intersection point mapP(xm, ym) between a straight line connecting the point P and the focal point of the hyperboloid mirror and each plane of the cube mapV. Therefore create an original image of the omni-vision camera by finding the mapping p (u, v) to the omni-vision camera using mapP, P′ (x, y, z), which is the coordinate in the space of mapP, and Eqs. (3) and (4).

Finally, a panoramic image is generated by expanding this to a distance-relative angle.

An original image created using the proposed method is shown in Fig. 10(a), and a panoramic image is shown in Fig. 10(b).

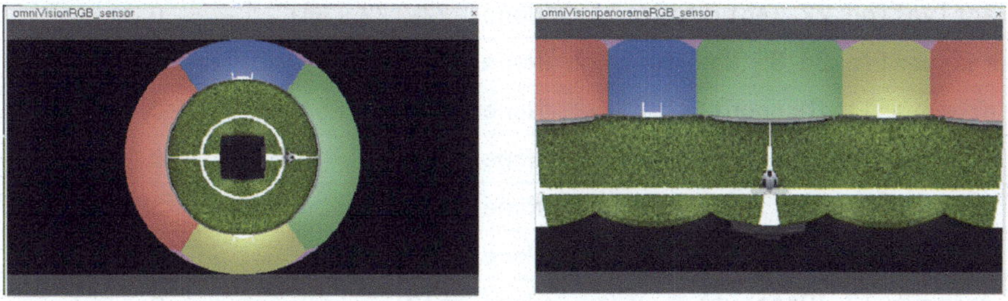

(a) (b)

Fig. 10. An original image created using the proposed method (a), and a panoramic image (b)

3 Algorithm

3.1 CNN

CNN is an abbreviation for Convolutional Neural Network, and well known NN technique often used for image recognition [6, 7]. By adding a convolution layer and a pooling layer, CNN perform convolution while judging the neighborhood area for example 3 × 3, It is known to withstand image distortion, rotation, and position changes. Table 1 shows our CNN model.

Table 1. Constitution of CNN

Layer (type)	Output Shape	Param #
conv2d_13 (Conv2D)	(None, 50, 50, 32)	896
activation_19 (Activation)	(None, 50, 50, 32)	0
conv2d_14 (Conv2D)	(None, 48, 48, 32)	9248
activation_20 (Activation)	(None, 48, 48, 32)	0
max_pooling2d_7 (MaxPooling2	(None, 24, 24, 32)	0
dropout_10 (Dropout)	(None, 24, 24, 32)	0
conv2d_15 (Conv2D)	(None, 24, 24, 64)	18496
activation_21 (Activation)	(None, 24, 24, 64)	0
conv2d_16 (Conv2D)	(None, 22, 22, 64)	36928
activation_22 (Activation)	(None, 22, 22, 64)	0
max_pooling2d_8 (MaxPooling2	(None, 11, 11, 64)	0
dropout_11 (Dropout)	(None, 11, 11, 64)	0
flatten_4 (Flatten)	(None, 7744)	0
dense_7 (Dense)	(None, 512)	3965440
activation_23 (Activation)	(None, 512)	0
dropout_12 (Dropout)	(None, 512)	0
dense_8 (Dense)	(None, 2)	1026
activation_24 (Activation)	(None, 2)	0

Total params: 4,032,034

Trainable params: 4,032,034

Non-trainable params: 0

3.2 Creating a Learning Data Set

In this experiment, as a simple localization task, let the robot learn so that it can judge whether it is inside or outside the center circle drawn in the field. In addition, the following environment was prepared to investigate the effect of wall color on learning and evaluation(see Figs. 11 and 12).

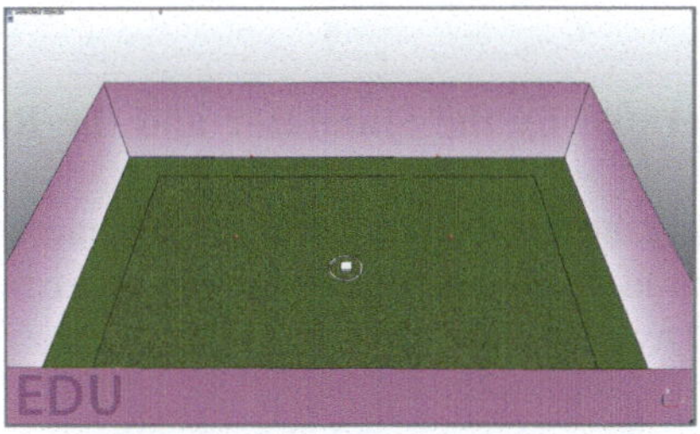

Fig. 11. Monochromatic wall environment

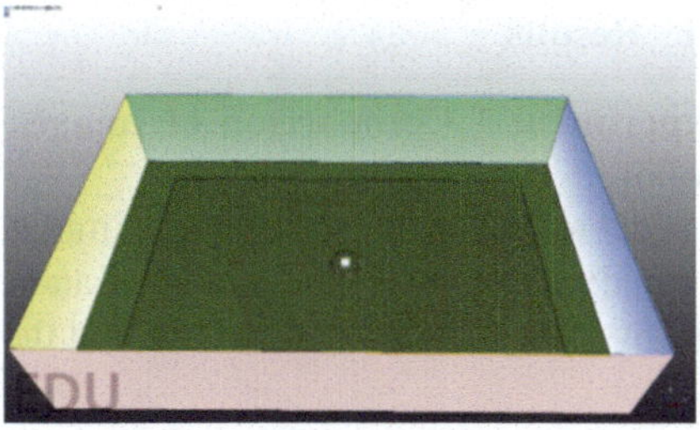

Fig. 12. Colorful walls environment

In this paper, a range where the circle can be stably visible by the omni-vision camera (see Fig. 13) is obtained in advance, and learning data was created by sampling panoramic images while randomly changing the position and direction of the robot in this range. In addition, based on the comparison result of the position of the robot and the size of the circle, the images were automatically classified into [IN of circle] and [OUT of circle].

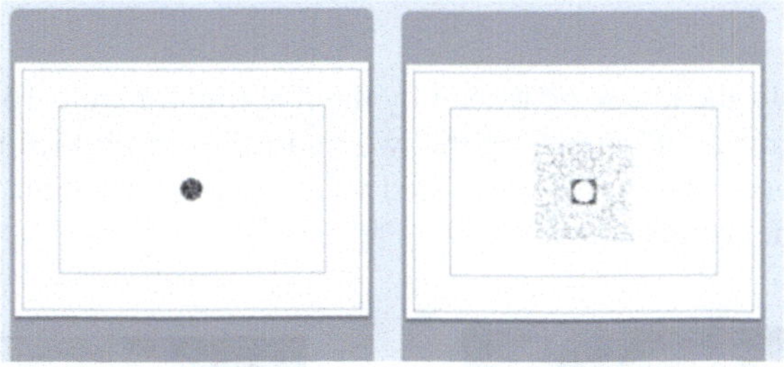

Fig. 13. Position and direction of image creation

4 Experiments

Here are three experiments. The first is to compare the simulated image using the proposed method and the image generated by the real robot to confirm whether the omni-vision camera can be simulated. The second is to confirm whether the task related to localization can be learned using the created simulated image and CNN. The third is to investigate suitable conditions as a pre-learning environment in transfer learning by investigating the influence of external unpredictable factors, like the color of the wall, on learning and evaluation. The fourth is an evaluation in a simulated real environment for a preliminary study of how pre-learning results are reflected in a real environment.

- Experiment 1 Comparison of images generated by the simulator and the real robot.
- Experiment 2 Learning and evaluating under each environment.
- Experiment 3 Evaluate another environment with the learning environment.
- Experiment 4 Evaluation in a simulated real environment using pre-learning results

4.1 Results

Experiment 1 Comparison of Images Generated by the Simulator and the Real Robot
In order to confirm that the image simulated using the proposed method can be generated correctly, the image was generated and compared at specific points in the soccer field shown in Fig. 14 in the simulator and the real environment.

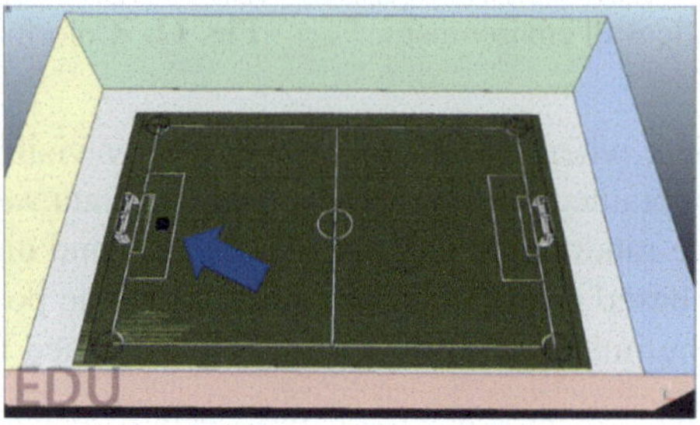

Fig. 14. Shooting location

Figure 15 shows the image generated by the real robot. Figure 16 shows the simulated image. As a result of comparing the two figures, we confirmed that the positions and refraction of each feature point matched well. Therefor, we think we succeeded in simulating an omni vision camera.

Fig. 15. The image generated by the real robot

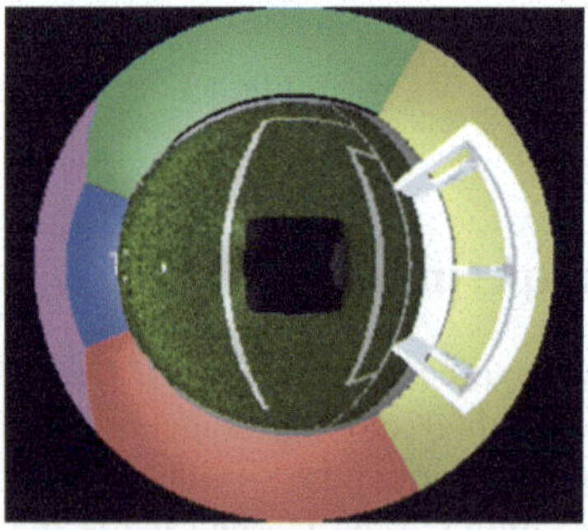

Fig. 16. The simulated image

Next, learning data was generated using a simulator. Figure 17 shows an example of learning data generated by the simulator.

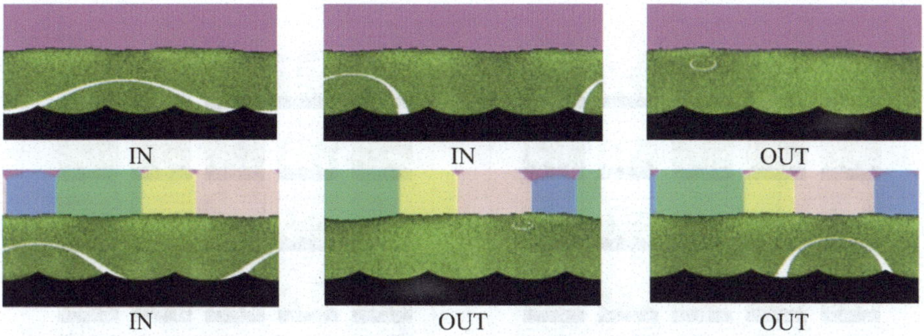

Fig. 17. Example of created images

Experiment 2 Learning and Evaluation in Each Environment
The learning progress is as shown in Fig. 18.

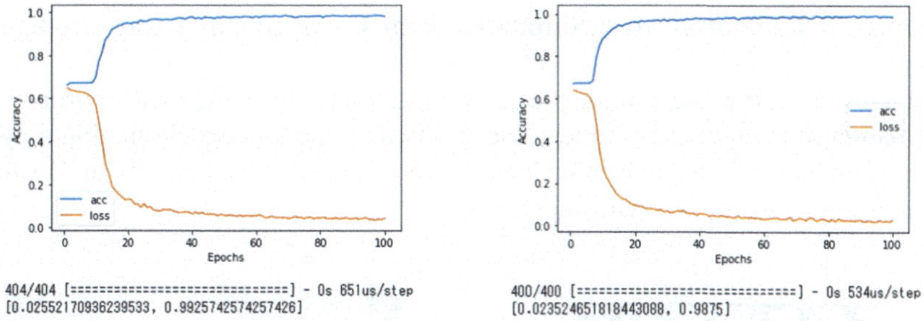

(a) Monochromatic wall environment (b) Colorful walls Environment

Fig. 18. The learning progress

The accuracy when learning with epoch50 was about 98.7% in the monochromatic wall environment and about 98.5% in the colorful walls environment. The accuracy when learning with epoch100 was about 99.2% in the monochromatic wall environment and about 98.7% in the colorful walls environment.

Experiment 3 Evaluation in Another Environment to the Learning
As a result of evaluating the colorful walls environment using the learning results in the monochromatic wall environment, the accuracy was about 95.7% (see Fig. 19).

As a result of evaluating the monochromatic wall environment using the learning results in the colorful walls Environment, the accuracy was about 92% (see Fig. 20).

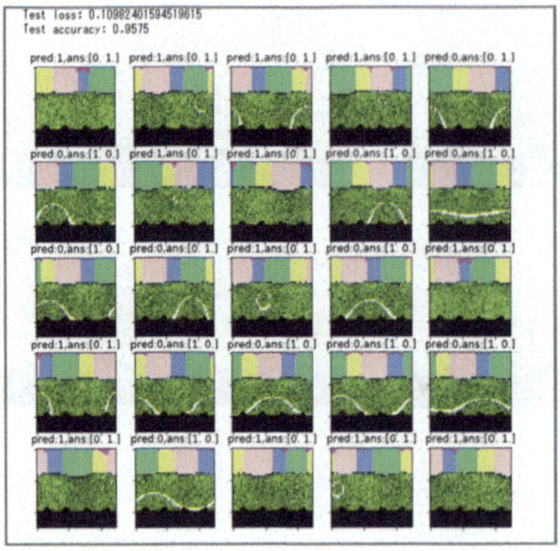

Fig. 19. Evaluation of the colorful walls environment based on learning results in the monochromatic wall environment

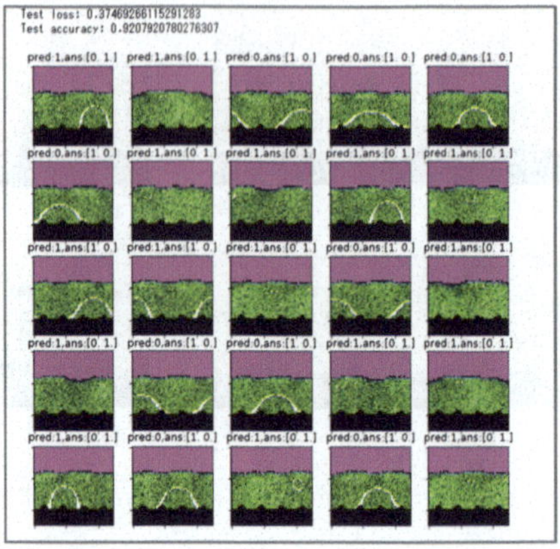

Fig. 20. Evaluation of the monochromatic wall environment based on learning results in the colorful walls environment

Experiment 4 Evaluation in a Simulated Real Environment Using Pre-learning Results

In order to evaluate the effect of the pre-learning results in the simulated environment on the evaluation in the real environment, we create the environment simulating a gymnasium (see Fig. 21) and the environment simulating open air (see Fig. 22), and evaluated each environment using pre-learning results.

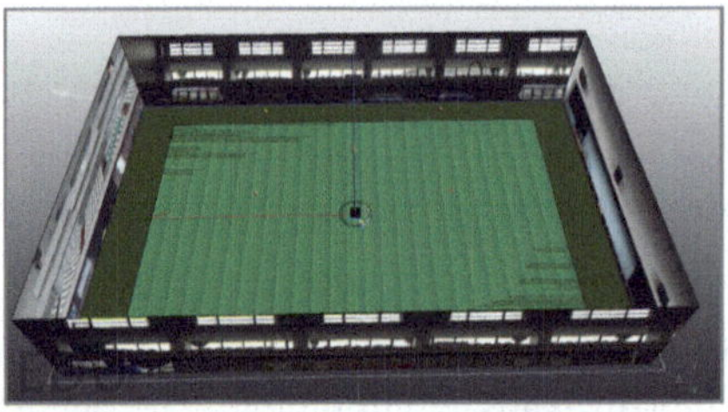

Fig. 21. The environment simulating a gymnasium

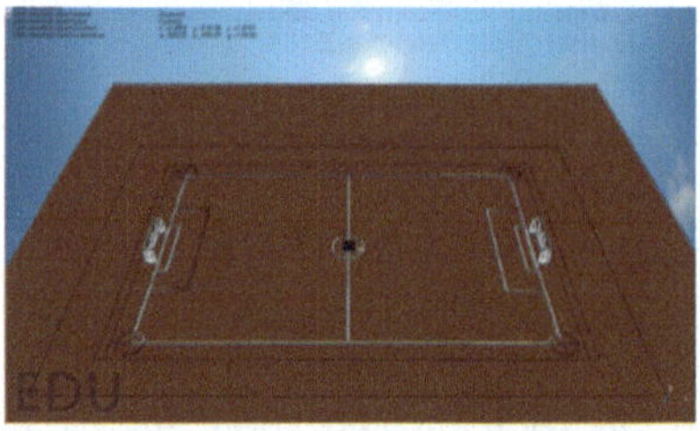

Fig. 22. The environment simulating open air

Table 2 shows evaluations, and numbers of epochs required for tuning, in each environment using pre-learning results. As a result of evaluating the environment simulating a gymnasium using the learning results in the monochromatic wall environment, the accuracy was about 98.7% and the number of epochs required for tuning was seven. As a result of evaluating the environment simulating a gymnasium using the learning results in the colorful walls Environment, the accuracy was about 92.7% and the number of epochs required for tuning was eight. As a result of evaluating the environment simulating open air using the learning results in the monochromatic wall environment, the accuracy was about 83.6% and the number of epochs required for tuning was six. As a result of evaluating the environment simulating open air using the learning results in the colorful walls Environment, the accuracy was about 93.3% and the number of epochs required for tuning was five.

Table 2. Evaluation results, and numbers of epochs required for tuning

	Monochromatic wall environment	Colorful walls Environment
Gymnasium	98.7%/99%(+7 epochs)	92.7%/99%(+6 epochs)
Open air	83.6%/99%(+8 epochs)	93.3%/99%(+5 epochs)

Figure 23(a) and (b) show examples of mistakes evaluating each environment using learning results in the monochromatic environment. Errors occurred even if it was completely outside the circle, a relatively simple identification problem.

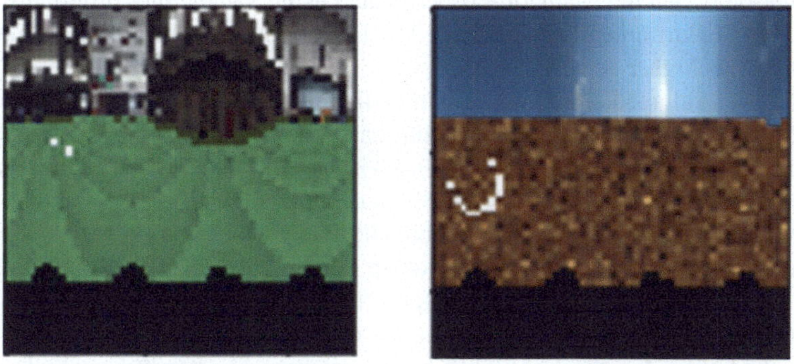

(a) **(b)**

Fig. 23. Examples of mistakes evaluating each environment using learning results in the monochromatic environment

4.2 Summary of the Experiments

The first was to confirm whether an omnidirectional image can be generated correctly using the proposed method. As shown in Figs. 15 and 16, the omni-vision camera could be simulated.

The second was confirmed whether learning can be performed using the created learning data. As epoch progresses, an increase in accuracy and a decrease in Loss function can be confirmed as shown in Fig. 18, and it is considered that learning has been successful. Regarding the difference in the learning results by epoch, it was confirmed that the learning with the monochromatic wall environment increased from about 98.7% to about 99.2% when learning with epoch50 and when learning with epoch100. On the other hand, learning in the colorful walls environment did not show much change between the case of learning with epoch50 and the case of learning with epoch100, and the result was lower than that of learning with the monochromatic wall environment. In learning in the colorful walls environment, The CNN used in this study does not have many layers, so it is presumed that the change in wall color could not be fully learned. Increasing the CNN layer may solve this problem. However, careful consideration is required as it consumes more robot resources. We need more experience this topic.

As the third, regarding the evaluation in another environment to the learning environment, the accuracy when the colorful walls environment is evaluated using the learning result of the monochromatic wall environment is about 95.7%, the accuracy when the monochromatic wall environment is evaluated using the learning results in the colorful walls environment is about 92%, the former was more accurate. Learning in the monochromatic wall environment does not change much in the wall color even if the position and orientation of robot change, so it is assumed that the weight of the change in the wall color on the judgment is low. On the other hand, learning in the colorful walls environment greatly changes the wall color depending on the position and orientation of the robot, so it is assumed that weight of the change in the wall color on the judgment becomes higher than in the case of learning in the monochromatic wall environment. For this reason, when the evaluation is performed using a wall color different from that of the learning environment, it is estimated that the learning in the monochromatic wall environment has higher accuracy.

Finally, we evaluated environments simulating real environments. The evaluation in the environment simulating the gymnasium has a very high accuracy of 98.7% in the evaluation using the monochromatic wall environment, whereas the evaluation in the environment simulating open air was 93.3% in the colorful wall environment, this result was higher than the result using the monochromatic wall environment. In addition, using the pre-learning results in the colorful walls environment, numbers of epochs required for tuning could be learned with fewer epochs, in both environments simulating the gymnasium and open air. From the above results, it was found that if information including the field color matches highly, higher accuracy can be obtained by learning in the monochromatic wall environment. On the other hand, in the environment where the field color is greatly different from the pre-learning environment, it was found that using the pre-learning result in the colorful wall environment gives higher accuracy. This is because the types of color information are larger in the colorful wall environment, and the weight of the field color information has decreased relatively, so it is assumed that a certain level of accuracy has been achieved even if the color of the field changes greatly.

Outside the field in the venue, there are things that can be predicted in advance and things that cannot be predicted in advance. How to use the knowledge about the color of walls and field is the future challenge. Also, we need to study on the accuracy, and the number of epochs required for learning in the real environment.

5 Conclusion

In this study, in order to reduce the cost of tuning a soccer robot, we examined prior learning with a simulator.

The aim is to reduce the tuning cost in the field by building an environment similar to the venue in the simulator and learning in advance with CNN, which is one of the deep learning. In this paper, as the first step, we performed a simple computer experiment on localization using omnidirectional images created in V-REP.

A soccer field can be regarded as a semi-structured environment with highly consistent structures in the environment. Here, we examined requirements suitable for tuning by changing the color of the wall at the boundary with the field. As a result, it was found that learning in the monochromatic wall environment can achieve high accuracy when the field color matches the pre-learning environment. However, no significant result was obtained for the tuning cost. In future research, we will use these survey results to study a method for tuning effectively using the learning results in the monochromatic wall environment and adapting to more complex problems.

References

1. Umeda, M., Date, H.: Spherical panoramic image-based localization by deep learning. Trans. Soc. Instrum. Control Eng. **54**(5), 483–493 (2018)
2. Koyasu, H., Miura, J., Shirai, Y.: Recognizing moving obstacles for robot navigation using real-time omnidirectional stereo vision. J. Robot. Mechatron. **14**(2), 147–156 (2002)
3. Yamazawa, K., Yagi, Y., Yachida, M.: Omnidirectional imaging with hyperboloidal projection. In: International Conference on Intelligent Robots and Systems (IROS 1993), Yokohama, vol. 2, pp. 1029–1034 (1993)
4. Yamada, S., Suzuki, H.: Real-time self-localization method for autonomous mobile robot. In: Japan Society for Fuzzy Theory and Intelligent Informatics Proceedings of the Fuzzy System Symposium, vol. 2015, no. 31, pp. 149–150 (2015)
5. Demura, K., Nakagawa, Y.: A Monte Carlo localization based on template matching using an omni-directional camera. J. Robot. Soc. Jpn. **27**(2), 249–257 (2009)
6. Fujiyoshi, H., Yamashita, T.: Image recognition by deep learning. J. Robot. Soc. Jpn. **35**(3), 180–185 (2017)
7. LeCun, Y., Bottou, L., Bengio, Y., Haffner, P.: Gradient based learning applied to document recognition. Proc. IEEE **86**(11), 2278–2324 (1998)

Examination of Analysis Method of Opinion Distribution in News Media Transferred on Web

Yasuko Kawahata[1,2](✉)

[1] Faculty of Social and Information Studies, Gunma University, 4-2 Aramaki-machi, Maebashi, Gunma 371-0044, Japan
kawahata@gunma-u.ac.jp

[2] Center for Spatial Information Science, University of Tokyo, 5-1-5 Kashiwanoha, Kashiwa-shi, Chiba 277-8568, Japan

Abstract. The global spread of public networks and social environment systems is not only a positive has some potential downsides aspect. It is transforming us into a society where information exchange, decision making, and consensus building are made across various boundaries. In particular, public opinion tends to decline due owing to the amplification of the echo chambers in unconscious consensus building, and it can be seen that online speech tends to have a major impact on mental health. In this research, the sentiments (Negative/Positive/Neutral), we are obtained from comments made from in response to videos transferred from various news media found on YouTube, which transmits media and information related to our daily lives, and is considered in the analysis of the distribution of opinions in various references.

Keywords: Opinion · Online social media · YouTube · Text mining

1 Introduction

Since public network devices have been spread all over the world, the Opportunities for asynchronous telecommunication beyond the space-time constraints have expanded enormously as a result of the worldwide spread of public network devices. In other words, it can be said that online communication is playing plays an important role as a physical place channel for communication related to our mental health care. In particular, with the arrival of large corporations with that use information and communication technology that to maintains a platforms that can acquire comprehensive daily lifelog data (e.g., "GAFA", "FANG", etc.), whether our daily data is unconscious or conscious, Cases that are such platforms have become commonly treated as a form of consensus have also become common [1]. Since Following the spread of public networks, research has been growing rapidly in the field of Web Science that to understands the mechanisms of social phenomena that have been difficult to visualize and quantify has been growing

H. Sato et al. (Eds.): IES 2019, PALO 12, pp. 156–166, 2020.
https://doi.org/10.1007/978-3-030-37442-6_14

rapidly [2]. In the process of research and development as described above, the data is shaped so that significant features can be extracted statistically and significantly. After processing, It the data is also becomes then used as model data for purposes after processing such as machine learning. In Through our daily online actions activity, personalized opinions functions, such as recommendation functions, can be automatically provided via public networks and APIs. In the future, the early stages of education we will need a lot of to establish defense, literacy, and care an awareness and caution for regarding such information on publicly available online networks where we live in the early stages of education. Also, as we will connect more diverse public networks will become available. We think that discussions on accessibility to devices and media will also accelerate. In that regard, local governments and educational institutions are can act as mental blocks for online social networks, where daily online access blocks do is not currently occur from children to adults everyday regulated regardless of age. It can be said that public opinion has shifted from owing to the amplification of echo chambers, especially in unconscious consensus building, and online speech has entered an era in which mental health is likely to be greatly affected. In this research, the sentiments (Negative/Positive) We were obtained through comments from videos transferred from various news media on YouTube, which transmits delivers media and information related to our daily lives, and is considered the analysis of distribution of opinions in various references.

2 Prior Research

AHP (Analytic Hierarchy Process (AHP) has been proposed since the 1980s to solve certain decision-making problems in groups. The feature of this model is the use of interval expressions to reach decisions throughout the group consensus building process that it canto approach the actual problem [3] and uses interval expressions to reach decisions throughout the group consensus building process. Research on consensus building as described above has been conducted in various ways, from the including theoryetical base to approaches to the practical solving of real problems [4]. In addition, even in the case of dissatisfaction in when the a group when not agreeing and is dissatisfied and does not agreeing on the an opinion, a pairwise comparison value of the group is can be calculated, and a quantitative tendency can be suggested. In the model constructed model, a dissatisfaction function is defined for the presented interval to minimize both the consistency and the dissatisfaction of the whole group. Research results corresponding to cases have also been devised [5]. While the above problem-solving methods related to consensus building have been devised, various opinions can be collected as concrete explicit texts on the Internet. In constructing a social framework, the public opinion, will, and bias on politics have always been considered to depend on the development and utilization of the media [6]. The media is also a symbol of advanced modernization that has emerged as a unique logical and independent institution that should respond to social systems. On the other hand, the media is an integrated part of other institutions such as "Politics, Work, Family and Religion". Many of the activities of these institutions are use both interactive and immediate social media and mass media.

It is being done through. In other words, it is an era when society itself becomes media, and opinions are constantly formed within our groups [6]. A number of techniques have emerged, such as concentrating the election results on the consensus building of opinions, and examining them from the resulting coefficients in of the explanatory variables resulting from the results [7–11]. The Analysis using a large-scale database is regarded as important, and has enabled the development of a method for clarifying the roles of forces that create a consensus, such as media,/external forces, public opinion, religion, region, and so on etc. that brought consensus has become possible. In this study, such a large-scale database is used in quantitative analysis using such a large-scale database is applied as a study approach for opinion distribution. In this research, we will explore the possibility of collaborative work of measurement and theoretical calculation for the a theory for quantitative analysis and prediction.

3 Dataset

In this section, we will discuss what the data was collected and used for to analyze people's the reactions of people to information transmitted by the media in society. At present, various studies on natural language processing have been carried out by expanding classification methods in machine learning and enriching the data that is the sources of dictionary data and learning data [12–15]. In other words, the results of these studies became made it possible to judge the strength of opinions from text collected by natural language processing. As a result, it became possible to This in turn has allowed the measurement of the distribution of opinions distribution in society as a continuous distribution, from positive references to negative references and including positive, negative, and neutral references, rather than as binary opinions alone. In this study, YouTube was selected because of differences in data acquisition legality between countries worldwide various data acquisition restrictions in the world. Various OSN s are available for real-time video distribution, but in the case of YouTube, news media with one of the world's leading TV and/newspaper networks has also published news reports on YouTube. Real-time distribution is Another feature of YouTube, is that real-time distribution is performed. It was also and it has been hypothesized that it could allow distribution analysis the distribution of opinions in the news media in each country, of the differences in coverage of the same news content, and the distribution of opinions on the differences in the user groups who made providing comments were possible. In addition, in order to grasp obtain the strength of the opinions under the same identical conditions, it data was limited to news videos in ch based on English-speaking language dictionary data. In this study, we obtained the calculated opinion distribution calculated by measuring the sentiment of comments in the top most frequently viewed videos plays of from major news media in each country (e.g., "CNN", "BBC", "The Sun", "Fox TV", etc.). There are cases where Though these YouTube videos involve English language (in English)., However, it is hypothesized that the video content of the video is globally transmitted., An analysis was performed based on expected and that comparisons regarding differences among media are expected, and that there are things that are as well as similarities to the transmissions on TV, such as parts that differences in receivinge otherwise the same identical news content differently depending on the reporting method. The analysis was performed.

3.1 Methods

The channels on YouTube for data acquisition were "CNN", "BBC", and "Fox TV", etc. The comment acquisition period for the each acquisition target video is set for each video individually due to restrictions on the comment data collection rules in the YouTube API. The acquired ch and acquisition periods per channel are described below.

(1) BBC (Upload date of each video-until April 30, 2019)
(2) Fox TV (Upload time of each video-until April 25, 2019)
(3) CNN (Upload time of each video-until April 17, 2019)

In this study, we collected comments on the videos with the highest number of playbacks within the acquisition period since the launch of each channel, and added negative, positive, and neutral scores for each comment. Scores on each of the three axes were defined with a range of [−1.0, −0.75, −0.5, −0.25, 0, 0.25, 0.5, 0.75, 1.0]. This time, the Processing is was omitted for scores of −1, 0, and 1, which is a are considered biased scores when considering the distribution. NLTK (Natural Language Toolkit (NLTK) [16] is used to analyze the input natural input language. Using vader VADER analyzer, feature values were extracted for each comment in each video, and finally, the distributions of negative, positive, and neutral scores in for all the comments of each video was were output on three axes. Additionally, when capturing the negative, positive, and neutral axis scores in a unified distribution, add 1 to the positive opinion score, negative opinions* (−1), and −1 without adding neutral opinions. A method scales of were convertinged to a scale of 2 was taken. Furthermore, standardization processing is was performed (CDF) in order to eliminate variations in the number of comments and sentiment scores in the acquisition period and the number of playbacks in each video.

4 Result

In previous research, we took up the to panalyzed the most frequently view of CNN videos [17]. At that time, it turned out that it was our results showed slightly inclined bias toward the animation related to the animals and the animation focused on the fun. In this paper, we have excluded cases where news and disasters with political intentions were clearly negative or positively biased as case studies.

4.1 Case (1)

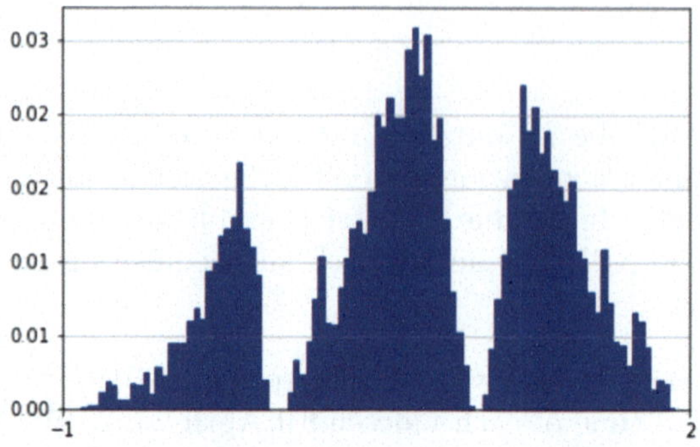

Fig. 1. Children interrupt BBC News interview - BBC News:Distribution of negative, positive and neutral comments (CDF)

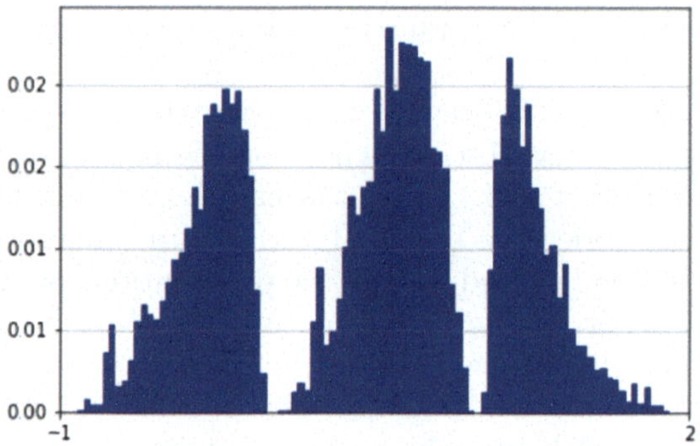

Fig. 2. "Internet meme ruined my career" BBC News:Distribution of negative, positive and neutral comments (CDF)

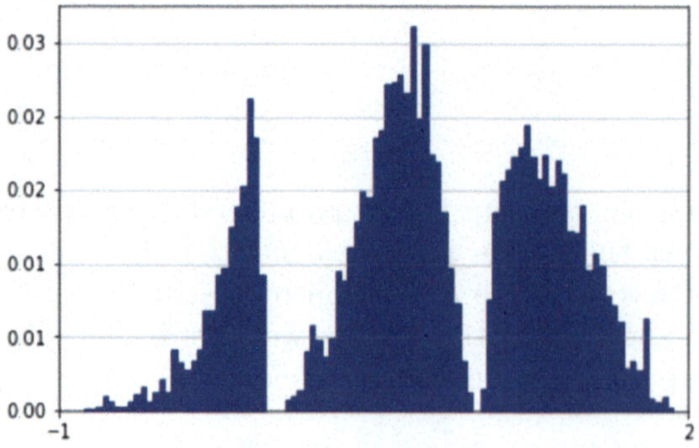

Fig. 3. FULL Interview: Prince Harry and Meghan Markle - BBC News:Distribution of negative, positive and neutral comments (CDF)

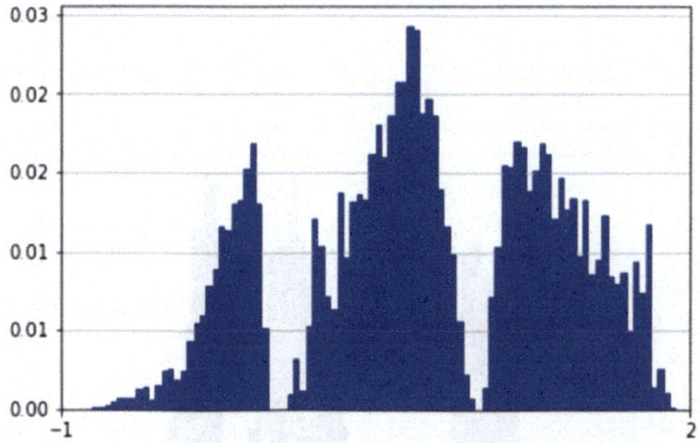

Fig. 4. Prof Robert Kelly is back & this time his wife & children are meant to be in shot! BBC News: Distribution of negative, positive and neutral comments (CDF)

4.2 Discussion (1)

Figure 1 shows a video that was caused by resulting from a live broadcast accident in a live broadcast, but it tends to be biased from toward a neutral opinion or positive opinion one rather than a negative distribution. However, as for the tragic news report and problem raising as shown in Fig. 2, it can be inferred that there is a concentration of neutral negative opinions. Also, in Fig. 3, the full interview of the imperial family gathered generated comments, and a very positive accumulation of opinions was observed, which was. Inferred to be due to people's general interest and people's positive interest. Also, in Fig. 4, the comments for video of the party of the broadcast accident in Fig. 1 was listed at the top of the number of views, but it is guessed we assume that this was also caused by a tendency to very produce laughter and positive emotions of in people. It was a tendency to. Regarding the BBC, it seems that the specific gravity tends to be biased towards video content, seems to gravitate around people's interests, positive or negative, and it can be seen that opinions are formed on the contents that are closer to the lives of ordinary people than the to contents of about animals and politics.

4.3 Case (2)

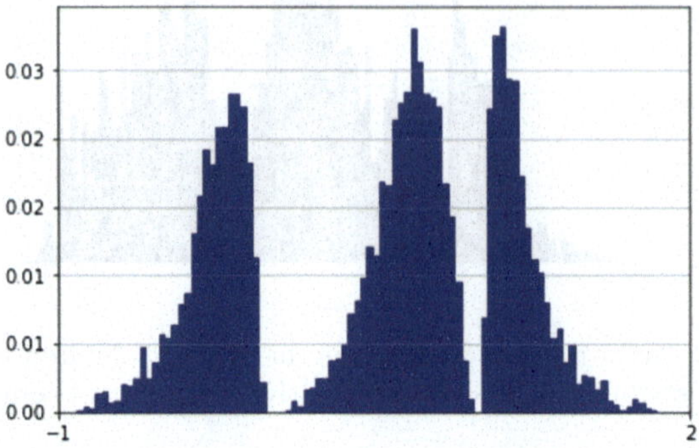

Fig. 5. Robert Shapiro reveals what OJ whispered after verdict:Distribution of negative, positive and neutral comments (CDF)

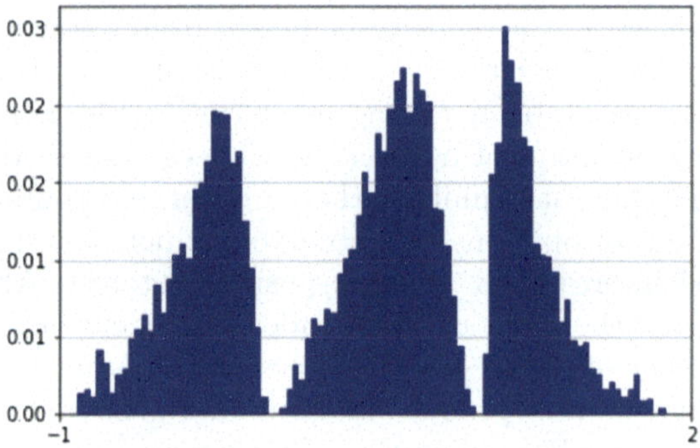

Fig. 6. Watch Jodi Arias' reaction as guilty verdict is read:Distribution of negative, positive and neutral comments (CDF)

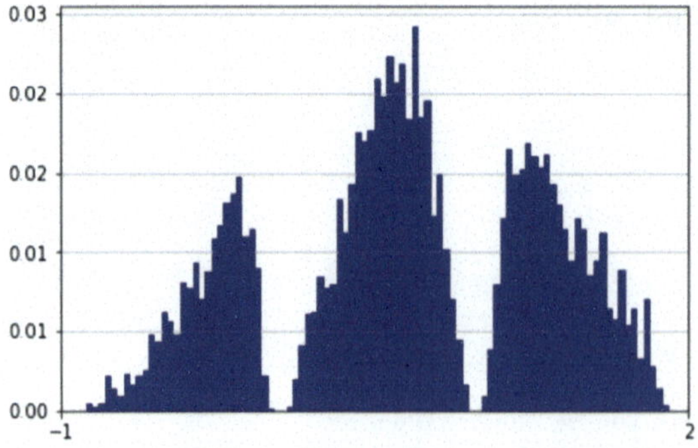

Fig. 7. Waterboy with Down Syndrome Scores Touchdown:Distribution of negative, positive and neutral comments (CDF)

4.4 Discussion (2)

Figure 5 shows the results from a video of various discussions on and post-trial comments by a former lawyer of O.J. Simpson on a political topic. This Topic can be inferred that Opinions on this topic are inferred to be divided because the opinions from both the negative and positive sides can be interpreted as being positive in the neutral perspective. Also, Fig. 6 is another video about the trial, and comments on regarding sensational cases are gathered were generated. In addition, In Fig. 7, videos listed at the top of the number of views tended to be related to sports activities including sports boys with intractable disabilities were listed at the top of the number of views. It was a tendency to be. Regarding Fox TV, videos that have risen to the top of the number of views focused on people active in politics, trials, and society have risen to the top of the number of views, and it is a which are topics that tends to can be divide opinions polarizing, so it they tended to be negative and positive rather than neutral.

4.5 Case (3)

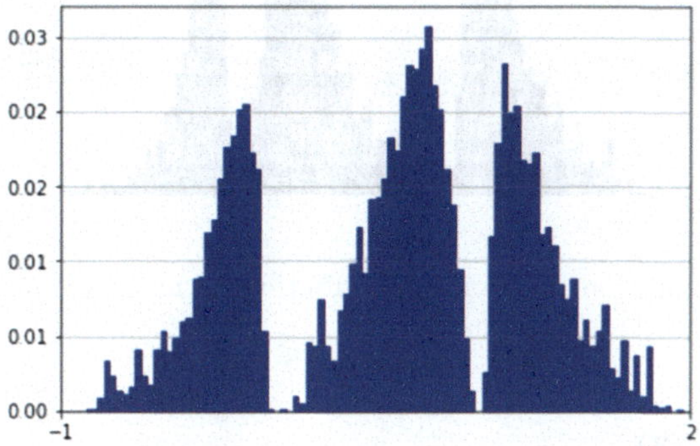

Fig. 8. Obama forgets to salute:Distribution of negative, positive and neutral comments (CDF)

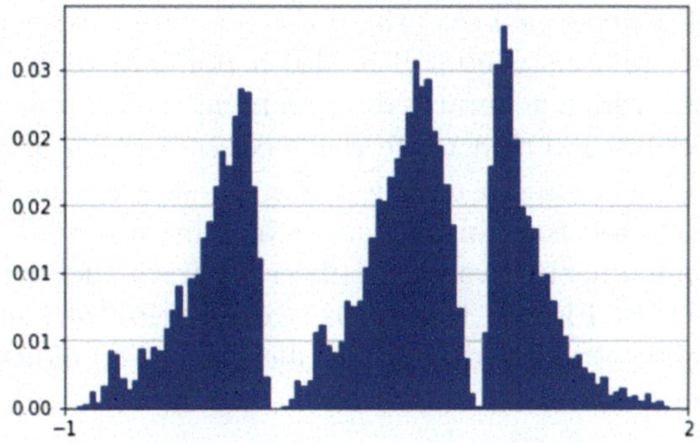

Fig. 9. Trump supporter leaves CNN anchor speechless: Distribution of negative, positive and neutral comments (CDF)

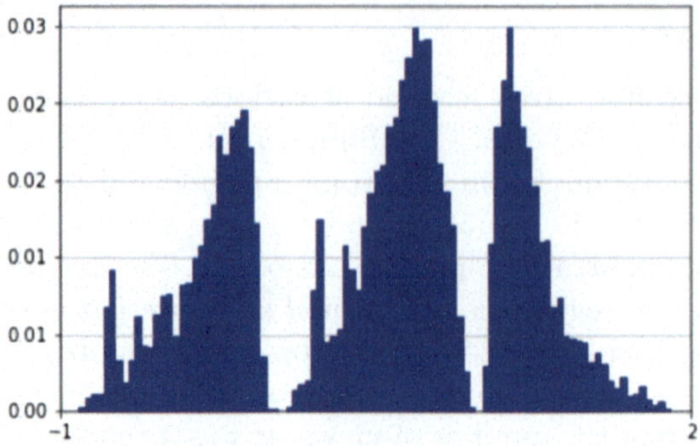

Fig. 10. Panther sneaks up behind man and then ...:Distribution of negative, positive and neutral comments (CDF)

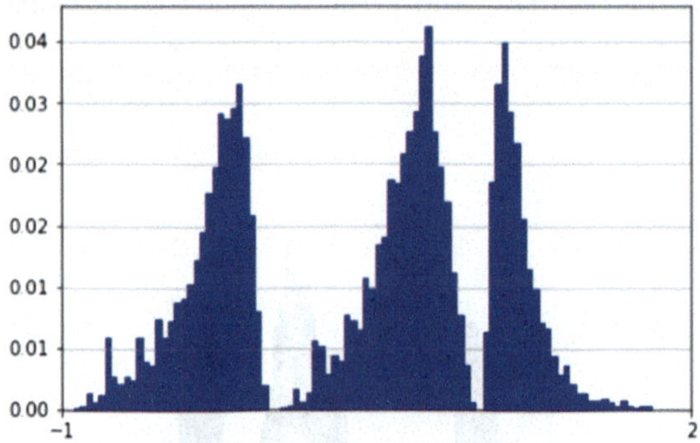

Fig. 11. Cop confrontation goes viral:Distribution of negative, positive and neutral comments (CDF)

4.6 Discussion (3)

Figure 8 is shows results from a political topic, but it the video was rich in American humor. This tended to be gathered in generate a neutral position near the positive side, but it is presumed that there was also some extreme negative gathering generation of opinions. Figure 9 is also a political topic, but it regarding election support, was which generates disagreement easily to disagree about election support. It was inferred that there was a concentration of negative opinions that were somewhat neutral. In addition, Fig. 10 shows a neutral collection of opinions on topics related to animals, but as in previous studies, negative opinions focused on condemnation was were observed from the viewpoint of animal welfare. In addition, in Fig. 11, there was a contention of conflicting opinions were made about police activities, but a collection of opinions urging negative antithesis was observed.

5 Summary

In this study, an attempt was made to understand the formation of social consensus from the negative-positive score distribution in comment text. In order to proceed with research in this direction in the future, positive, neutral, and negative joints are one of the issues. In addition, many cases were observed in the course of this study in which communication in the comment section on YouTube functions as if it were one independent SNS were observed in the course of this study. Therefore, it was also inferred from the content of videos and comment trends in each news media that quantitative knowledge such as in which cluster mental health problems occurred on social networks occurred. It is necessary to continuously consider continuously what kind of news content can be generated in a negative cluster and which multi-dimensional data is the causes the cluster.

Acknowledgements. We would like to also thank you for the support from the 2019 Scholarly Scholars Research Grant Program (Leading Initiative for Excellent Young Researchers (LEADER)), which provided support for this research in general.

References

1. French, J.R.P.: A formal theory of social power. Psychol. Rev. **63**, 181–194 (1956)
2. Harary, F.: A criterion for unanimity in French 's theory of social power. In: Cartwright, D. (ed.) Studies in Social Power. Institute for Social Research, Ann Arbor (1959)
3. Yoshiaki, Z., Sugiyama, M., Yamaki, N.: Group AHP using consensus building model. J. Oper. Res. **40**(2), 236–244 (2017)
4. Yahata, N., et al.: A population interval AHP method using dissatisfaction function. J. Oper. Res. Soc. Japan **45**(3), 268–284 (2002)
5. Yamaki, N., Sekiya, K.: Large-scale AHP with incomplete data (determination theory and related fields) (1998)
6. Hjarvard, S.: The mediatization of society. Nordicom Rev. **29**(2), 102–131 (2008)
7. Pak, A., Paroubek, P.: Twitter as a corpus for sentiment analysis and opinion mining. LREC **10**, 2010 (2010)
8. Agarwal, A., et al.: Sentiment analysis of Twitter data. In: Proceedings of the Workshop on Languages in Social Media. Association for Computational Linguistics (2011)
9. Siersdorfer, S., et al.: How useful are your comments?: analyzing and predicting youtube comments and comment ratings. In: Proceedings of the 19th International Conference on World Wide Web. ACM (2010)
10. Wilson, T., Wiebe, J., Hoffmann, P.: Recognizing contextual polarity in phrase-level sentiment analysis. In: Proceedings of the Conference on Human Language Technology and Empirical Methods in Natural Language Processing, pp. 347–354. Association for Computational Linguistics, October 2005
11. Sato, K., et al.: Correlation between web data and stock price fluctuations by negative-positive analysis (2015)
12. Lasswell, H.D.: The measurement of public opinion. Am. Political Sci. Rev. **25**(2), 311–326 (1931)
13. Lasswell, H.D.: Why be quantitative. Lang. Politics **40–52**, 1949 (1949)

14. Galam, S.: Rational group decision making: a random field Ising model at T = 0. Phys. A **238**(1–4), 66–80 (1997)
15. Sznajd-Weron, K., Sznajd, J.: Opinion evolution in closed community. Int. J. Mod. Phys. C **11**(6), 1157–1165 (2000)
16. Bird, S., Loper, E.: NLTK: the natural language toolkit. In: Proceedings of the ACL 2004 on Interactive Poster and Demonstration Sessions. Association for Computational Linguistics (2004)
17. Kawahata, Y., Ishii, A.: Consideration of simulation of consensus formation Online using Sociophysics method, GDN 2019 (2019)

Opinion Distribution Analysis of Video Media Related to Social Life

Yasuko Kawahata[1,2](✉)

[1] Faculty of Social and Information Studies, Gunma University, 4-2 Aramaki-machi, Maebashi, Gunma 371-0044, Japan
kawahata@gunma-u.ac.jp

[2] Center for Spatial Information Science, University of Tokyo, 5-1-5 Kashiwanoha, Kashiwa-shi, Chiba 277-8568, Japan

Abstract. The global spread of public networks and social media has transformed us into a society where information exchange, decision making, and consensus building are performed across various boundaries. In particular, public opinion tends to decline owing to the amplification of the echo chamber in unconscious consensus building. Furthermore, it can be seen that online speech has a major impact on mental health. In this study, comments on social media were analyzed and labeled as negative, positive, and neutral based on their sentiments. This was done using comments from various videos on YouTube, which is a social media platform that enables the sharing of videos and information related to our daily lives. Thus, we analyzed the distribution of opinions in various cases.

Keywords: Opinion · Social media · YouTube · Text mining

1 Introduction

With the advent of online communication, social media is playing an important role as a physical place for communication related to mental health [1,2]. Recently, research on social phenomena that have been difficult to interpret in quantitative terms has grown rapidly [3]. In the future, children will need to be educated regarding social media and dealing with such massively perceived information in the early stages of education. Because the information to be introduced has already learned the patterns that we have already acted on the web (machine learning models of consensus building through clicks, text input, etc.). The model data allows us to send and receive information based on recommendations fitted to each of our behavior patterns. This makes it easier for an individual to fall into a "filter bubble" that makes it easier for an individual to blindly believe in uniform information. The negative effects of this on adults and children, the sense of distance, and how information is deal with is closely analyzed by educational and research institutions. Both positive and negative public feedback can be seen in social media, particularly in unconscious consensus building. There is a need for a way of looking at this information from the

H. Sato et al. (Eds.): IES 2019, PALO 12, pp. 167–174, 2020.
https://doi.org/10.1007/978-3-030-37442-6_15

situation of enjoying the environment that is easily scalable. From this point of view, it can be said that we have entered an era where the diversity of online speech is likely to have a major impact on mental health. In this study, sentiments (negative/positive/neutral) are determined from videos related to food and daily life, and this information is discussed from various viewpoints. Furthermore, we examined a method to consider the distribution of opinions.

2 Dataset

This section explains the data that was collected and used to determine people's sentiments to information transmitted by social media based on past research methods [4]. It is now possible to judge the strength of opinions from collected text by using natural language processing. Therefore, we measure the sentiment distribution in society as a continuous distribution ranging from positive to negative and neutral references rather than as binary opinions [5–9]. In this study, YouTube was chosen because of various data acquisition restrictions around the world. Various OSNs are available for real-time video delivery; however, YouTube is unique in that leading TV/newspaper news media also publish news reports on YouTube [4]. Another feature is real-time video delivery. It is also hypothesized that it would be possible to distribute opinions based on the distribution of opinions in the news media of different countries, differences in the coverage of the same news content, and differences in user groups that have made the comments. In addition, to limit the variability of the conditions in which the data was obtained, we limited our database limited to videos based on English-speaking dictionary data. In this study, we determine the opinion distribution by measuring the sentiments from comments using the videos with the highest number of views on YouTube for queries involving "food." However, the videos can be viewed globally, and comparisons based on geographical differences are expected. The comment acquisition period of the acquisition target moving image is described in each case because of API restrictions on comment data collection. Within the acquisition period, we collected comments from the most watched videos since the inception of each channel.

2.1 Method

The comments were given either negative, positive, or neutral scores. The scores for each of the three axes ranged from [−1.0, −0.75, −0.5, −0.25, 0, 0.25, 0.5, 0.75, 1.0]. In this study, the processing of −1, 0, 1 was omitted. This is the bias score when considering the distribution of sentiments. Furthermore, natural language toolkit (NLTK) [10] was used for input natural language analysis. Moreover, feature values were extracted for each video comment using the Vader analyzer. Finally, the negative, positive, and neutral score distributions of all comments for each video were output on the three axes. In addition, when capturing negative, positive, and neutral axis scores as a unified distribution, positive opinion score, negative opinion * (−1), and 1 to −1 without adding neutral opinions, the "add A" method of converting to a scale of 2 was adopted. In addition, a standardization process (CDF) was used to eliminate variations in the number of comments and sentiment scores owing to the duration and number of views of each video .

3 Results

Top-viewed news videos have been analyzed in previous research [4]. In this paper, we first considered the trend of "Best Ever Food," which is the channel with the highest number of views in the video query that includes "food." The comment data acquisition period was from the upload date of each video to May 21, 2019.

3.1 Case (1)

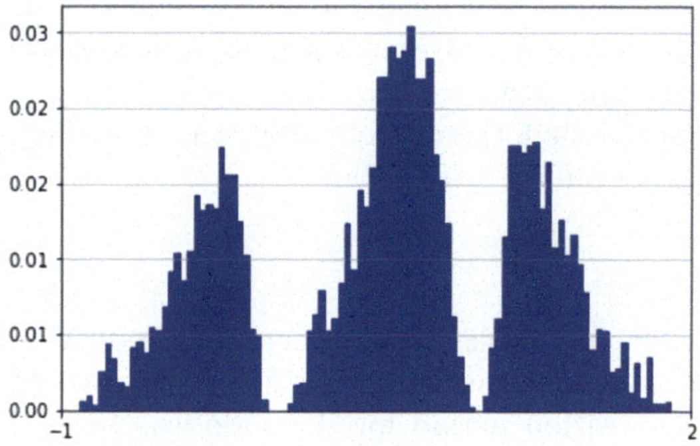

Fig. 1. Eating Japan's POISONOUS PufferFish!!!: Distribution of negative, positive and neutral comments (CDF)

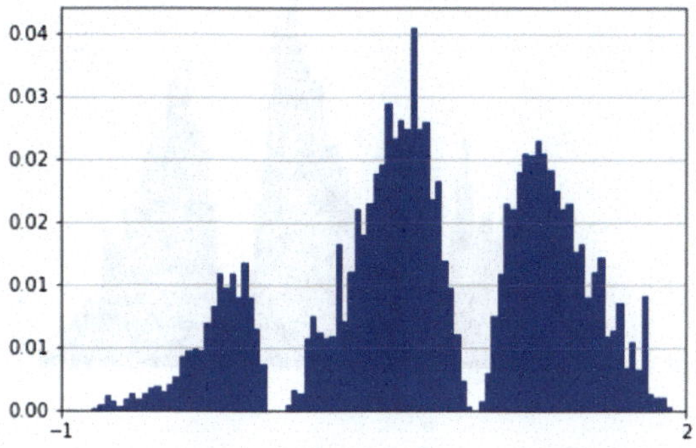

Fig. 2. Madagascar Street Food!!! Super RARE Malagasy Village Food!: Distribution of negative, positive and neutral comments (CDF)

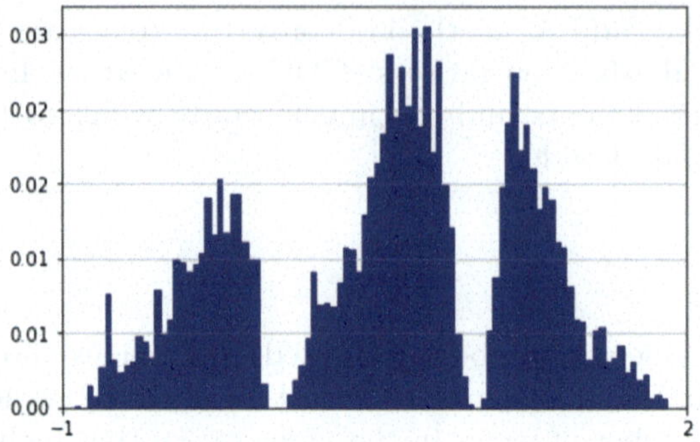

Fig. 3. EXOTIC Vietnamese food tour! SUPER RARE street food of Hanoi Vietnam: Distribution of negative, positive and neutral comments (CDF)

3.2 Discussion (1)

Figures 1, 2 and 3 show the videos with the unusual cuisines of each country with the most views. It was found that all of the videos tended to garner very neutral opinions and that there is no bias towards negative opinions. Furthermore, there was not much criticism or disaggregation in news such as previous research [4], and it was hypothesized that the content that was more entertaining tends to be a relatively neutral opinion distribution.

3.3 Case (2)

In this white paper, we first looked at the trend of "Food Wishes," which is the channel with the most views after Best Ever Food for video queries that include "food." The data acquisition period for this case was from the upload date of each video to May 23, 2019.

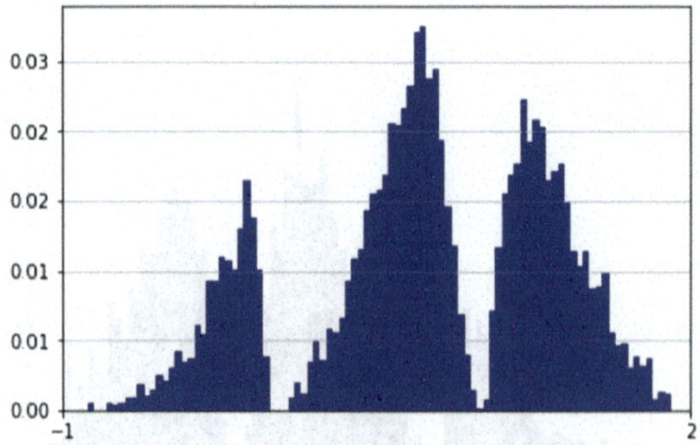

Fig. 4. Inside-Out Grilled Cheese Sandwich - Ultimate Cheese Sandwich : Distribution of negative, positive and neutral comments (CDF)

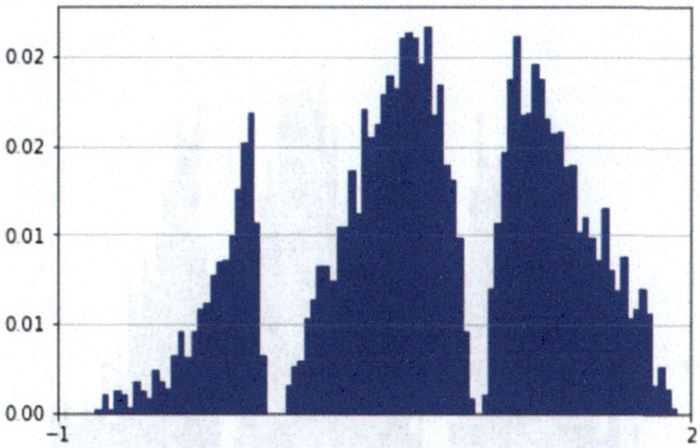

Fig. 5. Pita Bread - How to Make Pita Bread at Home - Grilled Flatbread: Distribution of negative, positive and neutral comments (CDF)

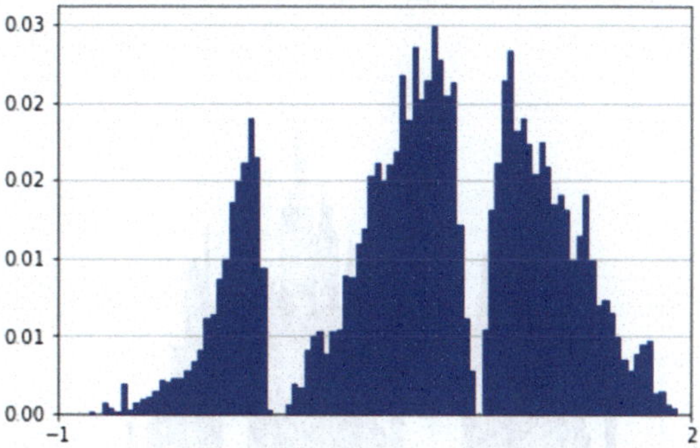

Fig. 6. Watch Jodi Arias' reaction as guilty verdict is read : Distribution of negative, positive and neutral comments (CDF)

3.4 Discussion (2)

The videos shown in Figs. 4, 5 and 6 introduce recipes that can be easily made at home. Furthermore, all videos tended to garner very neutral opinions, and there was no bias toward negative opinions. However, unlike in Case (1), there were cases where positive opinions were concentrated. This is highly practical and has a word of mouth (WOM) characteristic; therefore, this tendency was observed.

3.5 Case (3)

In this section, the videos with the highest number of views in the search results of the "Japanese food" search query are extracted and the results are described. The comment acquisition period is from the upload date of each video to June 23, 2019.

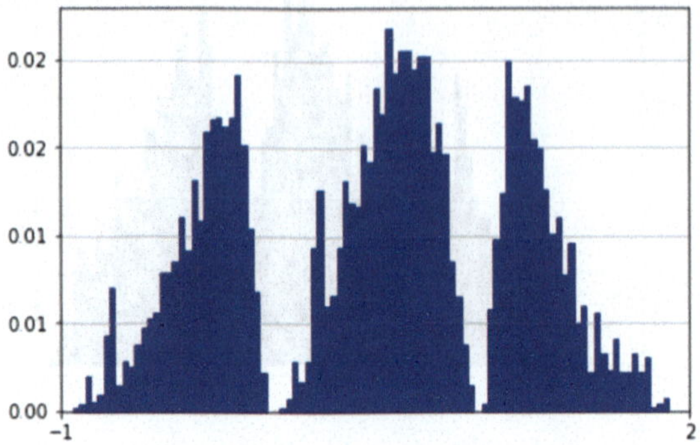

Fig. 7. Japanese street food-600 USD Giant rainbow-colored lobster seafood lobster: Distribution of negative, positive and neutral comments (CDF)

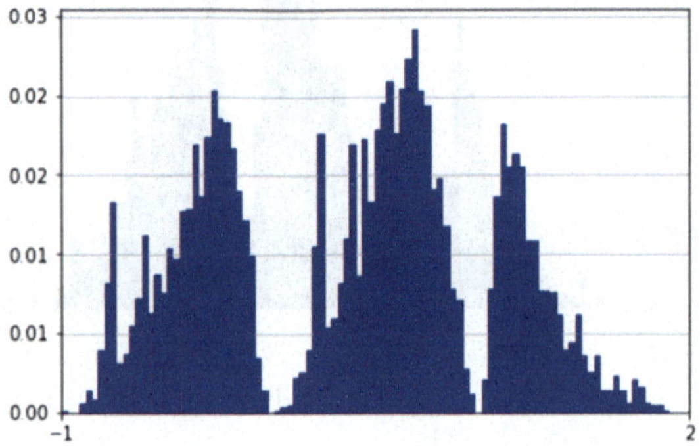

Fig. 8. Japanese: Japanese street food-Fugu: Distribution of negative, positive and neutral comments (CDF)

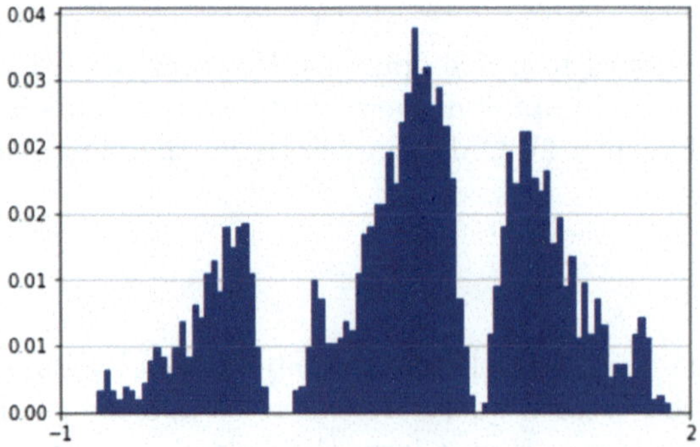

Fig. 9. Japanese: Japanese street food-Egg omlet "Tamagoyaki Tokyo"...: Distribution of negative, positive and neutral comments (CDF)

3.6 Discussion (3)

Figures 7, 8 and 9 show a popular video about Japanese food; however, unlike the trend of opinion distribution among the popular channels, critical comments and comments from the viewpoint of animal welfare tended to stand out. Regarding Fig. 9, there was a tendency similar to that in case (2) because of practical and helpful elements.

4 Summary

In this study, we examined the trend of opinion distribution on YouTube to understand the formation of social consensus from the distribution of negative and positive scores in social media comments. The main video media that was used for the analysis was related to food in daily life. There was a relatively positive opinion trend through cases (1) to (3). However, there are also themes that differ, as shown in Figs. 7 and 8. However, it was speculated that not only pleasantness but also convenience played a role in users browsing and making comments on food related videos on Youtube. One of the limitations of this study is the fact that the trend was analyzed only within the acquisition category "ch videos" (nearly 100). Furthermore, other search queries and popular videos in the theme and the trends that could not be obtained by only using ch were not considered. In addition, the detection of clusters of people that require social welfare and mental health care using information on the web and determining the right care for them is complex. The comment distribution of the videos related to "social welfare" was not obtained on another axis because of the significant amount of data and diversity of opinions. Therefore, we thought that observation and comparison of various social infrastructure themes were necessary. In addition, the distribution of media-inclusive care information based on large-scale data is also required. However, it should not be forced or unilaterally improved. Dissemination of media-inclusive information regarding each symptom, problem, individual activities, and companies that perform various social welfare activities are conspicuous. On the other hand, various trials and errors media-inclusive type using large-scale multidimensional data I think that are necessary for the care method.

Acknowledgements. We would like to acknowledge the support from the 2019 Scholarly Scholars Research Grant Program (Leading Initiative for Excellent Young Researchers (LEADER)).

References

1. Shepherd, A., et al.: Using social media for support and feedback by mental health service users: thematic analysis of a twitter conversation. BMC Psychiatry **15**(1), 29 (2015)
2. Doan, S., et al.: Extracting health-related causality from Twitter messages using natural language processing. BMC Med. Inform. Decis. Mak. **19**(3), 79 (2019)

3. French, J.R.P.: A formal theory of social power. Psychol. Rev. **63**, 181–194 (1956)
4. Kawahata, Y.: Examination of analysis method of opinion distribution in News media transferred on web IES2019, in press
5. Harary, F.: A criterion for unanimity in French 's theory of social power. In: Cartwright, D. (ed.) Studies in Social Power. Institute for Social Research, Ann Arbor (1959)
6. Yoshiaki, Z., Sugiyama, M., Yamaki, N.: Group AHP using consensus building model. J. Oper. Res. **40**(2), 236–244 (1997)
7. Yahata, N., et al.: A population interval AHP method using dissatisfaction function. J. Oper. Res. Soc. Japan **45**(3), 268–284 (2002)
8. Yamaki, N., Sekiya, K.: Large-scale AHP with incomplete data (determination theory and related fields) (1998)
9. Bird, S., Loper, E.: NLTK: the natural language toolkit. In: Proceedings of the ACL 2004 on Interactive Poster and Demonstration Sessions. Association for Computational Linguistics (2004)
10. NLTK Project, Natural Language Toolkit. https://www.nltk.org

Quantitative Consideration of Spatio-Temporal Information on Education and Culture in Japan Using Telephone Directory Archive Data

Yasuko Kawahata[1,2](✉)

[1] Faculty of Social and Information Studies, Gunma University, 4-2 Aramaki-machi, Maebashi, Gunma 371-0044, Japan
kawahata@gunma-u.ac.jp

[2] Center for Spatial Information Science, University of Tokyo, 5-1-5 Kashiwanoha, Kashiwa-shi, Chiba 277-8568, Japan

Abstract. In this study, an attempt was made to quantitatively understand the spatio-temporal situation with respect to the prosperity of educational and cultural facilities in municipalities nationwide. For this, Japanese telephone directory data, taken from the digital archive containing geospatial information (address and industry) of all the facilities/institutions in a given area, were used to analyze the time series changes and obtain the latest industrial trends. Multiple patterns were observed in the growth and decline of educational, cultural, and business fields, and a detailed understanding of the regional characteristics that lead to future trends, the growth of cities, and future tourism strategies in times of disaster, was achieved.

Keywords: Telepoint data · Digital archives · Spatio-temporal information

1 Introduction

At present, available geographic information is diverse, such as data sent from mobile base stations and information with real-time characteristics to grasp our behavior patterns. In the past, location information and travel directions were generally found by referring to symbolic information such as a route map using booklets and signs in the city. As for automobiles, car navigation systems have appeared since the 1980s, and research focused on the usage of such location information are in progress [1–3]. Furthermore, it is possible for a user to easily refer to real-time location information through personalized tools such as smartphone applications. There are a lot of various data that collected those data. There are several factors regarding these phenomena that require attention, such as leakage of personal information when referring to or transmitting personal location information within an application (locations can be disclosed

H. Sato et al. (Eds.): IES 2019, PALO 12, pp. 175–182, 2020.
https://doi.org/10.1007/978-3-030-37442-6_16

via services such as SNS). Therefore, there is a need for self-management and literacy when using smartphones to link personal and location information. However, despite these disadvantages, they allow for several tools that can improve quality of life, such as a map application displaying unrestricted (societally inclusive) facilities or information on evacuation shelter maps, public halls, welfare facilities, etc. that are installed voluntarily and registered and controlled by volunteers [4–6]. In addition to the map function, SNS and other communication tools can also be implemented, and game elements have been introduced to actively share information not only in Japan but around the world. In the future, the demand for such important information will continue to grow, and the circle of sharing will expand. Access to spatial information will be available under equal conditions in places where various adverse situations occur. Using such information, various methods of cartographic communication can be addressed by language, map symbols, signs, and road signs [6], changing each year in line with technological innovations. While various forms of spatio-temporal information are available, the telephone directory contains point data as geographical information that are regularly updated. These data include those registered for corporations in each industry that are periodically updated in units of latitude and longitude. Until now, the Ministry of Economy, Trade and Industry has led a wide range of research on "commercial statistics" regarding changes in shopping streets and commercial areas [7–9]. Approaches for surveying declining areas, such as a vacant house in a municipality, are also being investigated [10]. In addition, a variety of other approaches such as questionnaire surveys and fieldwork have been adopted as survey methods [11].

2 Purpose

Using the directory data, it is possible to identify the changes in the number of contractors in each region and industry from 2012 to 2018. In this paper, the above-mentioned data will be used to present a preliminary report on the changes and general conditions of education and culture throughout Japan. Notably, the data in this study considers trends within the range registered in the directory. However, in the case of education, there are several cases where corporations are registered with information such as compliance, cooperation with local governments, and so on. From this perspective, we believe that by analyzing these data, we can obtain important information that can help us in quantitatively and comprehensively understanding spatio-temporal features relevant to educational and cultural backgrounds.

3 Dataset

In this study, telephone directory data were used to obtain the latest industrial trends in municipalities nationwide. The analysis was carried out using the Japanese version of phonebook data containing industry information (22 data points from June 2012 to October 2018). The phonebook data contains geospatial information (address and industry) of all stores, institutions, and facilities, such as business establishments, hospitals, schools, parks, welfare facilities, and medical facilities. Therefore, it is possible to analyze detailed time-series changes of stores and facilities on a nationwide scale. Not only does the phonebook contain a wide range of data, from that of large corporations to individual, small-business owners, but the data are updated every four months, and therefore, detailed information about every store/facility can be obtained. The phonebook is essentially a large-scale digital archive with temporally high frequency and spatially accurate features. Each store/facility is classified into 25 major categories and 332 minor categories based on the type of industry. Previous studies have observed that these datasets can be scaled from the relationship between population and the number of stores and facilities in units of municipalities. The number of medical-related facilities is $\beta - 1$, and the number of facilities appears to increase linearly as the population increases. On the contrary, the facilities related to the real estate industry have $\beta \leq 1$, and the number of facilities increases beyond the population increase, and the accumulation progresses. In addition, facilities related to public institutions have $\beta \leq 1$ and the increase is lower than the rate of drought increase, although it is expected that efficiency will increase [1,2].

4 Overall Trends

First, in order to grasp detailed trends over the entire fiscal year, we analyzed the general situation regarding educational and cultural facilities from 2012 to 2017, when all the data for the fiscal year were obtained.

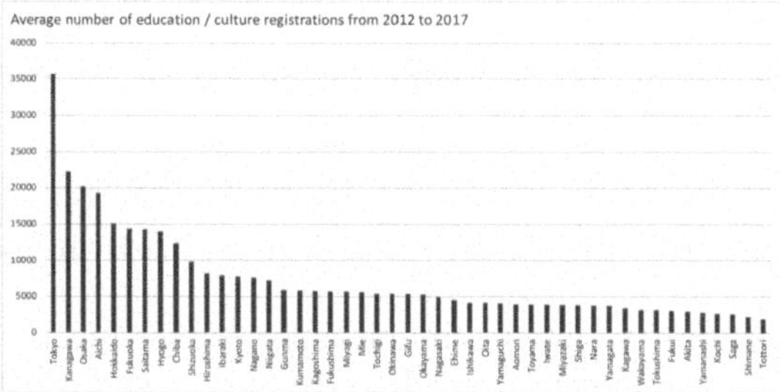

Fig. 1. Average number of educational and cultural facilities from 2012 to 2017

Looking at the average overall trend, as shown in Fig. 1, it can be seen that contractors are registered in densely populated areas such as Tokyo and Kanagawa, followed by major government-designated cities and the Kanto region.

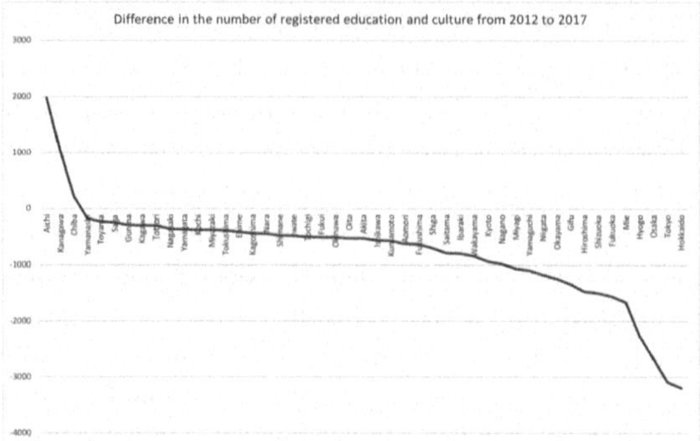

Fig. 2. Difference in the number of registered education and cultural facilities from 2012 to 2017

As seen in Fig. 2, on average, there is a tendency for the number of educational and cultural facilities in prefectures (Tokyo and Osaka), which existed at the top, to decrease significantly, whereas in Aichi and Kanagawa, it is observed to generally increase.

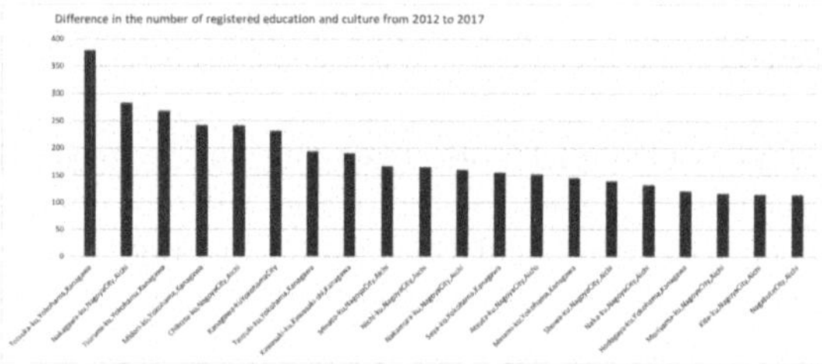

Fig. 3. Difference in the number of registered education and cultural facilities from 2012 to 2017 (by municipality top 20)

As can be observed in Fig. 3, the central areas in Aichi and Kanagawa are increasing.

Figure 4 depicts regions with a significantly decreasing trend from 2012 to 2017. Some cities in Yamaguchi tended to decrease significantly; however, new educational measures such as a scholarship system have been implemented since 2019, a point that should be observed to track the types of changes that have been observed continuously. In areas such as Fukuoka City (Fukuoka), Sapporo City (Hokkaido), and Shibuya-ku (Tokyo), where commercial development and other forms of urban development are progressing, there is a decline in the educational and cultural facilities. In addition, areas such as Kumamoto, Fukushima, and Miyagi, that once suffered major catastrophes, can also be observed to be on the decline. Once again, factors that should be considered regarding educational

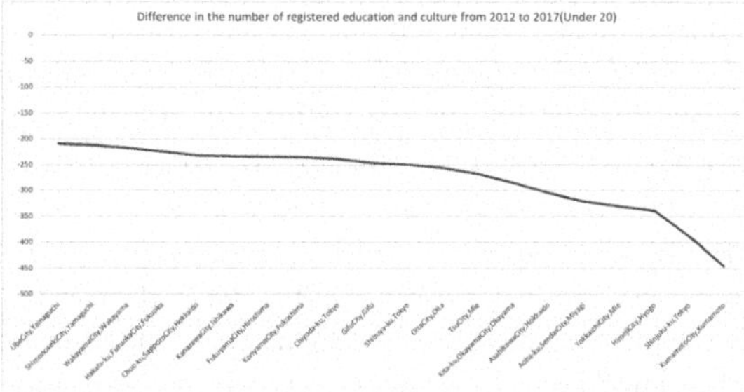

Fig. 4. Difference in the number of registered education and cultural facilities from 2012 to 2017 (by municipality under 20)

and cultural facilities and educational infrastructure after a major disaster were observed.

5 Discussion

Here, spatiotemporal archive data from 22 points from 2012–2018 were used in order to obtain a more detailed understanding.

5.1 Detailed Classification

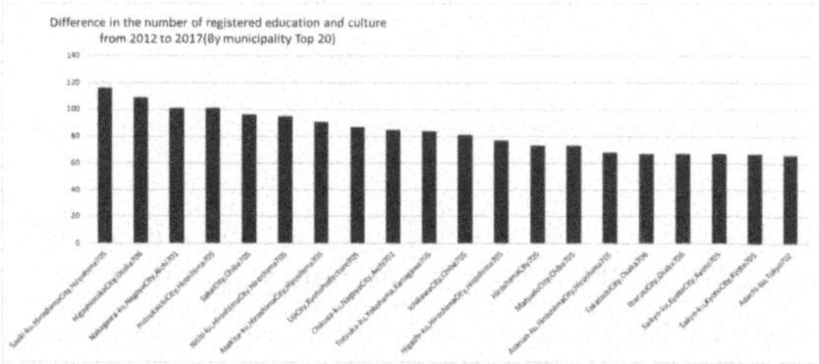

Fig. 5. Detailed classification of difference in the number of registered education and cultural facilities from 2012 to 2018 (by municipality top 20)

Industry numbering in Fig. 5

(A1) 701: Cram School/Preparatory School/Teacher
(B1) 702: Nursery/Kindergarten
(C1) 705: Business School/Classroom
(D1) 706: Musical Instruments Sales and Repair

As can be seen in Fig. 5, business-school-related trends are increasing in cities far from the center such as Hiroshima City (Hiroshima), Yokohama City (Yokohama), Uji City (Kyoto), Sakai City, and Matsudo City (Chiba). In addition, Nakagawa-ku (Nagoya) has shown the largest increase in nurseries since 2012, a sector of childcare and education whose insufficiency and operational failures have been condemned by the Japanese public in recent years [12]. Even when the conditions were narrowed down to only "(A1) 701," it was found that nursery schools in the six wards of Nagoya city (Nagoya) were on the rise in the top ranks nationwide, indicating that they have been actively working on countermeasures against the current situation.

In addition, it was observed that the number of corporations related to the sale and repair of musical instruments increased in areas far from the center of Higashi Osaka City, Takatsuki City, and Ibaraki City (Osaka). In addition, 17 cities in Osaka corresponded to the top 20 districts when the condition was extracted only for "(D1) 706" in the increase/decrease list of each industry in the municipalities nationwide. We tend to occupy more than a majority of the total, and we would like to continue to consider the types of trends that can be expected the future from the perspective of tourism and cultural beliefs for major events related to Japanese history such as the Olympics(2020) and Expo(2025).

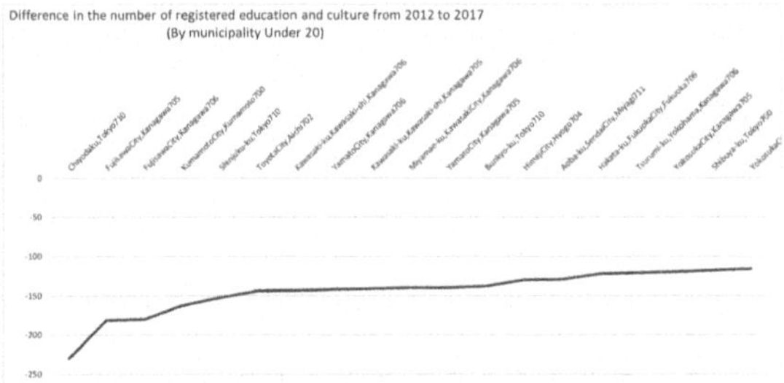

Fig. 6. Detailed classification of difference in the number of registered education and cultural facilities from 2012 to 2018 (by municipality under 20)

Industry numbering in Fig. 6

(A2) 700: Culture School
(B2) 702: Nursery/Kindergarten
(C2) 704: Elementary School
(D2) 705: Business School/Classroom
(E2) 706: Musical Instruments Sales and Repair
(F2) 710: Book Publishing
(G2) 711: University, Junior College, Graduate School

Figure 6 shows that the number of publishing-related businesses in the Tokyo metropolitan area in Chiyoda-ku, Shinjuku-ku, and Bunkyo-ku (Tokyo) has declined significantly. A tendency to be observed was observed. It was observed that these trends gradually decreased every 4 months since 2013. Furthermore, the number of cases has decreased by about 70 from January to April 2018. In addition, the number of business schools in Fujisawa City, Kawasaki City, Yamato City, Yokosuka City, and Yokohama City (Kanagawa) has been decreasing since 2013, while the number in Hakata-ku (Fukuoka) has been decreasing rapidly, particularly in 2015 and 2017 (an excessive trend was seen, such as one where there were over 100 cases in 4 months). In addition, the number of corporations related to musical instrument sales and repairs also showed a sharp decline after January 2015. The culture school in Shibuya-ku (Tokyo) has been on the decline since 2016.

6 Summary

In this study, an attempt was made to quantitatively grasp the spatio-temporal situation with respect to the prosperity of educational and cultural facilities in municipalities nationwide. In order to grasp the detailed trends, the overall trends were considered, and a detailed consideration was obtained to understand future tendencies.

(1) Certain areas such as Hakata Ward (Fukuoka), Sapporo Chuo Ward (Hokkaido), Shibuya Ward (Tokyo) were found to have excessive flow rates. However, it was observed that in such areas, where urban development was progressing in terms of commercial facilities, a downward trend was seen in terms of educational and cultural facilities.
(2) Kumamoto City, Fukushima Prefecture, Miyagi Prefecture and other areas where major disasters once occurred showed a declining trend in educational and cultural facilities.
(3) In 2013, the number of business schools in Fujisawa City, Kawasaki City, Yamato City, and Yokosuka City (around Kanagawa) also demonstrated a declining trend. There was a tendency to infer changes and globalization (transfer to overseas bases) in the Internet business area. In addition, the culture school in Shibuya-ku (Tokyo) has been on the decline since 2016, and it is speculated that the development of a public network that allows online learning and the distribution of devices contributed to this. Overall, (3) has been presumed to reflect phenomena that are unique to areas that are sensitive to trends such as online business.
(4) While Japan's lack of sufficient nursery schools has previously been condemned by the public [12], and it was found that Nagoya City (Aichi) has significantly increased its number of nursery schools since 2012. When similar conditions were narrowed down, it was found that nursery schools in the six wards of Nagoya City were on the rise in the top ranks across the country, indicating that they are actively working to address the aforementioned insufficiency.
(5) As an interesting consideration, a marked increase in musical instrument sales and repair in Osaka Prefecture was observed. In 2019, Hundred Birds

and Furuichi Burial Mounds in Sakai City (Osaka) were registered as UNESCO World Heritage Sites, and the Osaka Expo has been scheduled for 2025.

In order to gain insight into how such patterns develop in history, education, and culture, we hope to continue this study using the digital archive data in the phonebook.

Acknowledgements. We would like to extend our gratitude to the 2019 Scholarly Scholars Research Grant Program (Leading Initiative for Excellent Young Researchers (LEADER)), for providing us with support to conduct this research.

References

1. Bettencourt, L.M.A., et al.: Growth innovation, scaling, and the pace of life in cities. Proc. Nat. Acad. Sci. **104**(17), 7301–7306 (2007)
2. Enquist, B.J., Brown, J.H., West, G.B.: Allometric scaling of plant energetics and population density. Nature **395**(6698), 163 (1998)
3. Kawahata, Y., Mizuno, T., Ishii, A.: Measurement of human activity using velocity GPS data obtained from mobile phones. arXiv preprint (2017). arXiv:1706.04301
4. Bmaps. https://web.bmaps.world/ (Ref : 2019/8)
5. FunGoOut. https://fungoout.com/ (Ref : 2019/8)
6. Wheelog. https://www.wheelog.com/ (Ref : 2019/8)
7. Kashiwagi, T., Akiyama, Y., Shibasaki, R.: Monitoring method of three-dimensional distribution change of shops and offices by spatio-temporal integration of digital map and phone book data. Architectural Institute of Japan, 73 (626), 789-793 (2008)
8. Akiyama, Y., Sengoku, H., Tamura, K., Shibasaki, R.: Development of commercial statistical point data throughout Japan and verification of reliability of polygon data in commercial cluster areas (2011)
9. Sengoku, H., Akiyama, Y., Shibasaki, R.: Analysis of migratory behavior in commercial clusters using autologs of GPS mobile phones (2011)
10. Akiyama, Y., Sengoku, H., Shibasaki, R.: Development of time-series data sets of stores and offices using telephone directories and electronic maps, and analysis of the current state of commercial clusters. In: Proceedings of the 18th Geographic Information Systems Association Conference, pp. 233–238 (2009)
11. Kanamori, Y., Ariga, T., Matsuhashi, K.: Factor analysis and future estimation of vacant house rate. City Plan. Papers **50**(3), 1017–1024 (2015)
12. BBC NEWS. Is Japan's plan for more women in the workplace failing?. https://www.bbc.com/news/35888504 (Ref:2019/8)

Analyzing Urban Resilience Using Temporal and Spatial Archive Information Before and After a Disaster – Perspective of SDGs

Yasuko Kawahata[1,2](✉)

[1] Faculty of Social and Information Studies, Gunma University, 4-2 Aramaki-machi, Maebashi, Gunma 371-0044, Japan
kawahata@gunma-u.ac.jp

[2] Center for Spatial Information Science, University of Tokyo, 5-1-5 Kashiwanoha, Kashiwa-shi, Chiba 277-8568, Japan

Abstract. In this study, point data from a phone book was used as geographical information that was regularly updated to obtain insights into changes in the urban industrial structure before and after a disaster. From these data, a quantitative approach was taken for regional trends in the affected areas before and after a disaster. This includes the population before and after a disaster and the demand for mental health support. Local governments also make new city plans for reconstruction; however, issues remain concerning their sustainability. This paper considers the viewpoints of disaster recovery planning, sustainability, and urban resilience while referring to previous studies.

Keywords: Telepoint data · Digital archives · Spatio-temporal information · Resilience · Disaster recovery · SDGs

1 Introduction

The widespread use of various devices, such as smartphones, allows us to confirm our location and estimate the time to reach our destinations. Geographical information has various opportunities for creating several forms of non-verbal communication in our lives. In addition, the "Super City Initiative" and other proposals have called for the revision of the National Strategic Special Zones. Meanwhile, the urban use of microgeodata, such as GIS data, will be promoted further. Beyond the boundaries of industry-academia-government, smart city-related projects have been launched in various parts of the country aiming toward a "Smart city public-private partnership platform" [1]. Because interpolation is the subject of this paper, we highlight many prior investigations that addressed the relationships between individuals and a country. This was achieved by using SNS at the time of a disaster [2]. In addition, various support networks have been

H. Sato et al. (Eds.): IES 2019, PALO 12, pp. 183–193, 2020.
https://doi.org/10.1007/978-3-030-37442-6_17

established, such as providing information sharing platforms for mobile base stations such as "Disaster Message Boards". In the case of the 2011 Great East Japan Earthquake, various information links were made regarding the demand for goods and crisis information in the affected areas. Because Japan is an island country where unpredictable crises are likely to occur, discussions on how to raise resilience to disasters, urban regeneration, sustainable support, and mental care methods in disaster areas are necessary. These discussions and research can be used as effective knowledge for other countries. In this study, point data in a phone book was used as geographical information that is regularly updated to obtain insights into changes in the urban industrial composition before and after a disaster [4–6]. This data includes corporate registered data for each industry that is periodically updated in latitudinal and longitudinal coordinates. Trends in the number of contractors in each region and industry from 2012 to 2018 were identified. This paper focuses on the case of the 2011 Great East Japan Earthquake followed by a discussion of the surrounding areas, such as tsunami affected areas.

2 Purpose

In this study, point data in a phone book was used as geographic information that was regularly updated to obtain insights into changes in the urban industrial composition before and after a disaster [3–5]. This data includes corporate registered data of each industry that is periodically updated in latitudinal and longitudinal coordinates. Trends in the number of contractors in each region and industry from 2012 to 2018 were identified. This paper focuses on the case of the 2011 Great East Japan Earthquake followed by a discussion of the surrounding areas, such as the tsunami affected areas. There are too many uncertainties in earthquake predictions. As a result, there will be a need for continuous observation of industrial trends regarding countermeasures after earthquakes occur. Various factors such as regional characteristics and population form are necessary for reconstruction. Additionally, continuous mental health support from outside communities is required. Local governments also make new city plans for reconstruction; however, issues remain whether they are sustainable. This paper considers the perspective of disaster recovery plans, sustainability, and urban resilience.

3 Preview Research

Considerations for large-scale disasters can be addressed from various perspectives. Since the Great East Japan Earthquake, research on the awareness for the growing market demand for used cars, along with the contribution of ICT tools for communications during disasters, has been undertaken [6–8]. Furthermore, discussions about the transition of mobile tools in the event of a major disaster have been actively investigated from a transportation perspective. Additionally, research related to the redevelopment of disaster-stricken urban areas was undertaken due to the Great Hanshin-Awaji Earthquake [9]. In these research projects, annual business trends in the flooded areas were considered using telephone directory data. This study also discusses the importance of sustainability, issues, and past crisis comparisons as a reconstruction land readjustment project. Another approach using a phone was to identify how disasters have affected the diet that supports people's daily lives, which has also been investigated. Research was performed to determine what kind of tendencies were observed regarding food access difficulties that includes restaurants [10]. This survey extracted areas where damage was particularly severe along with changes in access regarding the rate of loss before and after a disaster. There is also research on how industrial forms have been reorganized, with a focus on Sendai City, the center of the disaster area [11–14]. Natural disasters recently caused by floods, such as torrential rain, have frequently occurred. These disasters are presumed to have been caused by global warming. As a result, there have been many problems with the operation of public facilities in municipalities, such as obtaining valuable documents and archives from libraries. Countermeasures have been developed in various regions; the damage status has been confirmed using research surveys [15]. Additionally, efforts are being made to assume the Nankai Trough for disaster simulations using human flow data. In the future, many evacuation routes using human flow data and telephone directory data will be used. This includes measures to improve reconstruction and resilience from familiar disasters to large-scale disasters. More diverse research teams are expected to require more case studies while using prediction models that undergo mathematical processing along with more proposals [16,17].

4 Dataset

In this study, telephone directory data was used to obtain the latest in industrial trends in municipalities across Japan. The analysis was performed using telephone book data with industry information in Japan between 22 June 2017 to 22 October 2018. The phone book data contains geospatial information (address and industry) of all stores and facilities (e.g. offices, hospitals, schools, parks, welfare facilities, medical facilities, etc.). This data allowed the analysis of a time series change. Moreover, the phone book data was updated every four months and also contains information for each store and facility. In addition, the phone book contains a wide range of data from large corporations to individual business owners who do not have any corporations. Stores and facilities were classified into 25 major categories and 332 minor categories according to the type of industry.

5 Trends Derived from Previous Studies

5.1 Relationship Between the Earthquake and Used Cars

Insights were obtained for companies that were registered as corporations that include used car dealerships before and after the Great East Japan Earthquake.

The municipalities along the coast of Miyagi, which were large-scale due to the tsunami damage, were increasing after 2012. When considering Miyagi as a whole, 53% of used cars were in demand.

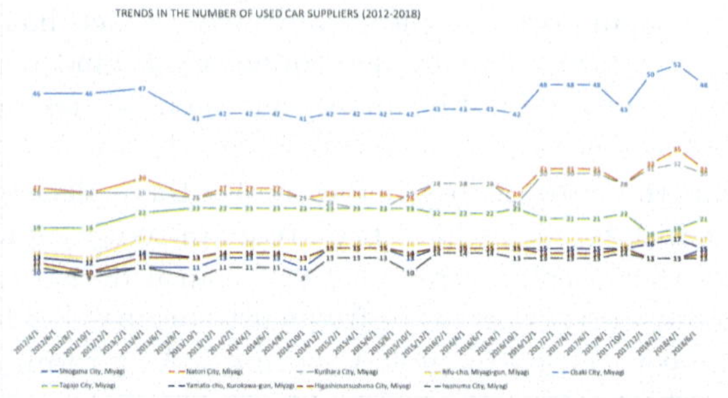

Fig. 1. Trends in the number of used car suppliers (2012–2018) (Top 10 in increasing trend)

As shown in Fig. 1, previous research revealed that the demand for used cars was very high immediately after the disaster. However, the demand remained flat from 2012 to 2018 because the tsunami damage was significant. Coastal cities (e.g. Shiogama City, Natori City) showed a slight, but increasing, trend in sales when analyzing the period between 2012 to 2018. This finding is one trend after a large-scale disaster in the coastal area where transportation means were cut off.

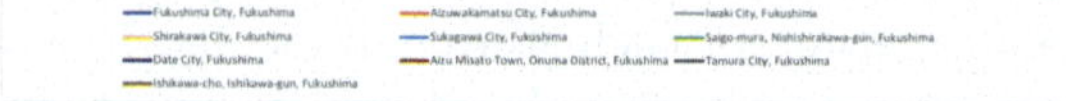

Fig. 2. Number of used car suppliers in the Fukushima Prefecture (Top 10 in increasing trend)

Figure 2 focuses on Fukushima, which had a similar large-scale impact. In the Fukushima Prefecture, there are some areas where the population decreased or was inaccessible due to the nuclear power plant disaster. However, the demand for used cars is higher than the Miyagi Prefecture.

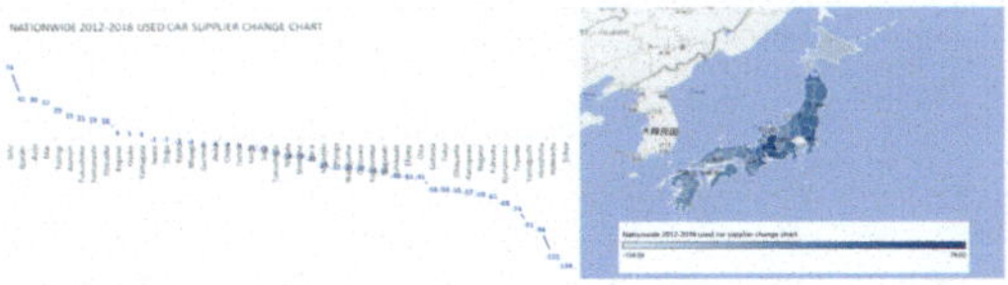

Fig. 3. Nationwide 2012–2018 used car supplier change chart

As seen from Fig. 3, the overall demand for used cars continued from Gifu and Ibaraki, followed by Aichi, Mie, and Tochigi. Fukushima was the top location for used car sales whereas Miyagi was in the middle for sales. Additionally, the demand for used car sales declined in Tokyo, Hokkaido, Hiroshima, Yamaguchi, Kumamoto and Fukuoka. Figure 3 shows that Kumamoto, which had a large-scale disaster in 2016 along with Fukuoka and Hiroshima, suffered from floods due to heavy rain. This corresponds to a disaster; yet, there was no demand for used cars. It was inferred that the demand for used cars increased after a tsunami. This is demonstrated in the case of Miyagi where the division of railways and mountain roads occurred.

Industrial classification in which there was a particularly increasing trend	Increase in 2012-2018	Increase in 2012-2014
Renovation (housing)	812	836
Kimono	198	220
Elderly welfare and care services	104	105
Job-specific construction business (construction industry)	87	105
General construction business (construction industry)	86	100
Safe / key (house)	83	100
Car maintenance	73	89
Household watering equipment	73	71
Household cleaning (housing)	58	40
Household cleaning (interior)	57	39
Exterminating pests and rats	57	38
Building materials handling	50	35
Rental / Rental	40	30
convenience store	38	21
Welfare / Child Welfare / Disability Services / Rehabilitation	37	20
Water, gas, power	34	16
Shoes / Bag / Accessories	32	14
Exterior construction	30	12
Used goods buying and selling	25	12
Administrative scrivener	21	11
Nursery school, kindergarten	16	10
Used car purchase sales	15	10

Fig. 4. Industrial classifications where there was an increasing trend for areas affected by large-scale tsunami damage

5.2 A Multifaceted Study on the Reconstruction Situation After a Large-Scale Earthquake Disaster

As shown in Fig. 1, when there was a large-scale crisis, such as a tsunami, there was a great demand for used cars. A characteristic trend for Tagajo City was obtained [18]. As shown in Fig. 4, there was an increase and decrease in business between 2012 to 2018. In particular, it can be seen that renovations, kimonos, welfare centers, and construction-related businesses had the largest increase immediately after the 2012–2014 earthquake. On the other hand, it was inferred that mice extinction, cleaning, rehabilitation, convenience stores, and gas water supply businesses contributed to the rise in kindergarten classes and

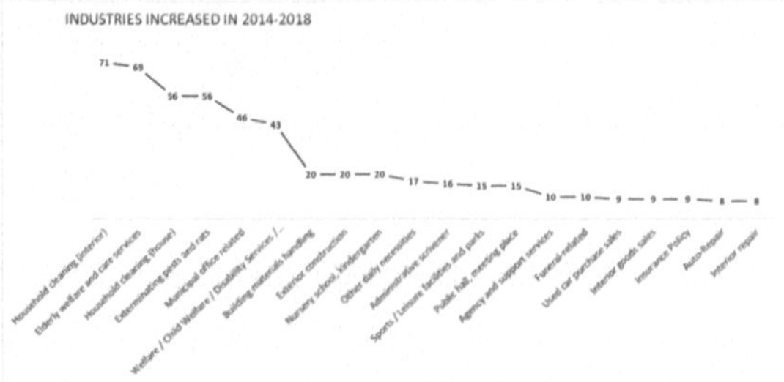

Fig. 5. Industries that increased in business between 2014–2018

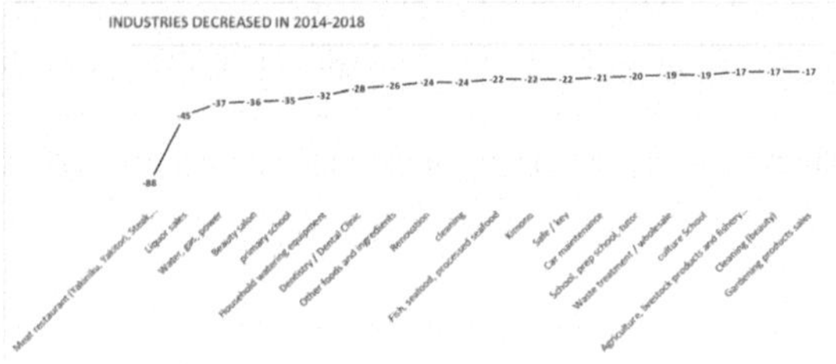

Fig. 6. Industries that decreased between 2014–2018

used car sales. Although demand for remodeling after the earthquake declined, it is speculated that cleaning-related issues (e.g. cleaning and mouse removal) were prominent and would continue to cause secondary damage due to tsunamis. In addition, the increase in convenience stores was presumed to be a good sign for the decline in elementary and junior high schools and the aging of welfare care-related facilities.

Figure 5 shows the number of contractors that increased, particularly between 2014 and 2018, in the tsunami-affected areas. House cleaning businesses saw the most growth whereas care-related and rat extermination businesses also increased. Additionally, leisure facilities such as sports, public facilities, and automobile repair centers also increased. Gradually, the infrastructure and transportation methods of local communities also increased. Cleaning and repair centers have been on the rise. Observing these signs is important as an indicator for resilience after a disaster.

Figure 6 lists the projects that decreased between 2014 to 2018 in the tsunami-affected areas. The number of restaurants, such as Yakiniku and Steak, were declining. In addition, liquor sales were on a downward trend. Additionally, the number of beauty salons, dentists, and renovation/cleaning businesses have decreased significantly in comparison to 2012 to 2014.

5.3 Characteristic Business in the Affected Areas

Figure 7 shows the time-series transition of the "Kimono" related business, which was listed as one of the top growing businesses. There has been a significant upward trend since 2013/4, and there has been an active effort for Japanese culture in the region. This includes the increase in the number of tourists in the NHK drama series, such as "Amachan", which is the theme of the earthquake disaster. It was observed that the number of business operators continued without any significant reductions in business.

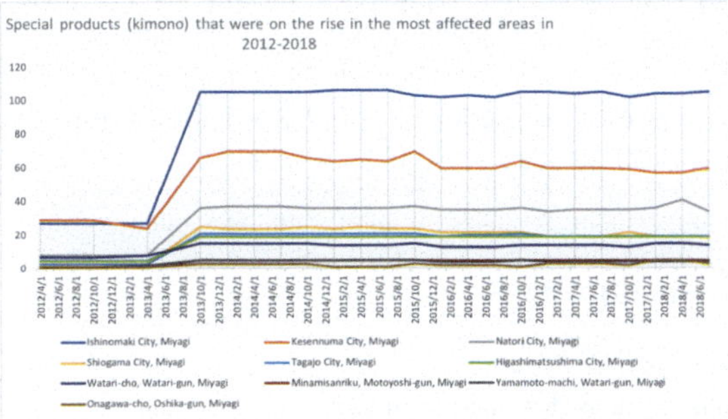

Fig. 7. Special products (Kimono) that were on the rise for the most affected areas in 2012 to 2018

6 Discussion

As shown in the analysis in Sect. 3, the businesses that featured a significant decline were education and culture. Human resource development in the community after a disaster affects the subsequent sustainability of the community [19,20]. A previous study on the formation of a city near a catastrophe also stated that new housing plans do not progress. This is due to the vacant housing problem; however, renovations still continue. From the above, development of towns while focusing on renovations are progressing; there is a tendency to infer this is due to aging. Construction of sustainable business support and regional management to promote residency is important. These trends were also observed in Figs. 4 and 5. However, there is a need for an educational and cultural environment for passing on to new generations from the perspective of inheriting memories from disaster-stricken areas and disaster recovery. Each region needs young leaders to conduct various educational activities using different activities, such as storytelling and illustrations, along with invitations to the creators. It can be inferred from the case study that the influx at an early stage makes the inheritance of local memories, along with the consensus building for the reconstruction of the affected areas, more fluid. This also has the potential of being a city that communicates new regional connections and how the city should be revitalized after a disaster. From this point of view, educational and cultural facilities were considered.

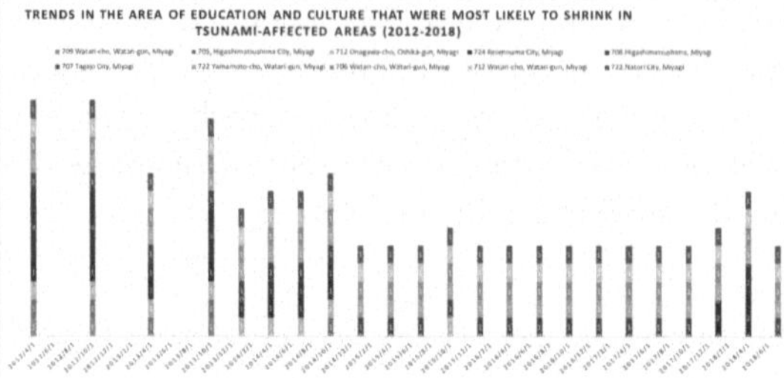

Fig. 8. Trends in the area of education and culture that were most likely to shrink in tsunami affected areas (2012–2018)

Industry numbering in Fig. 8

705: Business School/Classroom
706: Musical Instrument Sales and Repair
707: Language school
708: Junior high school
709: Sewing and handicraft materials and equipment sales
712: High school
722: Elementary School
724: Audio equipment and services

Figure 8 describes the time-series transition of businesses that are declining among educational and cultural businesses in the tsunami affected areas. A sharp downward trend has occurred in the above-mentioned industries since October 2013. This indicates that business schools and facilities related to manufacturing and hobbies such as music, sound, and handcraft are decreasing. A downward trend was also observed for junior and senior high schools.

Figure 9 describes the time-series transition of businesses that were on the rise in education and culture in the tsunami affected areas. As inferred from Fig. 9, here is a description of the post-disaster education and culture business. There was a significant decrease; however, business increased, which is a benchmark for post-disaster recovery.

Industry numbering in Fig. 9

702: Nursery/Kindergarten
709: Sewing, handicraft materials, and equipment sales
712: High school
717: Library

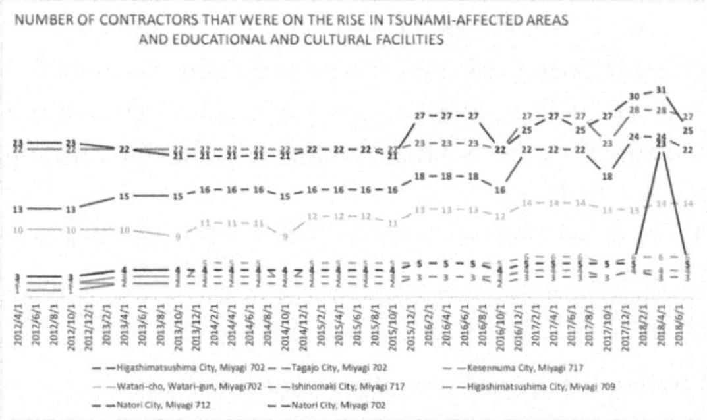

Fig. 9. Trends in the area of education and culture that were most likely to decline in tsunami affected areas (2012–2018)

From Fig. 9, it was observed that nursery schools and kindergartens have been increasing since 2016 in the three regions. This can be inferred as a positive sign that the development of children in the affected areas has begun. These signs will continue to be observed as a viewpoint of building a sustainable society after a large-scale disaster.

7 Summary

This paper focused on the case of the Great East Japan Earthquake in 2011 with the goal of obtaining quantitative considerations regarding the surrounding areas of the tsunami-affected areas using a telephone directory archive. There are too many uncertainties in earthquake predictions. There will also be a need for continuous observation of industrial trends regarding countermeasures after they occur. Various factors (e.g. characteristics of other regions, population form, etc.) are necessary for telephone directory data. Additionally, continuous external mental support is also required. Local governments also make new city plans for reconstruction; however, issues remain whether they are sustainable city plans. This study focused on quantitative considerations from the viewpoints of disaster recovery planning, sustainability, and urban resilience.

7.1 Considering on Characteristics of Tsunami-Affected Areas

(1) Previous research has shown that the demand for used cars was very high immediately after a disaster. Cities in the coastal areas that were damaged by tsunamis (e.g. Shiogama City and Natori City), saw their businesses as relatively flat with a modest increase from 2012 to 2018. However, these tendencies were adapted to the case of transportation, such as a tsunami, and the division of railways and mountain roads as described in the case of the Miyagi Prefecture.

(2) This describes the time-series transition of businesses that are declining among educational and cultural businesses in tsunami affected areas. A sharp downward trend has occurred in the above-mentioned industries since October 2013. This indicates that business schools and facilities related to manufacturing and hobbies (e.g. music, sound, and handcraft) are decreasing. This study also observed that there was a downward trend for junior and senior high schools.

7.2 Consideration of Resilience Perspective

(A) In Japan, a variety of businesses increased and decreased from 2012 to 2018 due to natural disasters. In particular, it can be seen that renovation, kimono, welfare care, and construction-related businesses had the largest increase immediately after the 2012–2014 earthquake.

(B) In particular, it is speculated that there was an increasing trend for several businesses. This includes: rat extermination; cleaning; rehabilitation; gas water supply; kindergarten; used car sales; and convenience stores. While the demand for remodeling after an earthquake has gradually declined, it was surmised that the conspicuous cleaning-related issues (e.g. cleaning and rat extermination) would yield secondary damage in comparison to the tsunami disaster. The increase in convenience stores was presumed to be a good sign for the decline in elementary and junior high schools and the aging of welfare care-related facilities.

(C) Looking at Fig. 9, nursery schools and kindergartens have been increasing since 2016 in the three regions. This can be inferred as a sign that redevelopment in the affected areas has begun. These signs need to be monitored continuously as a viewpoint of building a sustainable society after a large-scale disaster.

Acknowledgements. We would like to thank and acknowledge the support from the 2019 Scholarly Scholars Research Grant Program (Leading Initiative for Excellent Young Researchers (LEADER)), which provided support for this research.

References

1. Launching of Smart City Public-Private Partnership Platform, August 2019. https://archive.iges.or.jp/files/research/cty/pdf/20190328
2. NTT docomo Disaster Message Board, March 2011. http://dengon.docomo.ne.jp/Etop.cgi?es=0
3. Bettencourt, L.M.A., et al.: Growth, innovation, scaling, and the pace of life in cities. Proc. Natl. Acad. Sci. **104**(17), 7301–7306 (2003/2007)
4. Enquist, B.J., Brown, J.H., West, G.B.: Allometric scaling of plant energetics and population density. Nature **395**(6698), 163 (1998)
5. Kawahata, Y., Mizuno, T., Ishii, A.: Measurement of human activity using velocity GPS data obtained from mobile phones. arXiv preprint arXiv:1706.04301 (2017)

6. Shibuya, Y., Tanaka, H.: How does a large-scale disaster impact on the used-car market? A case study of the great East Japan Earthquake and Tsunami. Int. J. Econ. Policy Stud. **13**(1), 89–117 (2019)
7. Shibuya, Y., Tanaka, H.: Using social media to detect socio-economic disaster recovery. IEEE Intelligent Systems (2019)
8. Shibuya, Y.: Impact of ICT tools on disaster logistics issues: a case study of the great East Japan Earthquake and Tsunami of 2011. Int. J. Bus. Inf. **12**(3), 310–341 (2017)
9. Yamada, M., Yamada, M., Hada, K., Fujino, Y., Sakagami, K., Hayashida, T., et al.: Elucidation of damage mechanism in Mashiki-cho during the 2016 Kumamoto earthquake-Effect of ground characteristics-. Earthq. Eng. 73(4), $I_2$16–$I_2$24 (2017)
10. Masunabu, A., Kubota, A.: Research on the relationship between the redevelopment of the Tsunami-stricken urban areas and the trends of the affected businesses. City Plan. Proc. **53**(3), 905–912 (2018)
11. Kanamori, Y., Ariga, T., Matsuhashi, K.: Factor analysis and future estimation of vacant house rate. City Plan. Proc. **50**(3), 1017–1024 (2015)
12. Furuhashi, T., Tatana, Y., Sugaya, Y., Tamaki, T., Okumura, M.: Study on the estimation method of economic damage to the industrial sector due to the Great East Japan Earthquake. Proc. Jpn. Soc. Civ. Eng. D3 (Civ. Eng.) **70**(5), $I_1$97–$I_2$10 (2014)
13. Nagata, S., Nakatani, Y., Yano, K., Akiyama, Y.: Lost area of walking food access from 2010 to 2013. Abstr. Geogr. Soc. Jpn. **2015**(100276), (2015)
14. Keigo, M.: Relocation of office space in Sendai City due to recent rental office building location and tenant movement. In: Human Geographical Society Conference Abstracts Human Geographical Society (2013)
15. Kariya, R.: Record of water damage in Kanagawa Prefectural Library and its analysis and future disaster countermeasures. Bull. Kanagawa Prefect. Libr. = Kanagawa Prefect. Libr. Bull. **13**, 3-24 (2018)
16. Ogawa, Y., et al.: Personal damage estimation of Nankai Trough Earthquake Tsunami based on various scenarios using geo-big data-around Kochi city. E-j. GEO **13**(1), 140–155 (2018)
17. Mimura, S.: Pacific Island Countries and natural disasters-vulnerability and resilience (special issue: sustainable development and international relations in Pacific Island Countries). Ajiken World Trends **22**(2), 28-31 (2016)
18. Overview of the tsunami-affected areas (by municipality: municipal health record), August 2019. https://www.pref.miyagi.jp/pdf/kiki/12shiryou2.pdf
19. Koura, H.: A study on the living conditions of suburban housing estate and the sustainability of urban area. Proc. Urban Plan. **39**, 625–630 (2004)
20. Kawaguchi, T.: Sustainability of suburban residential areas in the age of population decline (2007)

Detecting Stock Market Bubbles Based on the Cross-Sectional Dispersion of Stock Prices

Takayuki Mizuno[1,2,3(✉)], Takaaki Ohnishi[3,4], and Tsutomu Watanabe[3,5]

[1] National Institute of Informatics, Hitotsubashi, Chiyoda-ku, Tokyo 101-8430, Japan
mizuno@nii.ac.jp

[2] Department of Informatics, SOKENDAI (The Graduate University for Advanced Studies), Hitotsubashi, Chiyoda-ku, Tokyo 101-8430, Japan

[3] The Canon Institute for Global Studies, Marunouchi, Chiyoda-ku, Tokyo 100-6511, Japan

[4] Graduate School of Information Science and Technology, The University of Tokyo, Hongo, Bunkyo-ku, Tokyo 113-8656, Japan

[5] Graduate School of Economics, The University of Tokyo, Hongo, Bunkyo-ku, Tokyo 113-0033, Japan

Abstract. A statistical method is proposed for detecting stock market bubbles that occur when speculative funds concentrate on a small set of stocks. The bubble is defined by stock price diverging from the fundamentals. A firm's financial standing is certainly a key fundamental attribute of that firm. The law of one price would dictate that firms of similar financial standing share similar fundamentals. We investigate the variation in market capitalization normalized by fundamentals that is estimated by Lasso regression of a firm's financial standing. The market capitalization distribution has a substantially heavier upper tail during bubble periods, namely, the market capitalization gap opens up in a small subset of firms with similar fundamentals. This phenomenon suggests that speculative funds concentrate in this subset. We demonstrated that this phenomenon could have been used to detect the dot-com bubble of 1998–2000 in different stock exchanges.

Keywords: Stock market · Financial bubble · Nowcast · Power law

1 Introduction

It is common knowledge in macroeconomics that, as Federal Reserve Board Chairman Alan Greenspan said in 2002, "...it is very difficult to identify a bubble until after the fact; that is, when its bursting confirms its existence." In other words, before a bubble bursts, there is no way to establish whether the economy is in a bubble or not. In economics, a stock bubble is defined as a state in which speculative investment flows into a firm in excess of the firm's fundamentals, so

H. Sato et al. (Eds.): IES 2019, PALO 12, pp. 194–202, 2020.
https://doi.org/10.1007/978-3-030-37442-6_18

the market capitalization (= stock price × number of shares issued) becomes excessively high compared to the fundamentals. Unfortunately, it is exceedingly difficult to precisely measure a firm's fundamentals and this has made it nearly impossible to detect a stock bubble by simply measuring the divergence between fundamentals and market capitalization [1–3]. On the other hand, we empirically know that market capitalization and PBR (= market capitalization/net assets) of some stocks increase during bubble periods [4–7]. However, they are also buoyed by rising fundamentals, so it is not always possible to figure out if increases can be attributed to an emerging bubble.

Recently it was reported that real estate bubbles can be detected from increased variability in the prices of houses with similar attributes: the location of the house, the size and floor plan of the house, and so on [8–10]. Similar houses can be regarded as having similar fundamental values. Therefore, increased variation in price among houses with similar attributes signify that speculative money beyond the fundamentals is flowing into a subset in the housing market triggering a real estate bubble.

By applying the real estate bubble detection approach to stock markets, this paper will propose a statistical method of detecting stock market bubbles from growing disparities in market capitalization between firms that are otherwise similar in industrial sector, size, and other attributes. There are similar firms in a stock market. The law of one price would dictate that the fundamentals of these firms are comparable. In other words, if the market capitalization of just few firms among similar firms is exorbitantly high, this suggests that the stocks of those firms are overvalued and that a bubble has formed.

A firm's financial standing is certainly a key fundamental attribute of that firm. Our approach will be to first identify the financial variables that most closely reflect fundamentals using random forest method. We investigate the variation in market capitalization normalized by fundamentals that is estimated by Lasso regression of a firm's financial standing. We will then observe the normalized market capitalization on each stock market from year to year. In non-bubble periods, the normalized market capitalization distribution is thin because market capitalization remains close to the firms' fundamentals. But if a bubble emerges, speculative funds flow into the stock of a small number of firms and this causes the upper tail of distribution to become fatter than during non-bubble periods. In this paper, by observing the upper tail of distribution we will detect the dot-com bubble of 1998–2000 on nine of the leading global exchanges: NASDAQ, NYSE, London SE, Tokyo SE, Paris SE, Korea SE, Shanghai SE, Hong Kong SE, and Taiwan SE. Market capitalization and other financial data were obtained from Thomson Reuters Quantitative Analytics. Classification of firms by industrial sector is according to Thomson Reuters two-digit SIC codes.

The rest of the paper is organized as follows. In Sect. 2, we identify the financial variables and weighting factors that most strongly reflect firms' fundamentals of different industries on the exchanges based on market capitalization and key financial variables during non-bubble periods. In Sect. 3, fundamentals roughly are estimated using Lasso regression of the key financial variables. We observe differences in market capitalization that is normalized by fundamentals. Section 4 demonstrates that stock market bubbles can be nowcast by observing the distribution of normalized market capitalization. And finally, Sect. 5 is a summary of the paper.

2 Financial Variables that Reflect Fundamentals

In definition of bubble, market capitalization remains close to the firms' fundamentals during non-bubble periods. In order to identify the financial variables that most closely reflect fundamentals, we focus on market capitalization in the years 1997 and 2004 when there were manifestly no stock market bubbles on world exchanges. We employ a regression tree and random forests to rank the importance of key financial variables—total assets, net assets, total revenue, operating income, net income, operating cash flow, and number of employees—that reflect the fundamentals.

We set logarithmic market capitalization and key financial variables during non-bubble periods to explain variables and explanatory variables. Figure 1 shows a regression tree for electronics firms listed on NASDAQ in 2004. Branching of the tree enable us to visualize net assets as a most important explanatory variable. Random forests are an ensemble learning method for regression that operates by constructing a forest or a multitude of regression trees [11]. Random forests quantitatively rank the importance of the explanatory variables. Figure 2(a) shows a time series of the importance of financial variables for electronics firms listed on NASDAQ from 1995 to 2005. One can see that net assets are critical in almost all of the years covered. The importance of net assets are around 0.5 in 1997 and 2004. Turning to Fig. 2(b), this shows the same time series for retailer firms listed on NYSE from 2004 to 2014. For this sector, one can immediately see that both operating and net incomes are predominantly reflected in market capitalization for these firms.

3 Variation in Market Capitalization that Is Normalized by Fundamentals

We construct Lasso model to estimate firms' fundamentals from financial variables by assuming that the market capitalization is close to the fundamentals during non-bubble periods. In general, Random forests are more accurate than Lasso regression in identifying important variables. However, the Random forests cannot extrapolate large fundamentals during bubble periods because large financial variables do not appear during non-bubble periods. We chose Lasso model to solve this extrapolation problem in this session.

Assuming that the relationship between fundamentals F and financial variables $\boldsymbol{X}$ follows a power function, we propose the regression as follows,

$$\begin{aligned}\log F = w_{A_T} logX_{A_T} + w_{A_N} logX_{A_N} + w_R logX_R + w_{I_O} logX_{I_O} \\ + w_{I_N} logX_{I_N} + w_C logX_C + w_N logX_N + \epsilon \end{aligned} \quad (1)$$

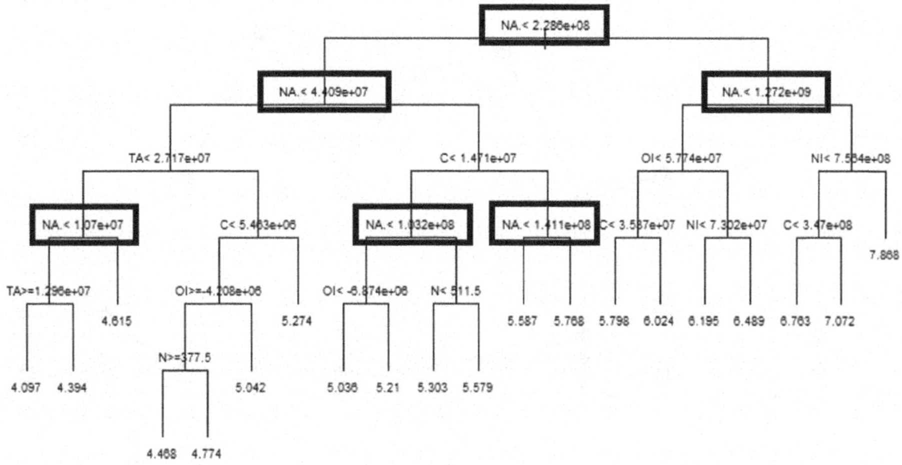

Fig. 1. Regression tree for electronics firms listed on NASDAQ in 2004. The non-explanatory variable is logarithmic market capitalization. Explanatory variables are (TA) total assets, (NA) net assets, (R) total revenue, (OI) operating income, (NI) net income, (C) operating cash flow, and (N) number of employees. Boxes represent branches subject to net assets.

where all of logarithmic variables are standardized, and the coefficients $\boldsymbol{w}$ depend on the market and industry. X_{A_T}, X_{A_N}, X_R, X_{I_O}, X_{I_N}, X_C, and X_N mean total assets, net assets, total revenue, operating income, net income, operating cash flow, and number of employees, respectively. In Lasso regression, we search the coefficients $\boldsymbol{w}$ that minimize the cost function as follows,

$$S_\lambda(\boldsymbol{w}) = ||logF - \boldsymbol{w}log\boldsymbol{X}||^2 + \lambda \sum_i |w_i| \tag{2}$$

while using the cross-validation. The hyperparameter λ, which maximizes the accuracy of Lasso model predictions (coefficient of determination), is chosen using the cross-validation.

Next, we estimate the coefficients and accuracy of Lasso regression for the electronics firms on NASDAQ and the retailer firms on NYSE during the non-bubble period in 2004. The $logF$ is set to the standardized logarithmic market capitalization. Data is randomly allocated 80% for learning and 20% for testing. Figure 3 displays the relationship between the hyperparameter λ and coefficients $\boldsymbol{w}$. The trend of the coefficients $\boldsymbol{w}$ is consistent with the importance of the random forest model. We obtain the optimal hyperparameter that maximizes the coefficient of determination.

We use the coefficients $\boldsymbol{w}$ during the non-bubble period in 2004 to estimate the fundamentals in different years. We normalize the market capitalization M at time t with the fundamentals F calculated by the Lasso regression (1) of financial variables at time t.

$$M'(t) = M(t)/F(t) \tag{3}$$

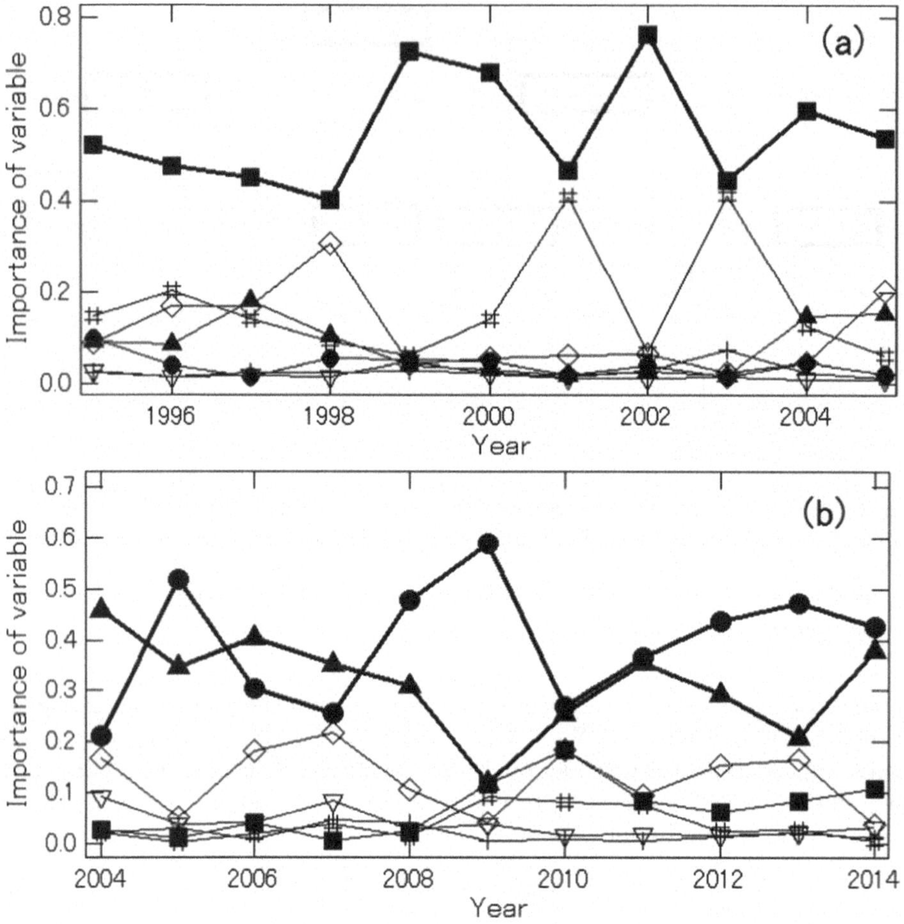

Fig. 2. Importance of financial variables: (a) electronics firms from 1995 to 2005 on NASDAQ, and (b) retailer firms from 2004 to 2014 on the NYSE. (#), (Black square), (+), (Black triangle), (•), (◇), and (∇) represent total assets, net assets, total revenue, operating income, net income, operating cash flow, and number of employees, respectively.

Figure 4 shows the cumulative distributions of normalized market capitalization $M'(t)$ on NASDAQ. The distribution function has always a fat-tail and depends on time t. When the tail is approximated by a power function, the slopes, β_t, are 1.5, 1.0, 1.5, and 1.3 for $t = 1997$, 1999, 2004, and 2014, respectively. Such time dependence is also observed in other stock markets.

4 Detecting Stock Market Bubbles: The Dot-com Bubble

The dot-com bubble reached its peak toward the end of 1999 as the NASDAQ composite index closed-in on the 5000 mark. Before the bubble in 1997 and

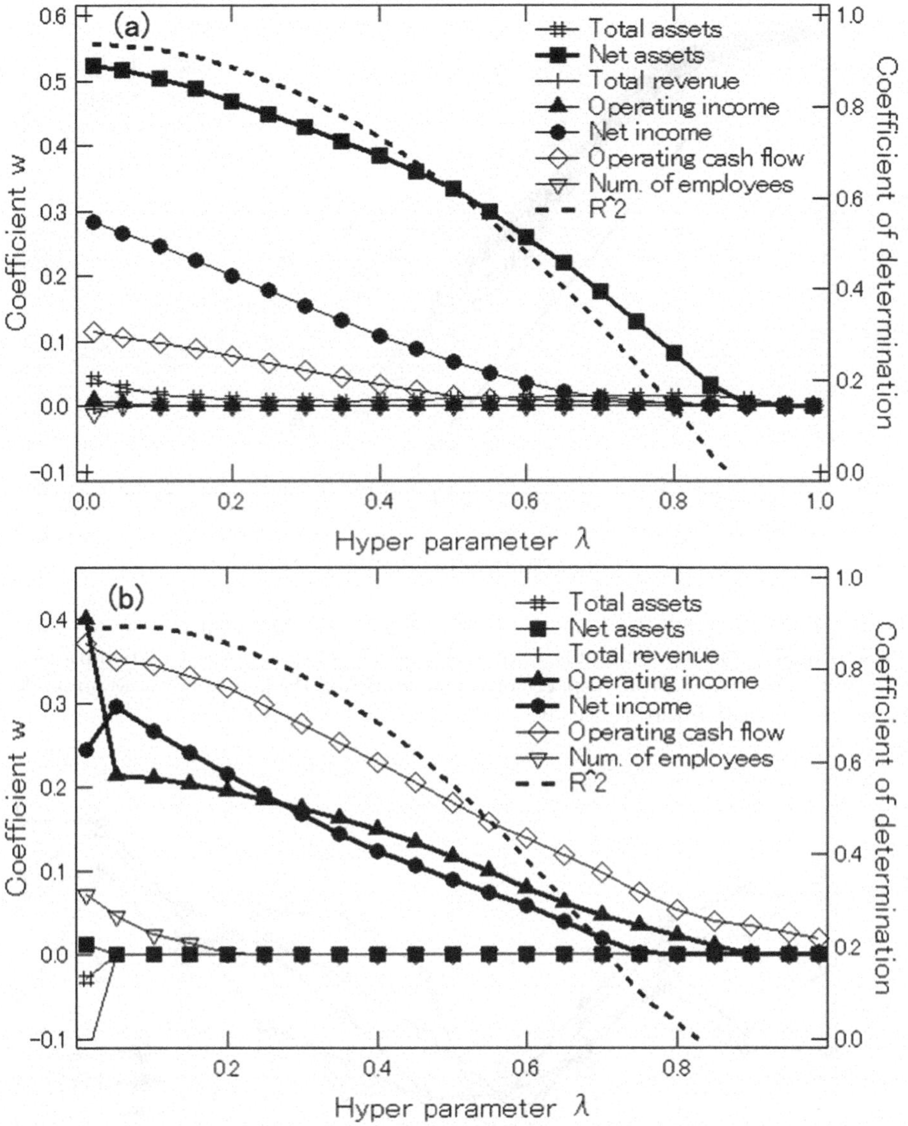

Fig. 3. Coefficient of financial variables in 2004: (a) electronics firms on NASDAQ, and (b) retailer firms on the NYSE. (#), (Black square), (+), (Black triangle), (•), (◇), and (∇) represent total assets, net assets, total revenue, operating income, net income, operating cash flow, and number of employees, respectively. The dashed lines show coefficient of determination.

after the bubble burst in 2004, the NASDAQ composite index hovered around 2000. Despite the collapse of Lehman Brothers in 2008, NASDAQ has come back and as of 2014 was again approaching 5000. Figure 4 shows distributions

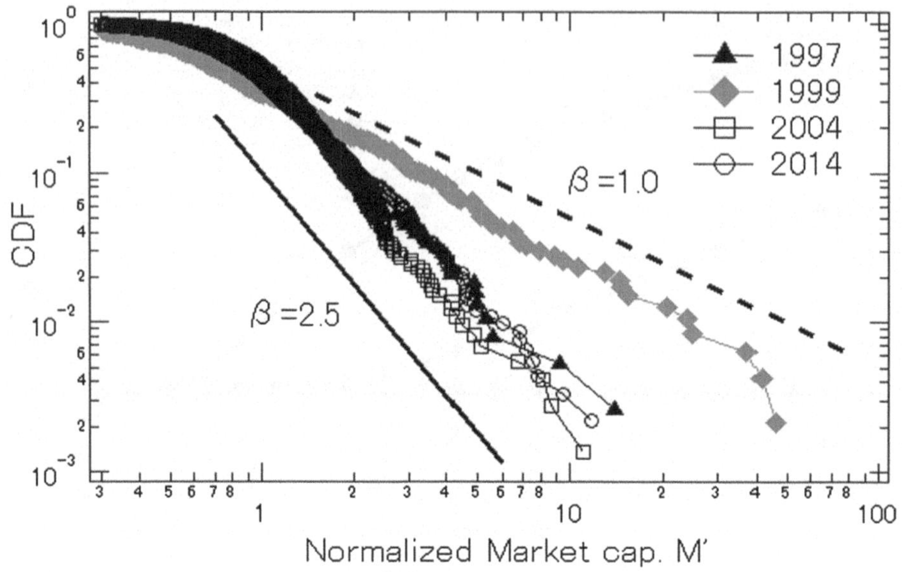

Fig. 4. Cumulative distributions of normalized market capitalization on NASDAQ. (Black triangle), (Black diamond), (Square), and (○) plots represent distributions for 1997, 1999, 2004, and 2014, respectively. The solid and dashed lines show power low slopes, $\beta = 2.5$ and 1.0, respectively.

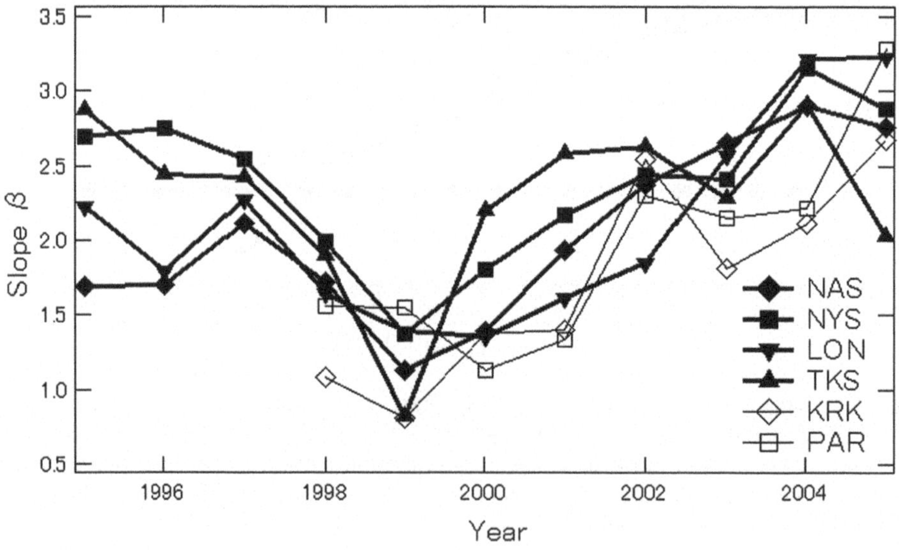

Fig. 5. Distribution's slopes, β_t, of normalized market capitalization. (Black diamond), (Black square), (Inverted triangle), (Triangle), (White diamond), and (White square) represent slopes for NASDAQ, NYSE, London SE, Tokyo SE, Korea SE, and Paris SE, respectively.

of normalized market capitalization in 1997, 1999, 2004, and 2014. The upper tail of distribution grew fat during the dot-com bubble but after the bubble burst, the distribution fell back to the same level as before the bubble. The fat upper tail is not present in 2014. This means that the stock prices increased in line with fundamentals. Market capitalization of firms in 2014 is firmly based on fundamentals.

Figure 5 shows time series of distribution's slopes, β_t, for the nine stock exchanges: NASDAQ (1995–2005), NYSE (1995–2005), London SE (1995–2005), Tokyo SE (1999–2005), Paris SE (1998–2005), Korea SE (1999–2005), Shanghai SE (1999–2005), Hong Kong SE (2000–2005), Taiwan SE (2000–2005). The slopes diminished during the dot-com bubble on all of the exchanges, then immediately steepened again after the bubble collapsed. Essentially, this means that market capitalization increases without any accompanying increase in fundamentals for a subset of stocks during bubble periods. Such an aberration can be readily detected by closely monitoring variations in normalized market capitalizations.

5 Conclusions

The defining characteristic of stock market bubbles is divergent between market capitalization (asset prices) and the fundamentals. In accordance with the definition of bubble, "market capitalization remains close to the firms' fundamentals during non-bubble periods", we identified firm's financial variables that most closely reflect fundamentals using machine learning. We observed considerable variability of market capitalization normalized by fundamentals that was estimated by Lasso regression model of financial variables. There is significantly greater divergence between market capitalization and fundamentals during bubbles. We have demonstrated that this phenomenon could have been used to detect the dot-com bubble of 1998–2000 across several stock markets. While it is difficult to determine if an individual firm or stock is caught up in a bubble, we can detect when a stock market is in the midst of a bubble.

We should note that there are types of bubbles that cannot be detected by the approach outlined here. The method we describe only works when identifying the type of bubble in which a concentration of speculative money flows into a small set of stocks. For example, our method would not work in a situation where speculative funds flow evenly into all stocks so that market capitalization diverges uniformly from fundamentals. In order to detect the formation of this kind of bubble, one would have to have knowledge of speculative money flows between financial markets. This is precisely the challenge that we intend to pursue next: detection of bubbles that arise from the concentration of speculative funds flowing into specific financial markets.

Acknowledgements. This work was supported by JSPS KAKENHI Grant Numbers 18H05217, 16H05904 and the Center of Advanced Research in Finance (CARF), University of Tokyo.

References

1. Brunnermeier, M.K.: Bubbles. In: Jones, G. (ed.) Banking Crises, pp. 28–36. Palgrave Macmillan, London (2016)
2. Palan, S.: A review of bubbles and crashes in experimental asset markets. J. Econ. Surv. **27**, 570–588 (2013)
3. Arshanapalli, B., Nelson, W.B.: Testing for stock price bubbles: a review of econometric tools. Int. J. Bus. Finan. Res. **10**, 29–42 (2016)
4. Mizuno, T., Ohnishi, T., Watanabe, T.: Power laws in market capitalization during the Dot-com and Shanghai bubble periods. Evol. Inst. Econ. Rev. **13**, 445–454 (2016)
5. Kaizoji, T.: Power laws and market crashes: empirical laws on bursting bubbles. Prog. Theor. Phys. Suppl. **162**, 165–172 (2006)
6. Kaizoji, T.: Precursors of market crashes: empirical laws of the Japan's internet bubbles. Eur. Phys. J. B **50**, 123–127 (2006)
7. Kaizoji, T., Miyano, M.: Why does the power law for stock price hold? Chaos Solitons Fractals **88**, 19–23 (2016)
8. Ohnishi, T., Mizuno, T., Shimizu, C., Watanabe, T.: Power laws in real estate prices during bubble periods. Int. J. Mod. Phys. Conf. Ser. **16**, 61–81 (2012)
9. Ohnishi, T., Mizuno, T., Watanabe, T.: House price dispersion in boom-bust cycles: evidence from Tokyo. Jpn. Econ. Rev. (2019, submitted)
10. Tay, D.J., Chou, C.I., Li, S.P., Tee, S.Y., Cheong, S.A.: Bubbles are departures from equilibrium housing markets: evidence from Singapore and Taiwan. PLoS ONE **11**, e0166004 (2016)
11. Breiman, L.: Random forests. Mach. Learn. **45**, 5–32 (2001)

Feasibility Study on Multi-agent Simulations of a Seasonal Influenza Epidemic in a Closed Space

Saori Iwanaga(✉), Hajime Yoshida, and Shigenori Kinjo

Japan Coast Guard Academy, Kure, Japan
s-iwanaga@jcga.ac.jp

Abstract. In a previous study, we traced the channel of infection at the Japan Coast Guard Academy as a closed space using real epidemic data. We introduced a refinement to a previous epidemic model to account for the incubation period and proposed a discrete-time epidemic model for seasonal influenza; in this manner, we obtained a reasonable incubation period and average infectivity ratio. In this study, to implement a realistic simulation, we investigate a simple method for calculating an accurate infectivity rate from real data. Therein, we consider a series of classes, including large classes attended by all students and classes for each grade, and multiplexed infection.

Keywords: Epidemic model · Seasonal influenza · Multi-agent simulation · Closed space

1 Introduction

In January 2017, an epidemic of influenza A, also known as seasonal influenza [1], occurred at the Japan Coast Guard Academy (JCGA). After three consecutive holidays from January 7 to January 9, all students returned to the dormitory and their classes. On Friday, January 13, two students were admitted to the academy infirmary. The epidemic spread rapidly, and the maximum number of infected patients was 27 on January 16 and 18, as shown in Fig. 1. In total, 37 (25%) students, including 33 (32%) freshmen and sophomores, were affected. However, the approximately 60 teachers who had contact with the students did not develop similar symptoms.

The CDC [2] has reported that the symptoms of seasonal influenza arise within one to four days after the virus enters the body. Thus, a person can transmit influenza to others before they know that they are sick, as well as while they are sick. In other words, seasonal influenza has an incubation period, including an exposed period and an infectious period in which no symptoms are observed. The SIR [3] and SEIR [4] models do not include this latter period and also do not apply to the JCGA case because the diseased students were isolated in the academy infirmary. We propose that an infectious period in which symptoms are not present is an important aspect to consider in an epidemic of seasonal influenza.

H. Sato et al. (Eds.): IES 2019, PALO 12, pp. 203–215, 2020.
https://doi.org/10.1007/978-3-030-37442-6_19

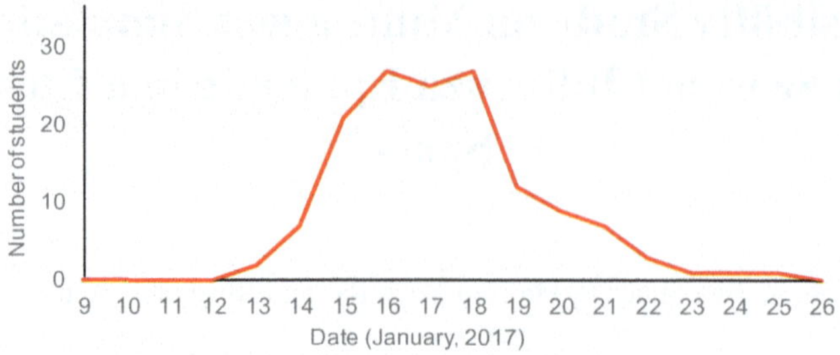

Fig. 1. Number of infected students in the JCGA case.

In a previous study [5], we traced the channel of infection at JCGA as a closed space using real epidemic data. We introduced a refinement to a previous epidemic model to account for the incubation period and proposed a discrete-time SEPIR model for seasonal influenza. In this manner, we obtained a reasonable incubation period and average infectivity ratio. We derived an incubation period of three days and students are able to infect other people beginning two day before symptoms [5], which was reasonable. But, our students may be more infectious than people. Another study [6] demonstrated that the incubation period of influenza is 3.6 days, with a standard deviation 1.6 days. Moreover, we calculated the average infection ratio for bedrooms and classrooms.

In this study, we focus on the infection rate and simulate a seasonal influenza epidemic using the SEPIR model with a multi-agent simulation (MAS) using real spatio-temporal model. We aimed to examine what type of countermeasures would be effective by simulating an epidemic of seasonal influenza, which is useful for managing groups of individuals. To perform a realistic simulation, an accurate infectivity rate must be obtained from real data by using a simple method.

Numerous epidemic simulations have been reported. EpiSimdemics [7] is a scalable parallel algorithm that simulates the spread of a contagion in large, realistic social contact networks using individual-based models. EpiSimdemics uses a SEIR-like model and is specifically designed to scale to social networks with 100 million students. EpiFast [8] is a large-scale realistic epidemic simulator based on a novel interpretation of stochastic disease propagation in a social contact network. This model implements the spread of disease via a social contact network and changes in the node SEIR states. FluTe [9] is an individual-based simulation of the spread of influenza across major metropolitan areas or the continental United States. Students in the model have realistic social contact networks, and transmission and infection are based on the current state of knowledge of the natural history of influenza. These three simulations use a large model with a social contact network rather than a real network, and they do not consider a preciously infectious period.

In this study, we simulate a seasonal influenza epidemic using the SEPIR model in a MAS with real spatio-temporal model. Using the JCGA data for 2017, we attempt to simulate the epidemic for future applications. In the next section, we present the SEPIR model for a seasonal influenza epidemic. We describe the simulation setting and the parameters in Sect. 3, and we report the simulation results and infectivity rates in

Sect. 4. We discuss the refinement of our approach in Sect. 5, and finally, we summarize the conclusions of this study.

2 Epidemic Model for Seasonal Influenza

We previously introduced a refinement to the SEIR model to account for an incubation period and proposed a discrete-time SEPIR model for seasonal influenza [5]. The incubation period [4] is divided into two periods, an exposed period and an infectious period with no symptoms. The state transition diagram of a student is shown in Fig. 2. We added the "P" status as the previously infectious period to the SEIR model [4], which was originally given as the infectious period. The exposed "E" student is infected but not yet infectious. Both the infectious "I" student and the infectious "P" student can infect the susceptible "S" student.

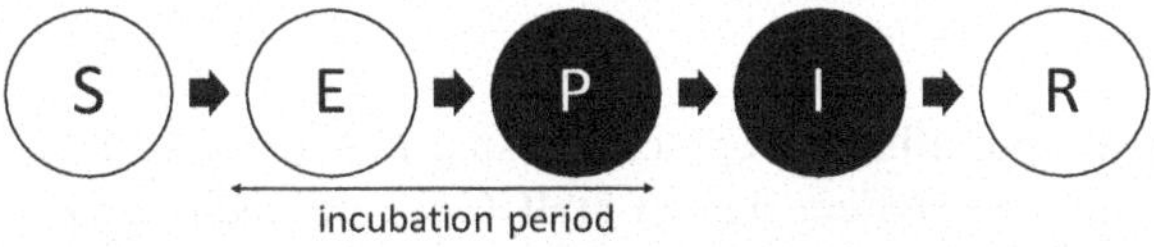

Fig. 2. A state change diagram of the epidemic model. The circles show the various states of the student. S: Susceptible state, E: Exposed state, P: Previously infectious state, I: Infected state; R: Recovered state. The black circles represent states in which the student can infect other students.

Considering the mean infectivity rates α and β (per day), the susceptible "S" student transits to the exposed "E" state, as given in Eq. (1), where t represents one day. P(t) is the number of previously infectious "P" students, S(t) is the number of susceptible "S" students, and I(t) is the number of infectious "I" students. The exposed "E" student transits to the previously infectious "P" state at a transition rate of σ, as given in Eq. (2), where E(t) is the number of exposed "E" students and R(t) is the number of recovered "R" students who have successfully cleared the infection. The previously infectious "P" student transits to the infectious "I" state at a transition rate of τ, as given in Eq. (3), and the infectious "I" student transits to the recovered "R" state at a transition rate of γ, as given in Eq. (4). The parameters σ, τ, and γ are transition rates (per day); γ is also denoted as the recovery rate. Similar to former models [3, 4], we set these rates as the inverse of the exposed "E" period, previously infectious "P" period, and infectious "I" period.

$$\Delta S = S(t+1) - S(t) = -\alpha S(t)P(t) - \beta S(t)I(t) \tag{1}$$

$$\Delta E = E(t+1) - E(t) = \alpha S(t)P(t) + \beta S(t)I(t) - \sigma E(t) \tag{2}$$

$$\Delta P = P(t+1) - P(t) = \sigma E(t) - \tau P(t) \tag{3}$$

$$\Delta I = I(t+1) - I(t) = \tau P(t) - \gamma I(t) \quad (4)$$

$$\Delta R = R(t+1) - R(t) = \gamma I(t) \quad (5)$$

In our previous study [5], we represented the bedroom, study room, and classroom as closed spaces. We showed that the bedroom and study room can be combined, and we utilized the bedroom to represent the dormitory. We also found that the exposed "E" period is one day, while the previously infectious "P" period is two days. Thus, the incubation period is three days, and the average infectious "I" period is 3.86 days. Because each infectious student is isolated in the academy infirmary, the infectivity rate $\beta = 0$ at JCGA. From Eq. (1), the average infectivity rate α is 0.0259 in the bedroom and 0.0206 in the classroom.

3 Simulation Preparation

3.1 Settings

We simulated a seasonal influenza epidemic using MAS, where each agent represents a student. There are 150 student agents at JCGA, and each agent follows the daily schedule presented in Table 1. Each day, the students have a sleeping time of 8 h in their assigned bedroom. On weekdays, the students have up to four classes of 90 min in the classroom. All classes are compulsory. The bedroom and classroom are considered as closed spaces with a corresponding risk of infection. Our simulation is performed using a spatio-temporal model with real data.

We first analyze an epidemic of influenza arising from two students because two students initially developed influenza on January 13. It was determined that the first two students were exposed on January 9. Thus, we set the first two agents as being exposed on January 9. We assumed that the exposed "E" period is one day, the previous infectious "P" period is two days, and the infectious "I" period is four days. We also set the average infectivity rate α as 0.0259 for the bedroom and 0.0206 for the classroom.

Table 1. Simulated schedule for the students.

Time	Weekday activity	Weekend activity	Room	Infectivity
base time 08:30				
08:45–10:15	class: first period		classroom	can be infected
10:30–12:00	class: second period		classroom	can be infected
13:00–14:30	class: third period		classroom	can be infected
14:45–16:15	class: fourth period		classroom	can be infected
22:30–06:30	sleep	sleep	bedroom	can be infected

3.2 Infection and State Change

Figure 3 shows an example of infection among the students in five classrooms (A–E) and three bedrooms (X–Z). Nine students are included, as identified by number, and the formatting of each number indicates the state of the student. A number that is not underlined represents a student that is susceptible "S," such as student 3. A bold number represents a student that is exposed "E," such as student 2 on day 1. An underlined number denotes a student is that is previously infectious "P," such as student 1.

In this model, the students in the class are the social contact network of the student. The roommate of each student is also included in the social contact network. During the day, the social network changes depending on the class schedule and the assignment of bedrooms. A previously infectious "P" student in a closed space such as a classroom or bedroom can infect susceptible "S" students in the same space. Each student changes state at the base time of 08:30.

At 08:30 on day 0, student 1 becomes previously infectious "P" for the first time. Student 1 infects student 2 at classroom A on day 0 (second period), as denoted by a circle. Then, student 2 changes from susceptible "S" to exposed "E" at 08:30 on day 1. Student 1 remains in the "P" state at 08:30 on the day 1 since student 1 has one more day in the incubation period.

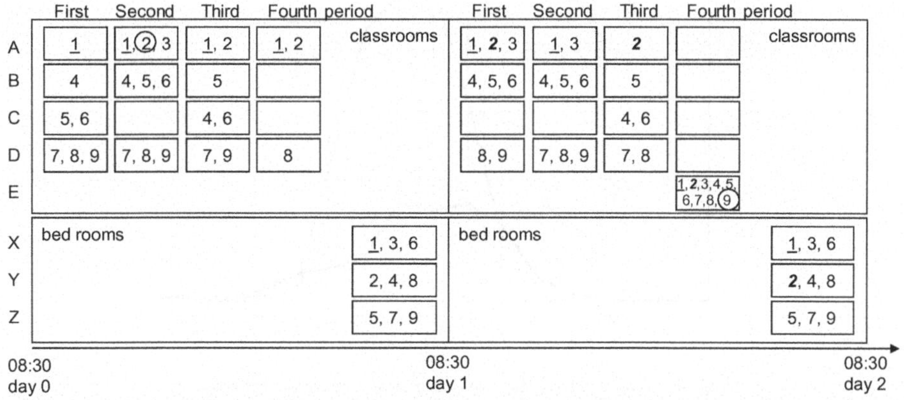

Fig. 3. Temporal and spatial states of the students. A number that is not underlined represents a student that is susceptible "S." A bold number denotes a student that is exposed "E." An underlined number represents a student that is previously infectious "P." Students 1–3, 4–6, and 7–9 are in the same grade.

3.3 Epidemic

We defined an epidemic as a case in which the infection spreads beyond the first two students.

4 Simulation Results

We simulated the period of January 9 to 25 as a trial, with 10,000 simulations. A seasonal influenza epidemic occurred in 9,959 of 10,000 trials. Figure 4 presents a time-series graph of the average number of "I" students, with the standard deviation denoted by an error bar. The red line presents the simulation results for an infectivity rate of 0.0259. The maximum number of infected students is observed on January 17 at 88.04, with a standard deviation of 27.22. The simulated "I" results are generally higher than the real data (purple line).

4.1 Review of the Average Infectivity Rate in the Classroom and Bedroom

A preliminary experiment was performed to confirm that the epidemic depends on the room type. For this purpose, we conducted two types of simulations. In the first case, students are infected only in classrooms, and in the second case, students are infected only in bedrooms. We found that the simulation results for the first case were nearly identical to those presented in Fig. 4. However, the simulation results for the second case are different from those in Fig. 4, with only a few students becoming infected, even though the infectivity rate (0.0259) for the bedroom is higher than that (0.0206) for the classroom.

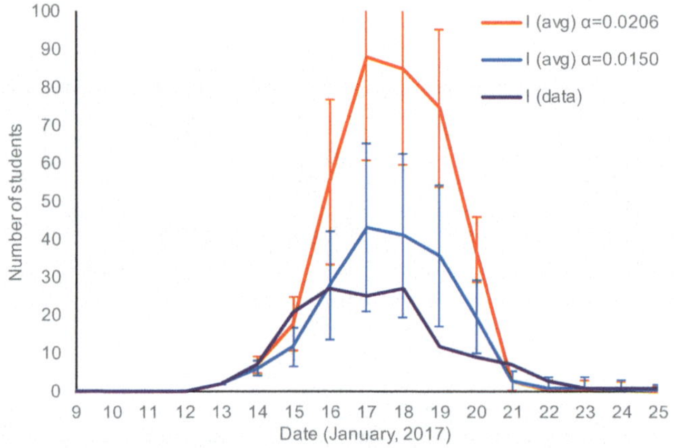

Fig. 4. Averages of the simulation results and real data.

4.2 Review of Class Schedule

Here, we consider the class schedule. There was a series of two classes for the freshmen's third and fourth period on January 11; some students were infected in these classes, and thus, we counted the "E" students twice. However, this approach is not appropriate for a daily infectivity rate [5]. Hence, we recalculated the infectivity rate α in classroom at 0.0150, which is approximately 70% of the previous rate.

In Fig. 4, the blue line presents a time-series graph of the average number of "I" students for an infectivity rate α of 0.0150, with the standard deviation indicated by error bars. A seasonal influenza epidemic occurred in 9,827 of 10,000 trials. The maximum number of infected students was 43.11, observed for January 17, with a standard deviation of 22.11. The "I" simulation results are closer to the real data; however, the real data are far outside the standard deviation for the first half, i.e., before January 18, with a particularly large deviation for January 15 in comparison to the previous results (α = 0.0206). The increasing rate of "I" is smaller than that for the real data because the infectivity rate α in the classroom is small.

4.3 Review of the Classroom Infectivity Rate

In the SEPIR model, an infected susceptible "S" student transits to state "E" and then "I" after an incubation period. We investigated the results for a single trial with regard to the number of "E" students, as presented in Table 2.

There are more "E" students on January 13 and 14, particularly among sophomores and juniors, in comparison to the real data. We found that it is better to consider the classroom infectivity rate for each grade to perform a more accurate simulation.

Table 2. Number of "E" students for each grade in one simulation trial and corresponding real data.

Number of "E" students		Date (January 2017)																
		9	10	11	12	13	14	15	16	17	18	19	20	21	22	23	24	25
One trial	Freshman	0	1	3	4	10	8	1	0	0	0	0	0	0	0	0	0	0
	Sophomore	0	1	0	0	9	14	0	0	0	0	0	0	0	0	0	0	0
	Junior	0	0	1	2	8	6	1	0	0	0	0	0	0	0	0	0	0
	Senior	0	0	0	0	0	0	2	0	0	0	0	0	0	0	0	0	0
Real data	Freshman	0	1	2	11	3	0	1	1	0	1	0	0	0	0	0	0	0
	Sophomore	0	1	3	3	2	0	3	0	1	0	0	0	0	0	0	0	0
	Junior	0	0	0	0	1	1	2	0	0	0	0	0	0	0	0	0	0
	Senior	0	0	0	0	0	0	0	0	0	0	0	0	0	0	0	0	0

4.4 Calculation of the Infectivity Rate

Before calculating the classroom infectivity rate for each grade, we must consider two issues.

- Multiplexed infection: Students are determined to be infected in a maximum of four classes per day.
- Large classes for all students: There are large classes that all students of all grades attend.

For multiplexed infection, as an example (Fig. 3), student 2 is infected in classroom A on day 0 (second, third, or fourth period) by student 1. Here we adopt the following cases and calculate the corresponding infectivity rate αm as follows.

1. Average the infectivity rates of each class per grade: Case M1
2. Distribute the infected students equally among the classes in which students are infected, and average the infectivity rates for each class per grade: Case M2
3. Distribute the infected students based on the number of students for each class in which students are infected, and average the infectivity rate of each class per grade: Case M3

For large classes attended by all students, the example (Fig. 3) includes real data except for the cases denoted by circle marks. This case includes a large class attended by all students in classroom E, where students of various grades can be infected. We adopt two cases, and the corresponding infectivity rate αb is defined as follows.

1. The classroom infectivity rate for each grade: Case B1
2. The classroom infectivity rate weighted by the number of infected students for each grade: Case B2

Case M1
The infectivity rate αm is 0.0292 for freshmen, 0.0069 for sophomores, and 0 for juniors and seniors, as shown in Table 3. For a large class attended by all students, the infectivity rate αb is 0.0066.

Case M2
The infectivity rate αm is 0.0192 for freshmen, 0.0042 for sophomores, and 0 for juniors and seniors, as shown in Table 3. For a large class attended by all students, the infectivity rate αb is 0.0036.

Case M3
The infectivity rate αm is 0.0172 for freshmen, 0.0032 for sophomores, and 0 for juniors and seniors, as shown in Table 3. For a large class attended by all students, the infectivity rate αb is 0.0066.

Table 3. Infectivity rates.

	Case M1	Case M2	Case M3	Case B1	Case B2
Infectivity rate	αm	αm	αm	–	–
Freshman	0.0292	0.00192	0.0172	αb	$\alpha b \times 3/6$
Sophomore	0.0069	0.0042	0.0032	αb	$\alpha b \times 2/6$
Junior	0	0	0	αb	$\alpha b \times 1/6$
Senior	0	0	0	αb	0
αb	0.0066	0.0036	0.0066	–	–

Case B1

The average infectivity rate αb for a large class attended by all students is αb.

Case B2

Three freshmen, two sophomores, and one junior are infected in a large class. We distribute the infected students with weighting based on the number of infected students for each grade, as shown in Table 3.

4.5 Simulation Results for a Large Class Attended by All Students

First, we compare Cases B1 and B2. Figure 5 displays a time-series graph of the average of "I," with the standard deviation denoted by error bars, for Case M1B1 (Case M1 & Case B1) and Case M1B2 (Case M1 & Case B2). A seasonal influenza epidemic occurred in 9,921 of 10,000 trials, and the maximum number of infected students was 43.83, occurring on January 17, with a standard deviation of 17.24 for Case M1B1, as shown by the gray line. A seasonal influenza epidemic occurred in 9,913 of 10,000 trials, and the maximum number of infected students was 37.48, observed on January 17, with a standard deviation of 14.93 for Case M1B2, as shown by the red line. The "I" results for the Case B1M2 simulation are close to the real data. We found that most of the real data points lie between $+\sigma$ and $-\sigma$ of the average for the first half, i.e., before January 18, for Case M1B2.

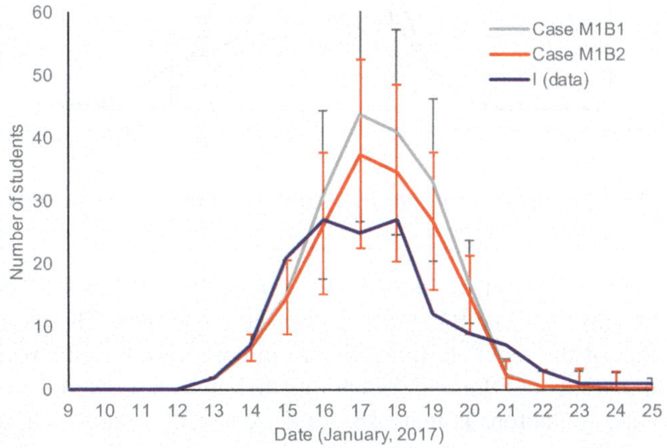

Fig. 5. Averaged simulation results for Cases B1 and B2 and real data.

4.6 Simulation Results for Multiplexed Infection

Here, we compare Cases M1, M2, and M3. Figure 6 presents a time-series graph of the average of "I," with the standard deviation indicated by error bars, for Cases M1B2, M2B2 (Case M2 & Case B2), and M3B2 (Case M3 & Case B2). The results for case M1B2 are identical to the previous results. A seasonal influenza epidemic occurred in 9,606 of 10,000 trials, and the maximum number of infected students was 16.32, observed on January 17, with a standard deviation of 10.09 for Case M2B2, as shown by the cyan line. A seasonal influenza epidemic occurred in 9,415 of 10,000 trials, and the maximum number of infected students was 13.79, observed on January 17, with a standard deviation of 9.20 for Case M3B2, as shown by the green line. The real data are far outside the standard deviation for Cases B2M2 and B3M2 during the first half, i.e., before January 18. The "I" results for Case B1M2 are close to the real data.

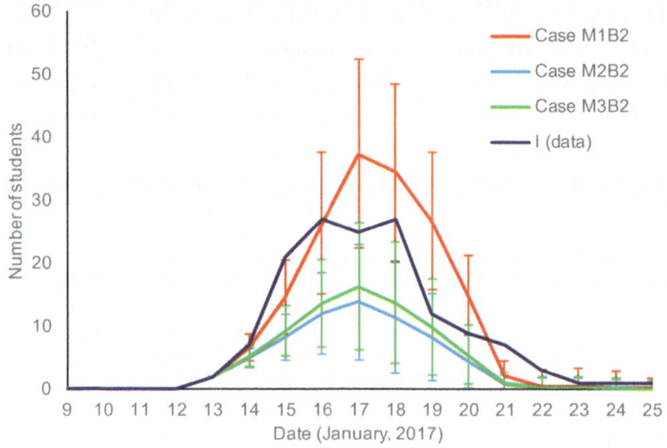

Fig. 6. Average simulation results fort Cases M1, M2, and M3 and real data.

Figure 7 presents a time-series graph of the average "E," "P," and "I" results for Case M1B2. The "I" results for Case B1M2 are close to the real data. The increasing rate of "I" is closer to that of the real data (purple line) in comparison to the results shown in Fig. 4. We found that most of the real data points lie between $+\sigma$ and $-\sigma$ of the average during the first half, i.e., before January 18, for Case M1B2. In this case, the number of "E" students is small.

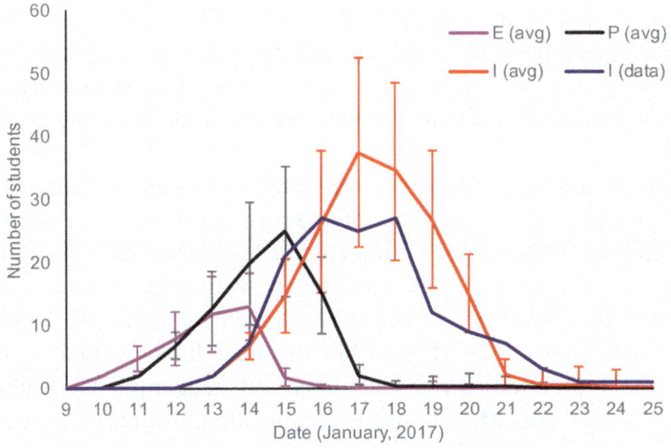

Fig. 7. Average simulation results for Case M1B2 and real data.

5 Discussion

Overall, we found that the simulation result is sensitive to the infectivity rate. To perform a realistic simulation, an accurate infectivity rate must be obtained from real data by a simple method.

Based on the infectivity rates for the bedroom and classroom, we found that the infection spread is greater in the classroom than in the bedroom, as reported for other studies. The infectivity rate of H1N1 influenza has been calculated as 0.01 for long-distance bus trips [10], where most of the passengers sleep. This condition corresponds to the bedroom in our model. In contrast, it has been reported that a school child will infect 2.4 other children due to daily contact within a school [11], corresponding to an infectivity rate of 0.016 for a classroom with 150 students. This condition corresponds to the classroom in our model. Thus, we reconfirmed that influenza is more infectious in the classroom than the bedroom.

Using the average infection rate, we simulated a seasonal influenza epidemic, as shown in Fig. 4. We found that the application of a simple average infective rate for the classrooms [5] is too rough and requires further attention. We must consider a series of classes, including large classes attended by all students and classes for each grade, and multiplexed infection.

First, we integrated the series of classes into one class. By considering a series of classes, the simulation results were closer to the real data, as shown in Fig. 4.

Second, in determining the infectivity rate for a large class attended by all students, the maximum number of infected students for Case B1M1 was found to be approximately 1.6 times the actual value. In addition, the maximum number of infected students for Case B1M2 was approximately 1.5 times the actual value. The "I" results for the Case B1M2 simulation were closer to the real data. We found that the increasing rate of "I" is closer to the actual result compared to the findings shown in Fig. 4 ($\alpha = 0.0150$) because the freshmen had a high infectivity rate.

Third, we calculated the infectivity rate for each grade in each classroom. By considering the infectivity rate for each grade, the simulation results became closer to the real data, as shown in Fig. 6. Children ages 5–18 years have a high influenza attack rate [11]. We found that the infectivity rate must be set for each grade since the grade is generally related to the age.

Finally, multiplexed infection is often ignored in simulation studies [7, 8]. From the JCGA data, we determined that the infected freshmen students are infected in 1.4 classes while infected sophomores are infected in 1.5 classes. Juniors and seniors were not infected in their classes. All infected students were infected in an average of 1.36 classes. We found that the overestimate was reduced for Case M1 when multiplexed infection is applied, while Cases M2 and M3 also gave lower estimates. We also found that the multiplexed infection can be calculated independently. The calculation of the infectivity rate αm for Cases M2 and M3 is complicated, whereas the calculation of αm for Case M1 is simple. We found that the calculation for Case M1 is reasonable.

Moreover, for multiplexed infection, a single student can be infected in both the classroom and the bedroom. In previous work, we found that 16 students were infected in classrooms [5] while 10 students were infected in bedrooms, with 5 students being infected in both rooms. Thus, double infection occurred, and we found that the multiplexed infection can be calculated independently.

6 Conclusion

By focusing on the infectivity rate, we simulated a seasonal influenza epidemic using the SEPIR model by performing a MAS with real spatio-temporal model. We aimed to examine what types of countermeasures are effective by simulating a seasonal influenza epidemic, which is useful for managing groups of individuals. To perform a realistic simulation, an accurate infectivity rate must be obtained from real data by a simple method. We found that the infectivity rate for the bedrooms can be calculated be simply averaging the infectivity rates for each bedroom. Moreover, we found that the classroom infectivity rate can be calculated as the average for each grade. In addition, we considered a series of classes, including a large class attended by all students and classes for each grade, and multiplexed infection.

In the future, we will examine the countermeasures that are effective by simulating a seasonal influenza epidemic. And we are going to apply the result to outside of JCGA.

References

1. CDC flu. https://www.cdc.gov/flu/keyfacts.htm
2. CDC spread. https://www.cdc.gov/flu/about/disease/spread.htm
3. Kermack, W.O., McKendrick, A.G.: A contribution to the mathematical theory of epidemics. Proc. R. Soc. Lond. A **115**, 700–721 (1927)
4. Keeling, M.J., Rohani, P.: Modeling Infectious Diseases in Humans and Animals. Princeton University Press, Princeton (2008)
5. Iwanaga, S., Kawaguchi, K.: Analysis of epidemic of seasonal influenza in closed space. In: The 22nd Asia Pacific Symposium on Intelligent and Evolutionary Systems, pp. 37–44 (2018)

6. Cowling, B.J., Fang, V.J., Riley, S., Peiris, J.S.M., Leung, G.M.: Estimation of the Serial Interval of Influenza. Epidemiology **20**(3), 344 (2009)
7. Barrett, C.L., Bisset, K.R., Eubank, S.G., Feng, X., Marathe, M.V.: EpiSimdemics: an efficient algorithm for simulating the spread of infectious disease over large realistic social networks. In: Proceedings of the 2008 ACM/IEEE Conference on Supercomputing, p. 37. IEEE Press (2008)
8. Bisset, K., Chen, J., Feng, X., Kumar, V., Marathe, M.: EpiFast: a fast algorithm for large scale realistic epidemic simulations on distributed memory systems. In: Proceedings of the 23rd International Conference on Supercomputing. ACM (2009)
9. Chao, D., Halloran, M., Obenchain, V., Longini, I.: FluTE, a publicly available stochastic influenza epidemic simulation model. PLoS Comput. Biol. **6**(1), e1000656 (2011)
10. Piso, R.J., Albrecht, Y., Handschin, P., Bassetti, S.: Low transmission rate of 2009 H1N1 Influenza during a long-distance bus trip. Infection **39**(2), 149–153 (2011)
11. Yang, Y., Sugimoto, J.D., Halloran, M.E., Basta, N.E., Chao, D.L., Matrajt, L., Potter, G., Kenah, E., Longini Jr., I.M.: The transmissibility and control of pandemic influenza a (H1N1) virus. Science **326**(5953), 729–733 (2009)

Simultaneous Local Adaptation for Different Local Properties

Ryota Kobayashi(✉), Ryo Takano(✉), Hiroyuki Sato(✉), and Keiki Takadama(✉)

The University of Electro-Communications, Tokyo, Japan
k1830046@uec.ac.jp

Abstract. This paper proposes a method of adaptation for a problem holds local optima with different properties around them. We employed neighborhood based method into Adaptive DE:JADE with a mechanism of local adaptation simultaneously at different local optima. Experimental results revealed (i) local search is effective to multiple properties problem around different local optima, (ii) a mechanism of LJADE ables to locate and explore local optima adaptively, (iii) LJADE perform slower but more stabled exploration than JADE.

Keywords: Adaptive Differential Evolution · Swarm intelligence · k-nearest computation

1 Introduction

Evolutionary Computation is getting attention as appliable algorithm to solve real-world problems. In order to operate computation for real-world problem, no mistakes should be made by any chance. Because of that reason, it is necessary to operate algorithm that performs better and stabled result.

One of the optimization problem we often face in real-life is a compositional problem that consists from multiple kinds of functions. Various evolutionary algorithms have been proposed to solve these problems. One of the algorithm called Adaptive Differential Evolution, well known as a strong evolutionary algorithm which controls its parameter adaptively. Examples of this class are jDE [4], SaDE [5], SHADE [6]. JADE [3] is also one of the class of Adaptive Differential Evolution aims avoiding to converge into local optima by archiving well solution through exploring.

However, these methods considers adapting to only one kind of properties at the same time. Because of that, JADE unable to locate global optimum property. To tackle this problem, we proposed Local JADE (:LJADE) which implemented a mechanism of Self-Adaptive Parameter Control with employing neighborhood based method into JADE. The control parameters are directory associated with individuals in this mechanism, the SPDE [7] and DESAP [8] also belong to this category. Our proposal method able to adapt different local properties at different local optima simultaneously and individually.

H. Sato et al. (Eds.): IES 2019, PALO 12, pp. 216–227, 2020.
https://doi.org/10.1007/978-3-030-37442-6_20

To validate our work, we had an experiment with Composition functions which was proposed in CEC2014 Benchmark Problems [9]. We compared the performance of JADE and LJADE with the mean and standard variance values of function error values for individual experiments. The experimental result has revealed (i) local search is effective to multiple properties problem around different local optima, (ii) a mechanism of LJADE ables to locate and explore local optima adaptively, (iii) LJADE perform slower but more stabled exploration than JADE. Note that this paper is solving a minimizing problem.

The organization of this paper is as follows. In Sects. 2 and 3 describes an algorithms of DE and JADE. Our proposal methods are shown in Sect. 4. Section 5 shows experimental settings and results with discussions. At last, Sect. 6 shows conclusion of this paper.

2 Differential Evolution

Differential Evolution [1] is one of meta-heuristic algorithm which is a population-based approach to function optimization. One of the reason that Differential Evolution is featured, is the simple algorithms, calculation, and few parameters yet powerful. In this section, we present the basic DE algorithm and the schemes that are used in this paper.

2.1 Basic Algorithm

DE begins by initializing entire population's positions $x_i^0 = \{x_{i,1}^0, x_{i,2}^0, \cdots, x_{i,D}^0\}, i = 1, 2, \cdots, NP$ with equation is shown (1).

$$f_t = f_{min} + rand(0,1) \times (f_{max} - f_{min}) \tag{1}$$

Where, i is a index number of population, NP represents number of entire population, D is dimension, t is iteration, f_{max}, f_{min} is maximum domain and minimum domain of the environment. For the next step, generate mutation vector $v_i^t = \{v_{i,1}^t, v_{i,2}^t, \cdots, v_{i,D}^t\}$ from current population P. To calculate v_i^t, apply strategy to entire population. Details of popular strategies is discussed in Sect. 2.2.

$$v_i^t = x_{r1}^t + F \times (x_{r2}^t - x_{r3}^t) \tag{2}$$

Where, r1, r2, r3 is random index number picked from $(1, 2, \ldots, NP)$ without any duplication. Parameter F controls mutation vector's size which was generated. There is another parameter for controlling crossover rate, CR. If this parameter gets smaller, more importance to child.

$$v_{i,j}^t = \begin{cases} v_{i,j}^t & If \;\; rand_j\,(0,1) \le CR \; or \; j = j_{rnd} \\ x_{i,j}^t & otherwise. \end{cases} \tag{3}$$

Where, $rand_j\,(0,1)$ is an uniform random number from the domain [0,1]. j_{rnd} is random number picked from $1, 2, \ldots, NP$. As the last step, compare fitness of child and parent and use one has better fitness as the next iteration parent.

$$x_i^{t+1} = \begin{cases} u_i^t & If \;\; f(u_i^t) < f(x_i^t) \\ x_i^t & otherwise. \end{cases} \tag{4}$$

DE algorithm repeats these steps until termination criteria is satisfied. The pseudo-code is shown in Algorithm 1.

2.2 Basic Strategies

At each generation g, DE creates a mutation vectors $v_{i,g}$ based on this operation. Various strategies have been proposed and this section describes basic strategies.

- rand/1
  $\boldsymbol{x_{r1}} + F \times (\boldsymbol{x_{r2}} - \boldsymbol{x_{r3}})$
  Most common strategy that creating mutation vector between two random individuals from population, and use it for another random individual. It is strong at global search by its randomness, but performance of convergence is lower than other strategy.
- best/1
  $\boldsymbol{x_b} + F \times (\boldsymbol{x_{r2}} - \boldsymbol{x_{r3}})$
  Creates mutation vector between two random individuals from population and use it for the best individual in the population. Convergence performance is higher than other strategy since it searches around the best individuals in the population. But it is easy to locate a local optimum as a global optimum.

Algorithm 1. The Algorithm of Differential Evolution(rand/1)

```
iter ← 0
initialize population x generating randomly.
Evaluate individual x_i, i = 1, 2, ··· , NP
for g = 1 to G do
    for i = 1 to NP do
        Randomly choose r1, r2, r3 and r1 ≠ r2 ≠ r3 ≠ i from P
        j_rnd = rand(1, D)
        for j = 1 to D do
            if rand_j(0, 1) < CR orj = j_rnd then
                v_{i,j,g} = x_{r1,j,g} + F (x_{r2,j,g} − x_{r3,j,g})
            end if
        end for
        Evaluate v_i
        if f (v_i) < f (x_i) then
            x_i ← v_i
        end if
    end for
end for
```

3 JADE

JADE is an algorithm that DE implements a mutation strategy "DE/current-to-p-best" with optional archive and controls parameters F and CR adaptively through searching spaces.

3.1 Adapting Parameter

At each generation, different value of F and CR is generated to every individuals x_i independently according to the Eqs. (5) and (6).

$$F_i \sim C(\mu_F, \phi_F) \tag{5}$$

$$CR_i \sim N(\mu_{CR}, \phi_{CR}^2) \tag{6}$$

Where $C(\mu_F, \phi_F)$ is a Cauchy distribution with location parameter μ_F and scale parameter $\phi_F = 0.1$. $N(\mu_{CR}, \phi_{CR}^2)$is a normal distribution with a mean parameter μ_{CR} and deviation parameter $\phi_{CR}^2 = 0.1$. Both of initial value for μ_{CR},μ_F is 0.5. This method assumes that individuals, that generated child that holds better fitness value, is a successful individuals. To update the adapting parameter μ_{CR},μ_F, it uses successful individual's parameter with following equation.

$$\mu_F = (1 - c) \cdot \mu_F + c \cdot mean_L(S_F) \tag{7}$$

$$\mu_{CR} = (1 - c) \cdot \mu_{CR} + c \cdot mean_A(S_{CR}) \tag{8}$$

S_F, S_{CR} is a set of a successful individual's parameters F,CR. $c \in [0, 1]$ manages the weight of updating parameters and $mean_A(\cdot)$ is a usual arithmetic mean. $mean_L(S_F)$ represents the Lehmer mean, which is described in Eq. (9).

$$mean_L(S_F) = \frac{\sum_{F \in S_F} F^2}{\sum_{F \in S_F} F} \tag{9}$$

3.2 DE/current-to-p-best

JADE creates mutation vector with a following equation with archived solution A. This method puts successful individual into the archive as a set of recently explored inferior solutions.

$$v_{i,g} = x_{i,g} + F_i \cdot (x_{best,g}^p - x_{i,g}) + F_i \cdot (x_{r1,g} - \tilde{x}_{r2,g}) \tag{10}$$

where, $x_{best,g}^p$ represents an individual that was randomly chosen from top $100p\%$ of current population with $p \in [0, 1)$. $x_{r1,g}$ is an individual from current population P and $x_{r2,g}$ is from P and archive A. The pseudo-code is shown in Algorithm 2.

Algorithm 2. The Algorithm of JADE

set $\mu_F = 0.5$; $\mu_{CR} = 0.5$
initialize population x generating randomly.
Evaluate individual $x_i, i = 1, 2, \cdots, NP$
for $g = 1$ to G **do**
  for i = 1 to NP **do**
    Generate CR_i and F_i
    Randomly choose $x^p_{best,g}$
    Randomly choose $x_{r1,g} \neq x_{i,g}$ from P
    Randomly choose $x_{r2,g} \neq x_{i,g} \neq x_{r1,g}$ from $P \cup A$
    $v_{i,g} = x_{i,g} + F_i \cdot (x^p_{best,g} - x_{i,g}) + F_i \cdot (x_{r1,g} - \tilde{x}_{r2,g})$
    $j_{rnd} = rand(1, D)$
    for $j = 1$ to D **do**
      if $rand_j(0, 1) < CR_i$ or$j = j_{rnd}$ **then**
        $u_{j,i,g} = v_{j,i,g}$
      else
        $u_{j,i,g} = x_{j,i,g}$
      end if
    end for
    if $f(x_{i,g}) \leq f(u_{i,g})$ **then**
      $x_{i,g+1} = x_{i,g}$
    else
      $x_{i,g+1} = u_{i,g}$; $x_{i,g} \rightarrow A$; $CR_i \rightarrow S_{CR}$; $F_i \rightarrow S_F$
    end if
  end for
  Randomly remove solutions from A to make size of A is NP
  $\mu_F = (1 - c) \cdot \mu_F + c \cdot mean_L(S_F)$
  $\mu_{CR} = (1 - c) \cdot \mu_{CR} + c \cdot mean_A(S_{CR})$
end for

Table 1. Parameter settings

Parameter	Value
G	200
D	2
NP	100
c(JADE)	0.1
c(LJADE)	0.5
k	5
p	0.05

4 LJADE

In this section, we propose LJADE which is the algorithm of JADE implementing neighborhood mechanism. Neighborhood was defined with k-nearest individuals calculated by Euclidean distance between individuals in the current population. Mechanism of adaptation was changed to adapt to local properties with each sub-population SP individually.

Algorithm 3. The Algorithm of LJADE

set $\mu_{F_i} = 0.5$; $\mu_{CR_i} = 0.5$ $i \in NP$
initialize population x generating randomly.
Evaluate individual $x_i, i = 1, 2, \cdots, NP$
for $g = 1$ to G **do**
    for i = 1 to NP **do**
        $FLAG_{S,i} = FALSE$
        Create SP_i with k-nearest from P
        Generate CR_i and F_i from $\mu_{F,i}$ and $\mu_{CR,i}$
        Randomly choose $x^p_{best,g}$ from SP_i
        Randomly choose $x_{r1,g} \neq x_{i,g}$ from SP_i
        Randomly choose $x_{r2,g} \neq x_{i,g} \neq x_{r1,g}$ from $SP_i \cup A_i$
        $v_{i,g} = x_{i,g} + F_i \cdot (x^p_{best,g} - x_{i,g}) + F_i \cdot (x_{r1,g} - \tilde{x}_{r2,g})$
        $j_{rnd} = rand(1, D)$
        for $j = 1$ to D **do**
            if $rand_j(0, 1) < CR_i$ or $j = j_{rnd}$ **then**
                $u_{j,i,g} = v_{j,i,g}$
            else
                $u_{j,i,g} = x_{j,i,g}$
            end if
        end for
        if $f(x_{i,g}) \leq f(u_{i,g})$ **then**
            $x_{i,g+1} = x_{i,g}$
        else
            $x_{i,g+1} = u_{i,g}$; $FLAG_{S,i} = TRUE$
        end if
    end for
    Randomly remove solutions from A_i to make size of A_i is k
    for $i = 1$ to NP **do**
        for $j = 1$ to k **do**
            if $FLAG_{S,j} = TRUE$ **then**
                $x_{j,g} \rightarrow A_i$; $CR_j \rightarrow S_{CR,i}$; $F_j \rightarrow S_{F,i}$
            end if
        end for
        $\mu_{F,i} = (1 - c) \cdot \mu_{F,i} + c \cdot mean_L(S_{F,i})$
        $\mu_{CR,i} = (1 - c) \cdot \mu_{CR,i} + c \cdot mean_A(S_{CR,i})$
    end for
end for

4.1 K-Nearest Neighborhood Based DE

Differential Evolution creates differential vectors from solutions of randomly chosen individuals which was distributed randomly in large search space to generate new solutions. This mechanism allows individuals to explore large space by and convergence at local optimum. Meanwhile, this method limits individuals to create the differential vectors among only k-nearest individuals calculated by Euclidean distance, so that individuals are allowed to explore only surrounding area of initial location. This mechanism was proposed in [2] aiming to locate multiple local optima simultaneously with DE algorithm.

4.2 Local Parameter Adaptation and Archive

In order to apply adaptation to each sub-population, μ_F and μ_{CR} needs to be calculated locally instead of globally as conventional method does. Since population in the sub-population made by k-nearest neighborhood changes in every iteration, we employ a mechanism of every individuals manage these adaptational value as their own value μ_{F_i} and μ_{CR_i}. Therefore, successful individuals parameter is collectable only among the SP and so does $x_{r1,g}$, $x^p_{best,g}$, $x_{r2,g}$ and A. The pseudo-code is shown in Algorithm 3.

5 Experimental Settings

In this section, an experiment is held to compare performance of LJADE with JADE. We applied these algorithms to Composition function F24 from CEC2014 Benchmark Problems [9]. The landscape of the function is shown in Fig. 1 with the location of global optimum marking with star mark. We evaluate algorithms by the error values achieved after G iteration in 31 runs. Details of parameters are used in this experiment shown in Table 1.

5.1 Performance Results

An overview of results are shown in Table 2. Mean error values achieved in 31 runs is shown with standard variance values which is surrounded by parentheses. In addition, the ratio of convergence at around global optimum after G iteration is shown. Since Evolutionary Algorithm performance highly depends on parameter NP, we also compared different value of NP with same parameters. The results of mean error values are shown in Fig. 2, standard variance values are shown in Fig. 3. Both of result's horizontal axis is NP, vertical axis insists Fitness value.

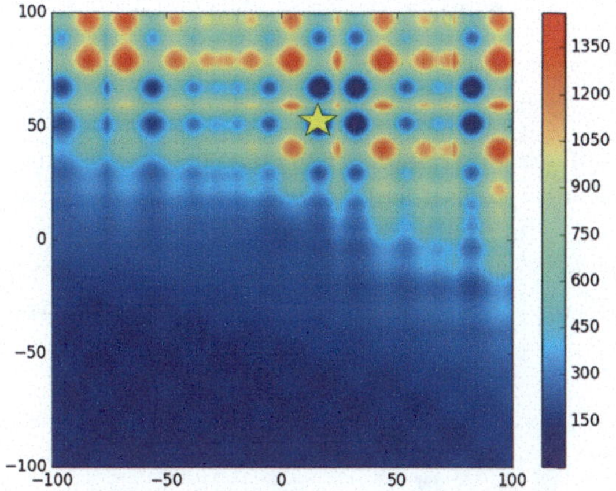

Fig. 1. Composition function F24

5.2 Algorithm Comparison

Figures 4 and 5 show JADEs convergence locations in the end of independent runs. White circle represents a position of each population on the generation. Note that these figures show JADE may lead into false convergence occasionally in Composition Function. On the other hand, Fig. 6 reveals LJADE ends up locating multiple local optima simultaneously. As the results shows, proposal method LJADE perform better than conventional method JADE in any NP value. Figure 3 shows standard variance values, which proves LJADE perform more stabled exploration through independent runs than JADE. This is also known from the ratio of locating optima around global optimum. LJADE is able to locate with $NP = 100$ in any case of random initial population placements, while JADE cannot constantly.

However, this result also reveals LJADE perform worse at locating around global optima with value of NP lower than 50, while LJADE's mean error value and standard variance value are lower than JADE. This can be explained by enhancement of LJADE's local search ability. Figure 7 shows the convergence location of LJADE with $NP = 25$ in the end of the run. It also converged into local optima same as JADE converge at, but at the same time, it locates way better local optimum above the figure. Mean while, JADE cannot. This causes JADE's exploration ability evaluation value in Composition Function extremely bad.

Both of Algorithms performance result gets better as population increases. However, While JADE increases 95% performance by increasing population from 100 to 400, LJADE increases 99%. This caused by lack of JADE's local search ability. JADE converge at local optima that is found in early exploration. According to F24's global optimum locates at far away from local optimum, JADE is suppose to adapt its parameter F much smaller. Since such a situation occurs

Table 2. Experimental result

	NP				
	25	50	100	200	400
JADE	6.497E+01	3.5717E+01	9.6774E+00	3.2274E+00	4.0843E-01
	(4.7247E+01)	(4.7681E+01)	(2.9565E+01)	(1.7668E+01)	(4.506E-01)
	7/31	14/31	28/31	30/31	31/31
LJADE	3.1147E+01	8.9811E+00	1.1853E+00	5.9900E-01	9.3146E-04
	(2.4585E+01)	(1.7851E+01)	(1.5919E+00)	(1.2563E+00)	(5.1017E-03)
	0/31	8/31	31/31	31/31	31/31

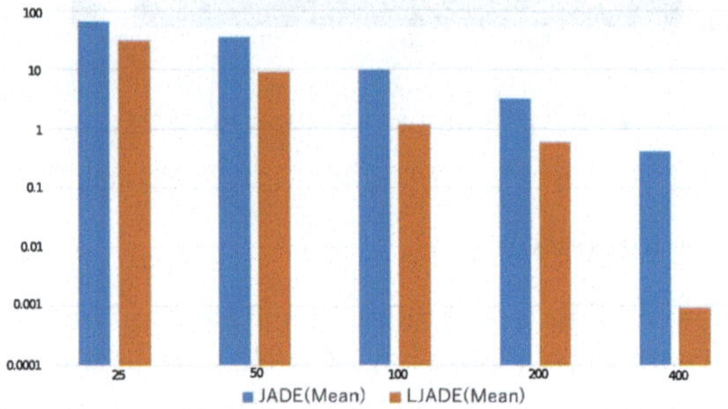

Fig. 2. Result mean comparison

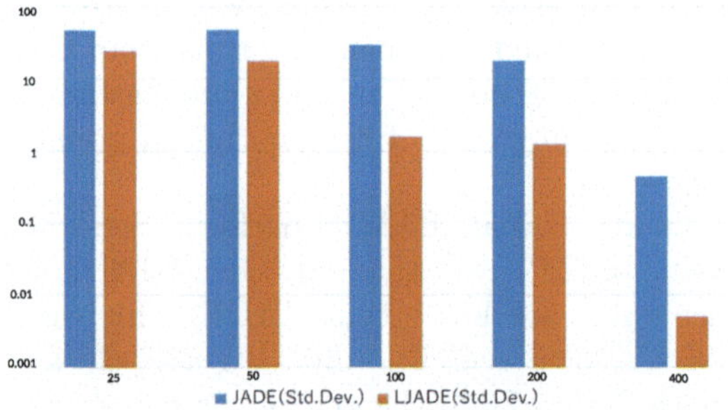

Fig. 3. Result standard variance comparison

after short stagnation moments, F is close to Max value 1, and cannot continue local exploration especially with too many population to explore.

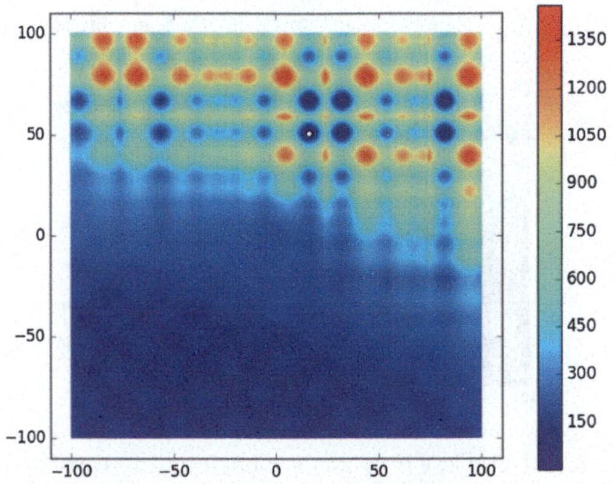

Fig. 4. JADE convergence at global optimum

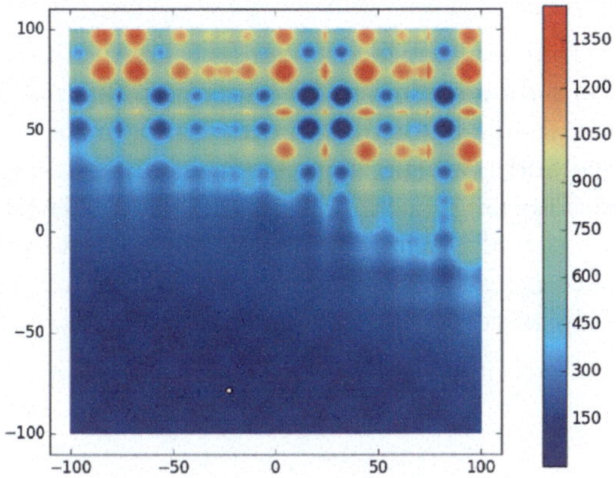

Fig. 5. JADE convergence at local optimum

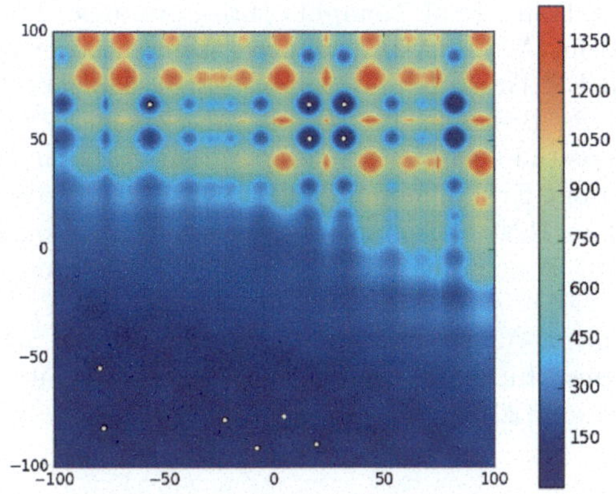

Fig. 6. LJADE convergence

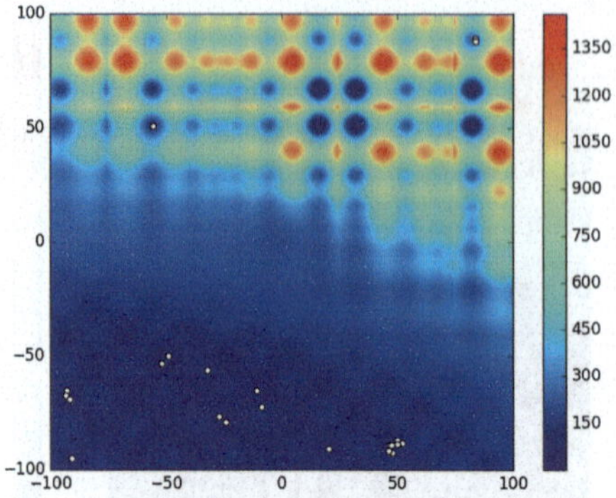

Fig. 7. LJADE($NP = 25$)'s locations of convergence

6 Conclusion

This paper assumes that real-world problem is a compositional problem which is a combinational environment of many kinds of problems. It is necessary to be adaptive and more stable algorithm in order to operate in real-world. JADE is known as adaptive and powerful algorithm, yet unstable to operate in compositional problem. The neighborhood based JADE(:LJADE) is proposed in this paper, to make the algorithm more stable and powerful. The results of experimental studies suggest that the proposed algorithms can provide better performance with stabled result in Composition problems.

References

1. Storn, R., Price, K.: A Simple and Efficient Heuristic for Global Optimization over Continuous Spaces. Kluwer Academic Publishers, Alphen aan den Rijn (1997)
2. Qu, B.Y., et al.: Differential evolution with neighborhood mutation for multimodal optimization. IEEE Trans. Evol. Comput. **16**(5), 601–614 (2012)
3. Zhang, J., Sanderson, A.C.: JADE: adaptive differential evolution with optional external archive. IEEE Trans. Evol. Comput. **13**(5), 945–958 (2009)
4. Brest, J., Greiner, S., Boskovic, B., Mernik, M., Zumer, V.: Selfadapting control parameters in differential evolution: A comparative study on numerical benchmark problems. IEEE Trans. Evol. Comput. **10**(6), 646–657 (2006)
5. Qin, A.K., Suganthan, P.N.: Self-adaptive differential evolution algorithm for numerical optimization. In: Proceedings IEEE Congress Evolutionary Computation, vol. 2, pp. 1785–1791 (2005)
6. Tanabe, R., Fukunaga, A.: Evaluating the performance of SHADE on CEC 2013 benchmark problems. In: 2013 IEEE Congress on Evolutionary Computation, Cancun, pp. 1952-1959 (2013)

7. Abbass, H.A.: The self-adaptive pareto differential evolution algorithm. In: Proceedings IEEE Congress Evolutionary Computation, Honolulu, HI, vol. 1, pp. 831–836 (2002)
8. Teo, J.: Exploring dynamic self-adaptive populations in differential evolution. Soft Comput. Fusion Found. Methodologies Appl. **10**(8), 673–686 (2006)
9. Liang, J., Qu, Y.B., Suganthan, P.: Problem definitions and evaluation criteria for the CEC 2014 special session and competition on single objective real-parameter numerical optimization (2013)

Evaluation of Search Performance of Evolutionary Computation by Transfer Entropy

Hiroshi Sato(✉)

1-10-20 Hashirimizu, Yokosuka, Kanagawa 2398686, Japan
hsato@nda.ac.jp

Abstract. In Evolutionary Computation, the search space made from genotype and the search space made from phenotype is usually quite different. This study tries to reveal the relation among genotype space, phenotype space, and fitness landscape using transfer entropy. The preliminary experiment shows a promising result.

Keywords: Evolutionary computation · Genetic algorithm · Genotype · Phenotype · Diversity · Search performance · Entropy · Transfer entropy

1 Introduction

Evolutionary Computation (EC) is a computational model which is inspired by biological evolution. Although there are many variant models in evolutionary computation, we choose the Genetic Algorithm (GA) as a reference model because GA has broad utility and is the most popular algorithm among them [1].

GA holds a population of solution candidates and uses several genetic operators such as crossover and mutation in order to evolve the population. Each individual in the population is typically represented by a gene string—usually binary string of zeros and ones. The individuals have fitness value which is the result of evaluation in the environment.

Here, there may be a problem caused by gene representation. For example, when GA represents two real numbers by binary strings, the distance between them in binary space is quite different from the distance in Euclidian space which is intrinsic for representing numbers. The genetic operators create new individuals called "children" around the original individuals called "parents." The children are close to the parents in binary space. However, they are not so close in Euclidian space. This situation is undesirable in view of the inheritance of characteristics.

There have been many models that give importance to the diversity of the population [2–4]. Notably, recent researches of the evolutionary computation directly manipulate the distribution of the population in the search space [5–7].

H. Sato et al. (Eds.): IES 2019, PALO 12, pp. 228–237, 2020.
https://doi.org/10.1007/978-3-030-37442-6_21

The distribution must be controlled carefully in order to find the optimal point in the space. It should be broad in the early stage of the search, but it should be concentrated in the last stage of the search. There are several techniques for controlling the distribution of the population. However, the maintenance of the diversity in the gene space is not sufficient because the diversity in the intrinsic representation space may be different.

The search stage is roughly divided into three stages: early stage, middle stage, and last stage. The time-series of the average fitness value of the population is a good indicator of the search stage. However, the time-series does not tell anything about how the population is distributed in the gene space.

Therefore, this study takes into account the other time-series such as the best fitness value, entropy of the gene distribution in order to evaluate the search progress. We introduced the transfer entropy for this purpose.

This paper consists of five sections. Section 2 introduces the concept of transfer entropy. We examine our evaluation through simple problems in Sect. 3 and discuss the experimental results in Sect. 4. Section 5 concludes this paper.

2 Transfer Entropy

Transfer entropy is the variation of the entropy, which represents the causality of two probabilistic variables [8]. The transfer entropy can be thought of as the extension of mutual information.

Let $X(t)$ and $Y(t)$ be two time-series. We represent the history of $X(t)$ from t to previous k steps:

$$X_t^k = (X(t), X(t-1), \ldots, X(t-k+1)). \tag{1}$$

If the X depends on its history, the probability of $X(t+1)$ can be written as

$$P(X(t+1)|(X_t^k). \tag{2}$$

Here, we think the conditional probability of $P(X(t+1))$ considering the history of Y from t to previous l steps:

$$P(X(t+1)|X_t^k, Y_t^l). \tag{3}$$

If X is determined only on its history free from Y, then Eqs. 2 and 3 are the same. However, Y usually affects X somehow. In this case, the transfer entropy from Y to X is defined as

$$T_{Y \to X} = \sum P(X_{t+1}, X_t^k, Y_t^l) log \frac{P(X(t+1)|X_t^k, Y_t^l)}{P(X(t+1)|X_t^k)} \tag{4}$$

Equation 4 represents the difference of information of these above two variables.

Contrary to the mutual information, the transfer entropy does not hold the similarity. The transfer entropy can be applied to the broader situation compared to the typical causality measurement such as Granger Causality [9] because the transfer entropy does not require anything about time-series.

3 Experiments

In this section, we examine the utility of the transfer entropy as the index of the performance.

3.1 Objective Functions

Many benchmark functions have been proposed for validation of optimization [10]. There is no standard, but they can be categorized. We pick up three categories: Multi-modal function, unimodal function with variable correlation, and ill-scale function. The followings are the typical function of these categories.

- Rastrigin's Function [10] (n-dimension)

$$f_1(x) = 10n + \sum_{i=1}^{n} [x_i^2 - 10\cos(2\pi x_i)] \tag{5}$$

  Rastrigin's function is a multimodal function with many local optimal. The global minimum of this function is $f^* = 0$ at $x^* = (0, 0, \ldots, 0)$ in the domain $-5.12 \le x \le 5.12$.
- Rosenbrock's Function [10] (n-dimension)

$$f_2(x) = \sum_{I=1}^{n-1} [100(x_{i+1} - x_i^2)^2 + (1 - x_i)^2] \tag{6}$$

  Rosenbrock's function is unimodal function whose global minimum $f^* = 0$ occurs at $x^* = (1, 1, \ldots, 1)$ in the domain $-5 \le x_i \le 5$.
- K-tablet Function [11] (n-dimension)

$$f_3(x) = \sum_{i=1}^{k} x_i + \sum_{i=k+1}^{n} (100x_i)^2 \tag{7}$$

  where, $k = 4/n$. K-tablet function is unimodal function, but it has ill-scale property. The global minimum of this function is $f^* = 0$ at $x^* = (0, 0, \ldots, 0)$ in the domain $-5.12 \le x \le 5.12$.

The fitness landscapes of these functions are depicted in Fig. 1.

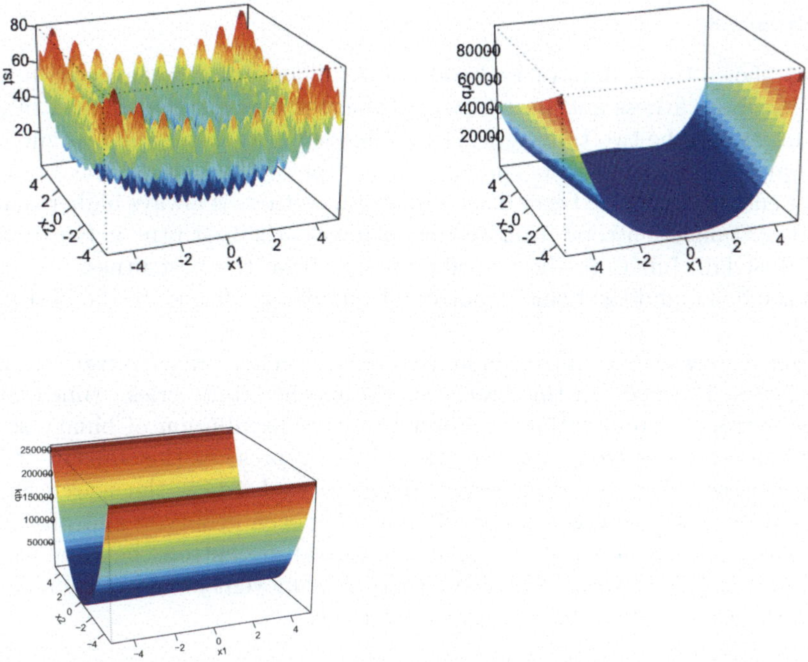

Fig. 1. Fitness Landscape in two dimension: (Upper left) Rastrigin Function, (Upper right) Rosenbrock Function, (Lower Left) K-tablet Function

3.2 Genetic Algorithm

We adopt Simple GA [1] because it is recognized as a benchmark algorithm in many literatures. The Simple GA is implemented in R using "GA" package [12]. Figure 2 is the flowchart of GA implemented in this package.

Table 1 shows the genetic operators and genetic parameters used in this experiments.

Table 1. Setting of the experiments

GA model	Simple GA	
Selection	Ranking Selection	
Coding	Binary String	Real Value
Crossover	One point Crossover	Arithmetic Crossover
Mutation	Random Bit Flip	Random Perturbation
Crossover rate	0.8	
Mutation rate	0.05	
Population size	50	
Objective function	Rastrigin Function, Rosenbrock Function	
Dimension of objective function	20	

3.3 Results

Figure 3 shows the results of real-coded GA. There are two time-series in the figure—average fitness and best fitness. Table 2 represents the transfer of information between the two time-series. In real-coded GA, the search was successful in all functions because genotype and phenotype are the same. The amount of transfer entropy from the beset fitness to average fitness is always higher than the amount of transfer entropy from average fitness to the best fitness. For example, in the Rastrigin function, the transfer entropy from the best fitness to average fitness is 0.0436, and the transfer entropy from average fitness to the best fitness is 0.0014.

Figure 4 shows the results of binary-coded GA. There are also two time-series in the figure. However, in this case, we add another time-series—time-series of average entropy of population calculated by the distribution of binary strings. Figure 5 shows the entropy of each case. Table 3 represents the transfer of information between the three time-series. In binary-coded GA, the search was not successful because there is a large difference between genotype and phenotype. In this case, there is no constant movement of transfer entropy between the best fitness and average fitness. However, there is a movement of transfer entropy from both fitness indices to entropy, respectively.

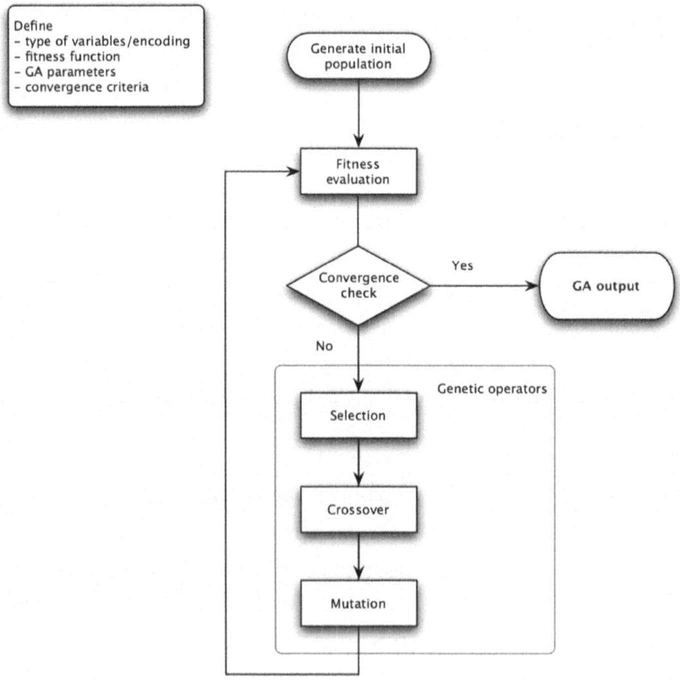

Fig. 2. Flow-chart of Genetic Algorithm (excerpt from [12])

Table 2. Transfer entropy between performance indices in real-coded GA

	Rastrigin		Rosenbrock		K-tablet	
X → Y	Best	Average	Best	Average	Best	Average
Best fitness	0	0.0436	0	0.0523	0	0.0196
Average fitness	0.0014	0	0.0127	0	0.0196	0

Table 3. Transfer entropy between performance indices in binary-coded GA

	Rastrigin			Rosenbrock		
X → Y	Best	Average	Entropy	Best	Average	Entropy
Best fitness	0	0.0343	0.0091	0	0.0095	0.0134
Average fitness	0.0023	0	0.0128	0.0476	0	0.0106
Entropy	0.0164	0.0069	0	0.0142	0.0042	0
	K-tablet					
X → Y	Best	Average	Entropy			
Best fitness	0	0.0032	0.0017			
Average fitness	0.0102	0	0.0020			
Entropy	0.0246	0.0017	0			

4 Discussion

The experimental result (Fig. 3 and Fig. 4) show that the performance of real-coded GA is much better than binary-coded GA. What can we say when we see the results from the viewpoint of transfer entropy?

First, we can find similarity in Table 2. Information on time-series flows from the best fitness to average fitness in all cases. The amount of transfer information can be thought of as the difficulties of the optimization function. These three functions can be sorted by the difficulties: K-tablet function > Rastrigin's function > Rosenbrock's function. This order corresponds to the order of the difference of transferred entropy between time-series of performance index—best fitness and average fitness.

On the other hand, there is no apparent similarities between the best fitness and average fitness in Table 3. We may say that the transfer of information the search will cause failure when the search is not successful. We can find the similarity between entropy and the best fitness. There are always directed information flow from entropy to the best fitness. The interesting thing is that there is no tendency between entropy and average fitness.

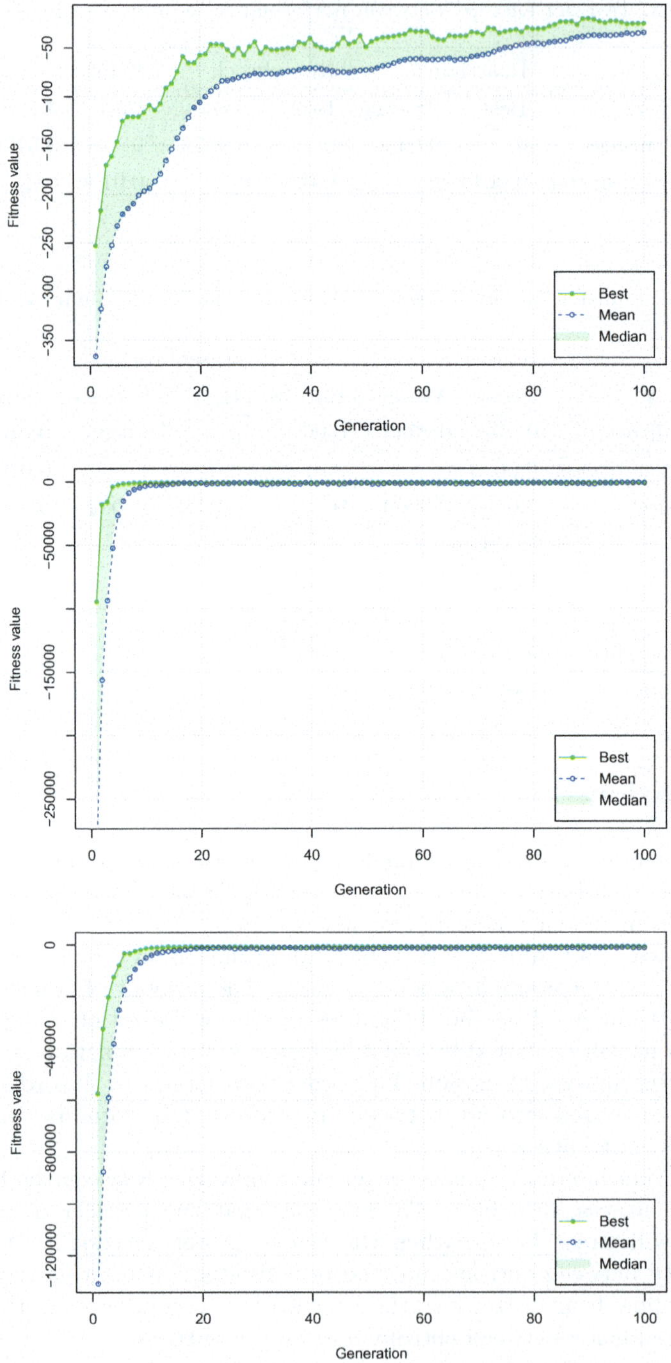

Fig. 3. Optimization of benchmark functions by real-coded GA: (Upper left) Rastrigin's function, (Upper right) Rosenbrock's function, (Lower left) k-tablet function

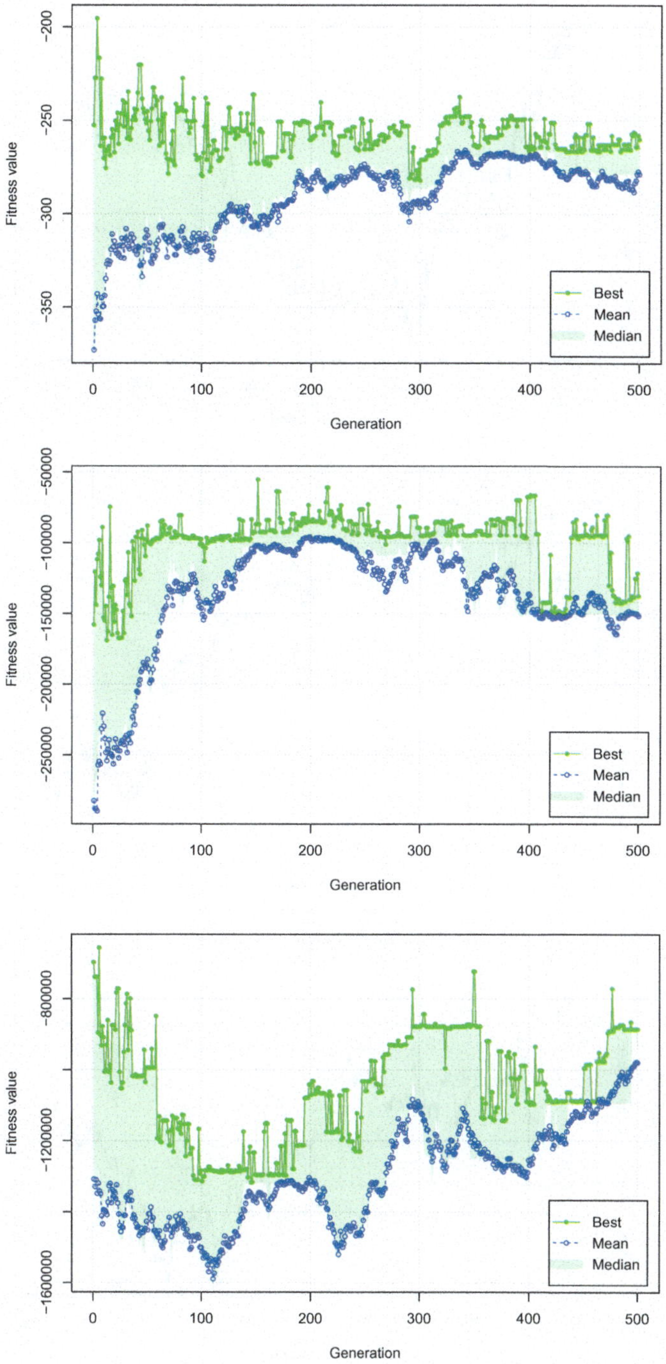

Fig. 4. Optimization of benchmark functions by binary-coded GA: (Upper left) Rastrigin's function, (Upper right) Rosenbrock's function, (Lower left) k-tablet function

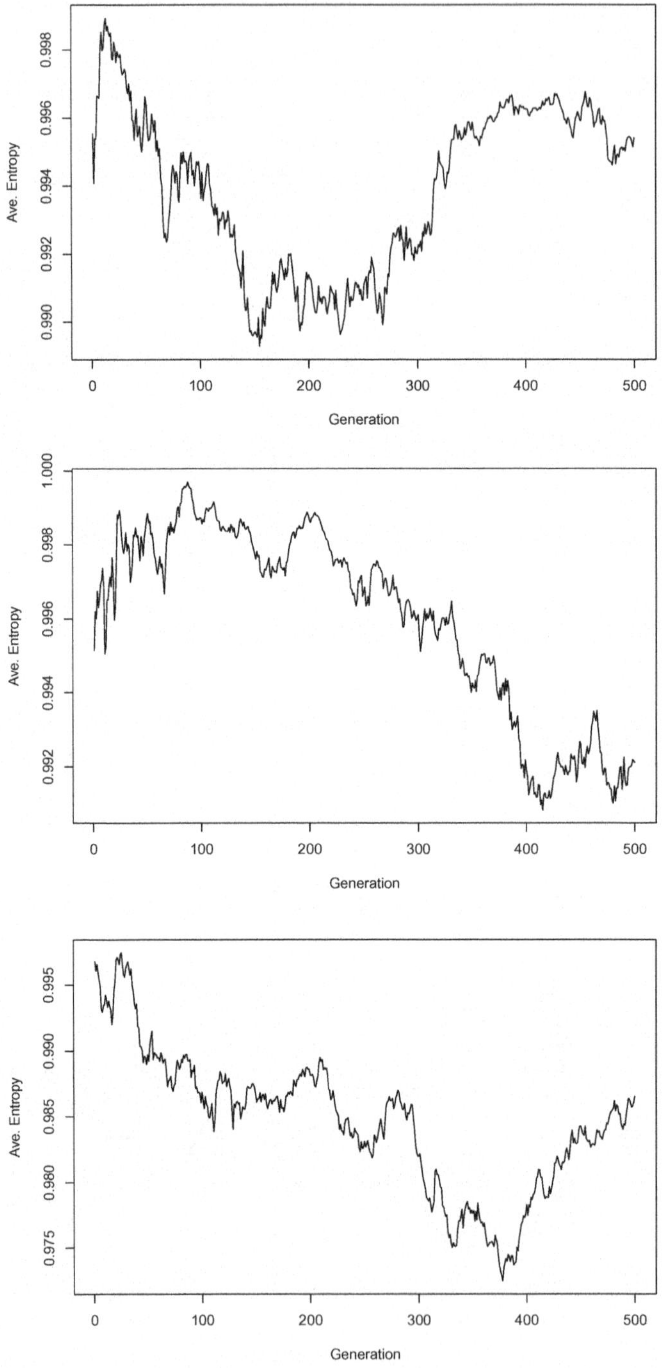

Fig. 5. Entrpry of population represented by binary strings: (Upper left) Rastrigin's function, (Upper right) Rosenbrock's function, (Lower left) k-tablet Function

5 Conclusion

The difference between genotype and phenotype makes the problem difficult for evolutionary computation to search and optimize. The difference also makes it hard to analyze the search process of evolutionary computation. In this paper, we focus on the time-series of several performance indices such as average fitness value and entropy and investigate the information flow between these indices. Through the experiments of function optimization, we show transfer entropy can be a useful tool for analysis of the search process of evolutionary computation. Applying this method to other types of problems such as combinatorial optimization and multi-objective function optimization is our future work.

References

1. Goldberg, D.E.: Genetic Algorithms in Search Optimization and Machine Learning. Addison-Wesley, Reading (1989)
2. Mengshoel, O.J., Goldberg, D.E.: The crowding approach to niching in genetic algorithms. Evol. Comput. **16**(3), 315–354 (2008)
3. Mori, N., Yoshida, J., Tamaki, H., Nishikawa, H.: A thermodynamical selection rule for the genetic algorithm. In: Proceedings of 1995 IEEE International Conference on Evolutionary Computation, pp. 188–192 (1995)
4. Satoh, H., Yamamura, M., Kobayashi, S.: Minimal generation gap model for GAs considering both exploration and exploitation. In: Proceedings of the IIZUKA 1996, pp. 494–497 (1996)
5. Mühlenbein, H.: The equation for response to selection and its use for prediction. J. Evol. Comput. **5**(3), 303–346 (1997)
6. Baluja, S.: Population-based incremental learning: a method for integrating genetic search based function optimization and competitive learning, Technical Report, Carnegie Mellon University (1994)
7. Harik, G.R., Lobo, F.G., Goldberg, D.E.: The compact genetic algorithm. IEEE Trans. Evol. Comput. **3**(4), 287–297 (1999)
8. Schreiber, T.: Measuring information transfer. Phys. Rev. Lett. **85**, 461–464 (2000)
9. Granger, G.W.J.: Investigating causal relations by econometric models and cross-spectral methods. Econometrica **37**(3), 424–438 (1969)
10. Yang, X.S.: Test problems in optimization. In: Xin-She, Y. (ed.) Engineering Optimization: An Introduction with Metaheuristic Applications. Wiley, Hoboken (2010)
11. Koabayashi, S.: The frontiers of real-coded genetic algorithms. J. Japanese Soc. Artif. Intell. **24**(1), 128–143 (2009). (in Japanese)
12. Scrucca, L.: GA: a package for genetic algorithms in R. J. Stat. Softw. **53**(4), 1–37 (2013)

Sociophysics Approach of Simulation of Charismatic Person and Distrusted People in Society Using Opinion Dynamics

Application of New Opinion Dynamics Theory Including both Trust and Distrust Between People

Nozomi Okano and Akira Ishii(✉)

Department of Applied Mathematics and Physics,
Tottori University Koyama, Tottori 680-8552, Japan
ishii@tottori-u.ac.jp

Abstract. The problem of socially isolated people who are not trusted by the people around them is considered using a new opinion dynamics theory that can handle both trust and distrust among people. This can be not only in the real world, but also in human relations on SNS. Using the new theory of opinion dynamics, we perform simulations related to isolated persons. We found that a charismatic popular person could be saved by having a relationship with an isolated person. However, on the other hand, the charismatic population is weakened. The best way is to resolve distrust of themselves by everyone before using the popularity of the charismatic person.

Keywords: Opinion dynamics · Trust · Distrust · Charismatic person

1 Introduction

The issue of socially isolated people is important. There are various reasons for isolation, but in particular, consider the problem of those who are not trusted by the people around them. This can be not only in the real world, but also in human relations on SNS. How to save the people who lose their credibility and isolation on SNS is important, but it is not easy to talk to every person in society and restore credibility.

As noted frequently in the news, books and other mass media, the clash of extreme opinions expressed online leads to increased social risks in everyday life. Online exchanges remain part of the public record, and an era where the media literacy of third parties will be even more important is quickly approaching. It is possible to analyze opinion exchanges by analyzing online logs. The process of exchanging opinions and using this to form intent is a research field known as opinion dynamics.

H. Sato et al. (Eds.): IES 2019, PALO 12, pp. 238–252, 2020.
https://doi.org/10.1007/978-3-030-37442-6_22

Of the approaches of opinion dynamics, rather than the binary approach which asks whether someone agree or disagree, we will try to tackle this problem by taking an approach that deals with the transition of opinions continuously. Unfortunately, a representative of the theory of opinion dynamics that handles the transition of opinions continuously is the Bounded Confidence Model, which is a theory aimed at consensus building [1–5]. Using these theories, we can handle only the trust relationship between persons in the simulation. Recently, Ishii et al. proposed a new theory of opinion dynamics that deals with human relationships of both trust and distrust [6,7]. Using this theory, for example, we can calculate the effect of charismatic person in the simulation [8]. The charismatic person in ref.8 [8] is defined that he/she is trusted by everyone around the person. Thus, in this paper, we apply this idea to simulate a person who were not trusted by everyone around the person.

2 Theory

A theory for analyzing the processes that aim at reaching agreement within society (or within a small group), opinion dynamics has been studied from various perspectives for many years [9–14]. Attempts have also been made to apply game theory, [15] which is a kind of opinion dynamics, to agreements as well. One can also refer to comprehensive reports on opinion dynamics theory [5]. However, to analyze opinion exchanges remaining in the vast log data of modern society, it goes without saying that what is needed is a theory that can be used for quantitative analysis focused on integration with analysis for large-scale data.

Theories of opinion dynamics suitable for quantitative analysis can be broken down roughly into two types. One method involves a theory that treats contradictory conditions and discrete opinions as 1 and 0, or 1 and −1. Because presidential elections in the US and France, and national voting as seen with Brexit and the like involve voting where there is one clear winner, it is easy to apply this binary theory to such cases. The other method involves a theory that considers opinion as a one-dimensional (or multidimensional) continuous value. Consensus building and similar phenomena are often thought of in this way.

The representative example of discrete binary theory is the theory that applies the Ising model of Galam's magnetic physics theory [16–18]. In this theory, each persons opinion can be represented by I_i, with interaction by opinion exchange expressed as

$$-J\sum_i\sum_j I_i I_j \tag{1}$$

Lets say that this involves social conflict, not energy. Because the dynamics of society act to restrict conflict, in the case of this formula, when $J > 0$, conflict is decreased when one shares the opinions of ones peers.

Moreover, a local majority decision model using the concept of renormalization groups from theoretical physics is also binary, consisting of approval and disapproval [19,20]. Utilizing this theory, Galam published a study on the analysis of the referendum of Britain's EU withdrawal (BREXIT) [21] and the election of US President Trump [22].

Meanwhile, representative examples of the theory of opinions as consecutive values include the Deffuant-Weisbuch Model [1,2] and the Bounded Confidence Model known as the Heselmann-Krause Model [3]. Although there have been attempts to deal with bipolarization resulting from conflicting opinions as an expansion of the Deffuant-Weisbuch Model [23–25], the Hegselmann-Krause Model considers only consensus building. Opinions take continuous values from 0 to 1, but it is worth noting that 1 denotes agreement and 0 indifference; there is no assumption of a dissenting opinion. In the extension of the Deffuant-Weisbuch Model [23–25], the manner in which N persons reciprocate each others influence is constant, with this determined by the degree of disagreement between each other and no evidence of such elements as each persons inherently unique trust or distrust.

In the present research, based on the model of Hegselmann-Krause, which makes it easier to devise settings for exchanging the mutual opinions of N persons more theoretically, this is expanded to include the factors that lead people to disagree and the influence of the mass media.

Our model is based on the original bounded confidence model of Hegselmann-Krause [3]. For a fixed agent, say i, where $1 \leq i \leq N$, we denote the agents opinion at time t by $I_i(t)$. As shown in Fig. 1, person i can be affected by surrounding people. According to Hegselmann-Krause [3], opinion formation of agent i can be described as follows.

$$I_i(t+1) = \sum_{j=1}^{N} D_{ij} I_j(t) \tag{2}$$

This can be written in the following form.

$$\Delta I_i(t) = \sum_{j=1}^{N} D_{ij} I_j(t) \Delta t \tag{3}$$

where it is assumed that $D_{ij} \geq 0$ for all i, j in the model of Hegselmann-Krause. Based on this definition, $D_{ij} = 0$ means that the opinion of agent i is not affected by the opinion of agent j. In this theory, it is expected implicitly that the final goal of the negotiation among people is the formation of consensus.

However, in the real society in the world, the formation of consensus among people is sometimes very difficult. We can find many such examples in the international politics in the world history. Even in domestic problems, the opinions between people pursuing economic development and people claiming nature conservation are not compatible and agreement is difficult. Since it is not possible to define the payoff matrix for such serious political conflict, application of game theory may be difficult. Thus, in order to deal with problems that are difficult to form consensus among these people, it is necessary to include the lack of trust between people in our opinion dynamics theory. Here, as a result of exchanging opinions, consider the possibility that the opinions of two people with different opinions change move in different directions. We consider the distribution of opinions with in the positive and negative directions of a one-dimensional axis. In this case, the value range of $I_i(t)$ is $-\infty \le I_i(t) \le +\infty$. Here, we assume that $I_i(t) > 0$ means positive opinion and $I_i(t) < 0$ means negative opinion. In the limitation of Hegselmann-Krause model, one can assume that $1/2 \le I_i(t) \le 1$ corresponds to positive opinion and $0 \le I_i(t) \le 1/2$ corresponds to negative opinion. However, our definition is intuitive, easy to understand and easy to apply to various examples.

We modify the meaning of the coefficient D_{ij} as the coefficient of trust. We assume here that $D_{ij} > 0$ if there is a trust relationship between the two persons, and $D_{ij} < 0$ if there is distrust relationship between the two persons.

According to our previous theory [7], we consider here that people disregard the opinion far removed from their opinions without agreeing or repelling. Also, opinions that are very close to himself/herself will not be particularly affected. To include the two effects, we use the following function instead of $D_{ij}I_j(t)$ as follows,

$$D_{ij}\Phi(I_i, I_j)(I_j(t) - I_i(t)) \tag{4}$$

where

$$\Phi(I_i, I_j) = \frac{1}{1 + exp(\beta(|I_i - I_j| - b))} \tag{5}$$

This function is Sigmoid function and it works as a smooth cut-off function at $|I_i - I_j| = b$. The typical graph of this function is shown in Fig. 1. Using this Sigmoid function, we assume that if the opinions of the two are too far apart, they will not be totally influenced by each other's opinion. Moreover, because of the factor $I_j(t) - I_i(t)$, the opinion $I_i(t)$ is not affected by the opinion $I_j(t)$ if the opinion $I_j(t)$ is almost same as the opinion $I_i(t)$.

For the factor $I_j(t) - I_i(t)$, we consider that $I_j(t) - I_i(t)$ gives the same effect whether $I_i(t)$ and $I_j(t)$ are both positive or both negative or either positive or negative. This is very natural that, for example, even between conservatives, there are intense debates between those with moderate conservatives and those with radical conservatives.

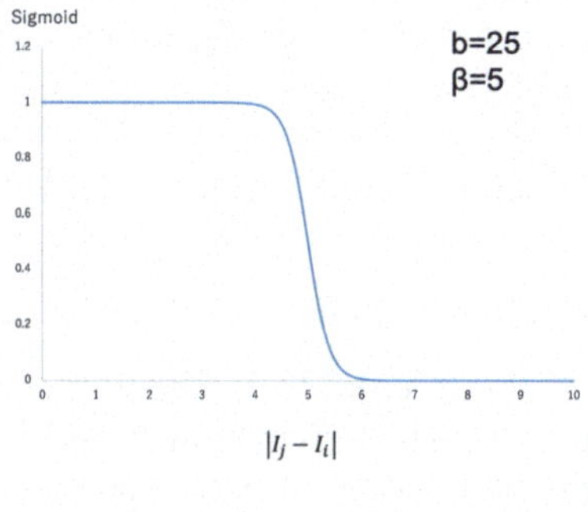

Fig. 1. The typical graph of the Sigmoid function (4) as smooth cut-off function.

Influences of mass media and government statements can not be ignored in the formation of public opinion. Such mass media effect can also work even for negotiations of small size group. Since formula of our theory above is similar to the model of hit phenomena [26] where the popularity of certain topic is analyzed using the sociophysics model, we introduce here the effects of mass media similar to the way of ref. [26]. Let $A(t)$ be the pressure at time t from the outside and denote the reaction difference for each agent is denoted by the coefficient c_i. The coefficient c_i can have different values for each person and c_i can be positive or negative. If the coefficient c_i is positive, the person i moves the opinion toward the direction of the mass media. On the contrary, if the coefficient c_i is negative, the opinion of the person change against the mass media direction.

Therefore, including such mass media effects, the change in opinion of the agent can be expressed as follows.

$$\Delta I_i(t) = c_i A(t)\Delta t + \sum_{j=1}^{N} D_{ij}\Phi(I_i(t), I_j(t))(I_j(t) - I_i(t))\Delta t \tag{6}$$

We assume here that D_{ij} is an asymmetric matrix; D_{ij} and D_{ji}, $D_{ij} \neq D_{ji}$ and D_{ij} and D_{ji} can have different signs.

Long-term behavior requires attenuation, which means that topics will be forgotten over time. Here we introduce exponential attenuation. The expression is as follows.

$$\Delta I_i(t) = -\alpha I_i(t)\Delta t + c_i A(t)\Delta t + \sum_{j=1}^{N} D_{ij}\Phi(I_i(t), I_j(t))(I_j(t) - I_i(t))\Delta t \tag{7}$$

2.1 Opinion Dynamics for Two Agents

Let us first consider the case where the opinions of the two agents are the same. In the calculation below, we set $A(t)$ to be zero. All calculations in this paper, we assume that $A(t)$ is zero. In the actual simulation of the real society behaviors, the time-dependence of the external effect $A(t)$ is also significant.

Using the model of the previous paper [7], we can write down the equations for the case of N = 2 in the following way,

$$\Delta I_A(t) = c_A A(t)\Delta t + D_{AB}\Phi(I_A(t), I_B(t))(I_B(t) - I_A(t))\Delta t \tag{8}$$

$$\Delta I_B(t) = c_B A(t)\Delta t + D_{BA}\Phi(I_B(t), I_A(t))(I_A(t) - I_B(t))\Delta t \tag{9}$$

We name here the two agents A and B. In the first case, we set both D_{AB} and D_{BA} are positive. It means that the two person A and B trust each other. Thus, the opinion $I_i(t)$ moves in the positive direction as shown in Fig. 2. This means that by having a conversation with an agent of the same positive opinion, the two agents A and B will change its opinion to be more and more positive. Similarly, if

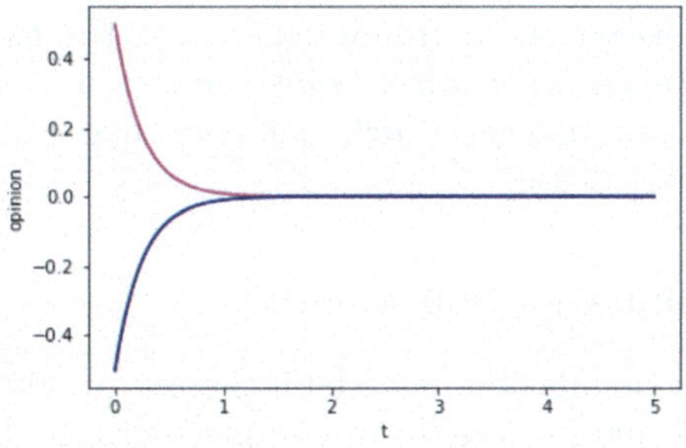

Fig. 2. Calculation result for N = 2. Adv = 0, $D_{AB} = 1.0$, $D_{BA} = 1.0$. The initial value is $I_A(0) = 0.005$, $I_B(0) = 0.2$.

the opinions of both agents are the same negative opinion, the opinions become more and more negative.

In the revised model [7], people is not influenced each other if their opinion is greatly distant. Also, it is not affected by opinions similar to his/her own opinion. Furthermore, it is assumed that the opinion will be influenced in proportion to the difference in strength of opinion between the two, $I_j(t) - I_i(t)$. Then, the calculation is corrected as shown in Fig. 2. As can be seen in Figure, in this theory, the two opinions converge on one opinion, even if they have similar positive opinions.

Next, we consider the case where the opinions of the two agents are opposite $I_A(t) > 0$ and $I_B(t) < 0$ where $D_{AB} < 0$ and $D_{BA} < 0$. Thus, the opinion of agent A changes to more a positive one. It means that, in a discussion with an agent in disagreement, as there is no trust relationship with that agent, we consider that agent A held his/her opinion more firmly. Similarly, the agent B held his/her opinion more firmly, too. For the case of $I_A(t) < 0$ and $I_B(t) > 0$ with $D_{AB} < 0$, the result is same. This result shows that the dialogue of people who do not trust each other never leads to an agreement.

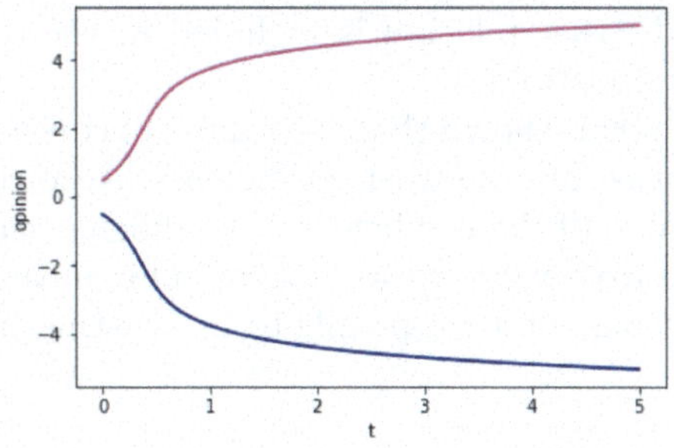

Fig. 3. Calculation result for N = 2. Adv = 0, $D_{AB} = -1.0$, $D_{BA} = -1.0$. The initial value is $I_A(0) = 0.005$, $I_B(0) = 0.2$.

In the calculation based on the modified theory, the two who repelled with different opinion from each other, after being separated to some extent, become parallel lines and do not diverge. Such behavior like Fig. 3 shown using the present theory is more realistic.

2.2 Opinion Dynamics for 300 Agents

We show here a simple calculation using Ishii theory for 300 persons. We assume that the connection of 300 persons form complete graph. The case of the trust relationship among 300 persons is shown in Fig. 4. This result means the consensus building. This opinion trajectory showed the left side of the figure is, in principle, very similar to the simulation results of the original Bounded Confidence Model [1–3].

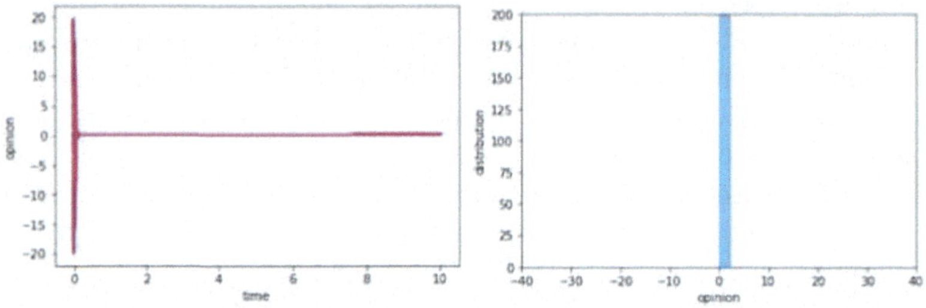

Fig. 4. Calculation result for N = 300 [27]. The human network is assumed to be complete graph. The left is the trajectories of opinions. The right is the distribution of opinions at the final time of this calculation. D_{ij} are set to be 0 to 1 randomly so that every person trust for everyone.

In Fig. 5, we show the result of simulation where positive D_{ij} and negative D_{ij} are mixed in half and half. In this case, we found that there are no consensus building. As we can see from the figures, the calculated results are very different. The negative D_{ij} or distrust relation is included in the society, the consensus building seems to be impossible.

As Ishii-Kawahata [28] is shown that the aspect of consensus building changes with the ratio of the trust and distrust coefficients. According to the study, the ratio of the positive value of the coefficient of confidence changes just like phase transition near the positive value ratio of 55%. This means that it is sufficient for 55% of the connections between people to be trusting in order for society to reach consensus.

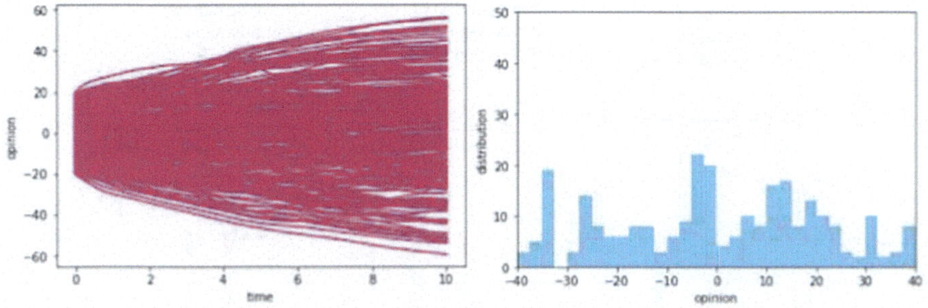

Fig. 5. Calculation result for N = 300 [27]. The human network is assumed to be complete graph. The left is the trajectories of opinions. The right is the distribution of opinions at the final time of this calculation. D_{ij} are set to be -1 to 1 randomly so that every person trust or distrust for everyone.

3 Results and Discussions

3.1 Model for Calculation

In this work, we consider one charismatic person and several persons who are not trusted by people in society. The charismatic person is defined as who is trusted by all people in the society. It means that D_{ij} to the charismatic person from the people in the society is positive value. The person who is distrusted by all people in the society is defined as D_{ij} to the person from the people in the society is negative value.

First, we consider a charismatic person. We assume that the charismatic person high trust from everyone. In Fig. 6, we show examples of the calculation including one charismatic person. As you can see in the figure, there are more people who have similar opinions to those who have strong trust from all. There are two peaks corresponding to the two charismatic persons. The will of the two charismatic person is defined here to be 10. According to the result, we find intermediate clusters between the two peaks in the case of two apart opinions for the two charismatic persons. In this paper, we define here that strongly trusted means strong popularity, but, in principle, trust is not equal to popularity. To treat trust and popularity separately will be a problem in future.

We also calculate a person who is not trusted by everyone. All D_{ij} directed to that person is a negative number, which makes it a setting that is not trusted by all. D_{ij} for the person from everyone is set to be -10. As you can see in Fig. 7, we can see that those who are not trusted by all are avoided from all in the calculation and are isolated even on the opinion distribution. Thus, using the new opinion dynamics theory [7], it turns out that the isolation of unreliable people can be well expressed from calculations.

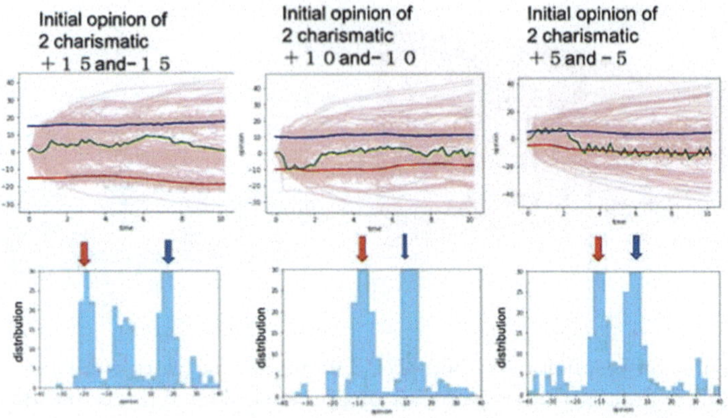

Fig. 6. Calculation result for N = 300. The human network is assumed to be complete graph. The upper is the trajectories of opinions. The lower is the distribution of opinions at the final of this calculation. D_{ij} are set to be -1 to 1 randomly so that every person trust or distrust for everyone. Two persons (blue and red lines) is set to be charismatic person. D_{ij} for the charismatic person from everyone is set to be 10. The green line is one of the ordinary persons. The arrow means the position of the opinion of the charismatic person.

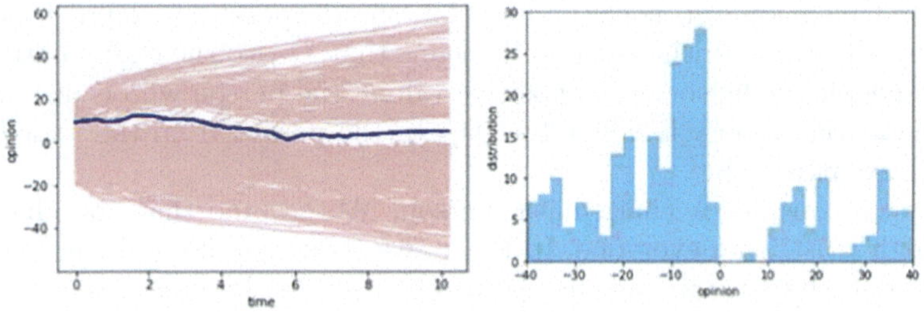

Fig. 7. Calculation result for N = 300. The human network is assumed to be complete graph. The left is the trajectories of opinions. The right is the distribution of opinions at the final of this calculation. D_{ij} are set to be -1 to 1 randomly so that every person trust or distrust for everyone. One person (blue line) is set to be the person who is not trusted by everyone. D_{ij} for the person from everyone is set to be -5.

3.2 One Charismatic Person and One Distrusted Person

Next, we consider the case of coexistence of the charismatic person and the person who is not trusted by everyone. In the following calculations, we set $D_{ij} = 10$ from all for the charismatic person and $D_{ij} = -10$ from everyone for the person who is not trusted by all is set to be -10. D_{ij} between the other people is set to be between -1 to 1 using random number.

In Fig. 8, we show a typical result that the charismatic person and the untrusted person exist independently. In the calculation of Fig. 8, we set the trust of the charismatic person to the untrusted person to be +10 and the trust of the untrusted person to the charismatic person to be −10. This setting means that the charismatic person considers the untrusted person seriously, but the untrusted person does not trust the charismatic person. For the case that both the charismatic person and the untrusted person distrust each other, we obtain the similar result as Fig. 8.

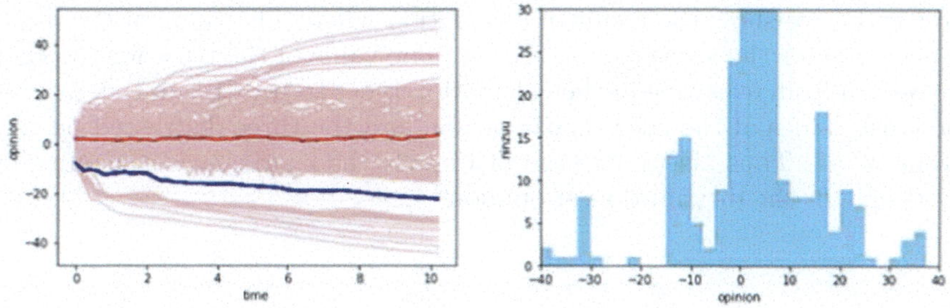

Fig. 8. Calculation result for N = 300. The human network is assumed to be complete graph. The left is the trajectories of opinions. The right is the distribution of opinions at the final of this calculation. D_{ij} are set to be −1 to 1 randomly so that every person trust or distrust for everyone. One person (red lines) is set to be charismatic person. D_{ij} for the charismatic person from everyone is set to be 10. One person (blue line) is set to be the person who is not trusted by everyone. D_{ij} for the person from everyone is set to be −10.

3.3 One Charismatic Person and Three Distrusted Persons

We consider here the problem of one charismatic person and three persons who are distrusted by everyone in the society. We can consider several cases for this problem. First, we consider the case of the charismatic person and three distrusted persons are trusted each other strongly.

We set that the charismatic person is trusted by everyone in the society. The value of the trust for the charismatic person from everyone in the calculation is set to be +10. The charismatic person and the three distrusted persons are trusted each other with the value of trust +10. The value of the distrust from everyone for the three person are −10. As we see in the left of Fig. 9, the charismatic person collect many people in the distribution of opinion. The three distrusted persons form a cluster of distrusted person. The distrusted person cluster is avoided by everyone.

For the result in the right of Fig. 9, because of the strong trust relation between the charismatic person and the three distrusted persons, the four form a cluster but the cluster is avoided by everyone. Since the Eq. (7) is linear, the value of the trust of the cluster is, roughly, sum of the trust values of the four people to everyone. Thus, it is quite natural that the cluster is avoided by everyone.

In Fig. 10, we show the result of calculation where The value of the distrust from everyone for the three person are −3. The value of the trust for the charismatic person from everyone in the calculation is set to be +10. In this case, sum of the trust values of the charismatic person and the three distrusted people to everyone is positive, +1. Thus, as we can see in Fig. 10, the cluster of the charismatic person with the two distrusted person collect many people in the distribution of opinion.

In Fig. 11, we show the result of calculation where The value of the distrust from everyone for the three person are −1. The value of the trust for the charismatic person from everyone in the calculation is set to be +10. In this case, sum of the trust values of the charismatic person and the three distrusted people to everyone is +7. Thus, as we can see in Fig. 11, the charismatic person collects many people in the distribution of opinion.

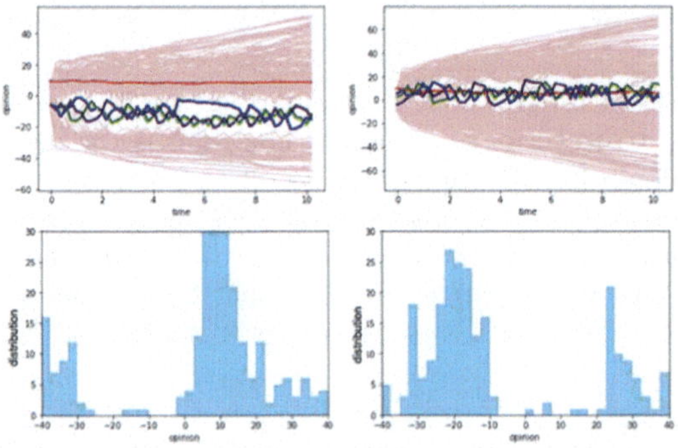

Fig. 9. Calculation of human relation between the charismatic person and three distrusted persons in 300 peoples. The red means the charismatic person and the blue, green and indigo means the persons distrusted by everyone.

For the calculation of Fig. 12, we set that the three distrusted persons distrust the charismatic person with the trust value −10. The trust value of three distrusted person by everyone is set to be −10. The three distrusted persons trust each other with the value of trust +10. In this case, a cluster of the three distrusted persons and the charismatic person is independent. The cluster is avoided by everyone and the charismatic person collect many people in the distribution of opinion.

For the similar calculation is shown in Fig. 13 where the trust value of the three distrusted persons by everyone is −1. We set that the three distrusted persons distrust the charismatic person with the trust value −10. The three distrusted persons trust each other with the value of trust +10. A shown in Fig. 13, the charismatic person collect many people in the distribution of opinion and the three distrusted person do not form cluster.

In the calculation of Fig. 14, we set that the charismatic person distrust the three distrusted person with the value −10 and the three distrusted person

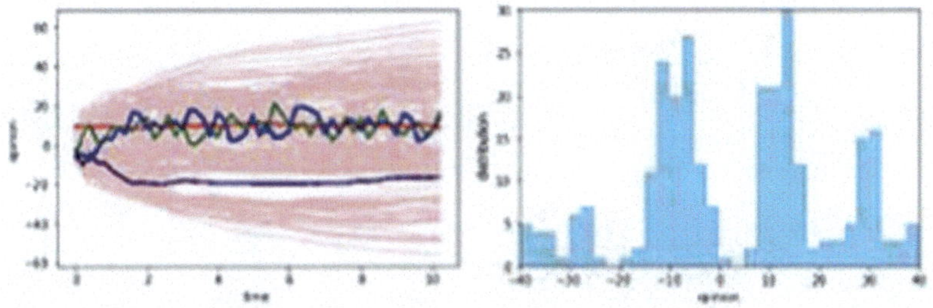

Fig. 10. Calculation of human relation between the charismatic person and three distrusted persons in 300 peoples. The red means the charismatic person and the blue, green and indigo means the persons distrusted by everyone.

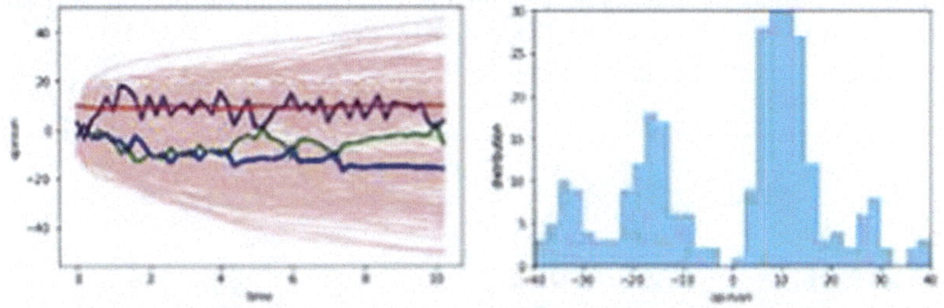

Fig. 11. Calculation of human relation between the charismatic person and three distrusted persons in 300 peoples. The red means the charismatic person and the blue, green and indigo means the persons distrusted by everyone.

distrust the charismatic person with the value -10. The three distrusted persons trust each other with the value of trust $+10$. In the calculation, the charismatic person and the three distrusted persons are avoided each other. The charismatic person collect many people in the distribution of opinion and the three distrusted persons form a cluster which is avoided by everyone.

3.4 Discussion

The results show that the popularity of charismatic people can be countered if people with strong distrust are close to charismatic people. On the other hand, if the distrust of three people is weak, the popularity of charismatic people is unaffected by their distrust. Therefore, trusting a charismatic person who is distrusted by the public is risky and can reduce the popularity of the charismatic person. People who are distrusted by the public will be able to eliminate the distrust they have from the public before relying on the popularity of charismatic people.

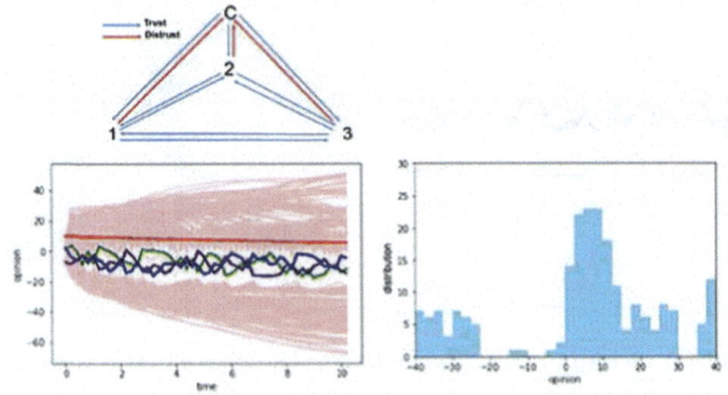

Fig. 12. Calculation of human relation between the charismatic person and three distrusted persons in 300 peoples. The red means the charismatic person and the blue, green and indigo means the persons distrusted by everyone.

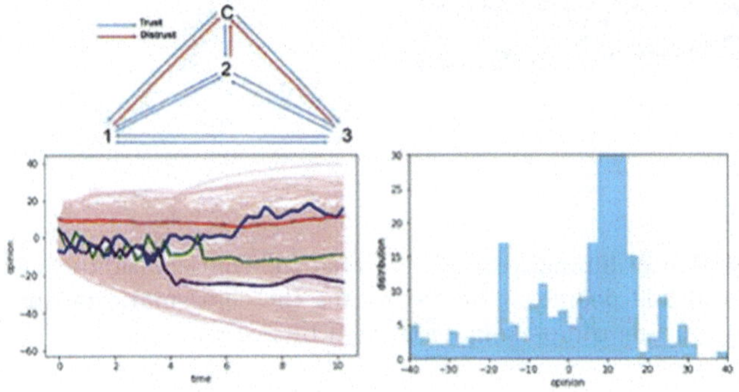

Fig. 13. Calculation of human relation between the charismatic person and three distrusted persons in 300 peoples. The red means the charismatic person and the blue, green and indigo means the persons distrusted by everyone.

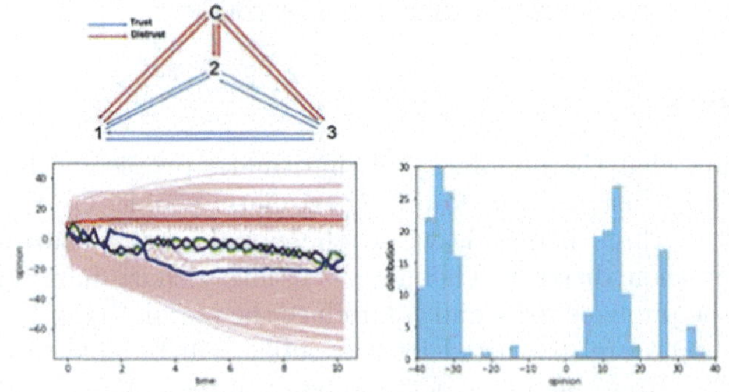

Fig. 14. Calculation of human relation between the charismatic person and three distrusted persons in 300 peoples. The red means the charismatic person and the blue, green and indigo means the persons distrusted by everyone.

4 Conclusion

In this paper, we apply our new opinion dynamics theory to the problem of the isolation of people who is distrusted by everyone in the society. Our new opinion dynamics theory is an extension of the bounded confidence model by Hegselmann-Krause but we include not only trust but also distrust in the human relationship.

The case of one charismatic person with three person who are distrusted by everyone in the society. People who are distrusted by the public cannot rely on the popularity of charismatic people to eliminate isolation. Efforts to resolve the distrust that oneself has from the public will be effective in eliminating isolation.

Since the calculation results are very native, our new opinion dynamics theory can be useful to resolve problems in society.

Acknowledgement. This work was supported by JSPS KAKENHI Grant Number JP19K04881.

References

1. Deffuant, G., Neau, D., Amblard, F., Weisbuch, G.: Mixing beliefs among interacting agents. Adv. Complex Syst. **3**, 87–98 (2000)
2. Weisbuch, G., Deffuant, G., Amblard, F., Nadal, J.-P.: Meet, discuss and segregate! Complexity **7**(3), 55–63 (2002)
3. Hegselmann, R., Krause, U.: Opinion dynamics and bounded confidence models, analysis, and simulation. J. Artif. Soc. Soc. Simul. **5**(3), 1–33 (2002)
4. Castellano, C., Fortunato, S., Loreto, V.: Statistical physics of social dynamics. Rev. Modern Phys. **81**, 591–646 (2009)
5. Sîrbu, A., Loreto, V., Servedio, V.D.P., Tria, F.: Opinion dynamics: models, extensions and external effects. In: Loreto, V., et al. (eds.) Participatory Sensing, Opinions and Collective Awareness. Understanding Complex Systems, pp. 363–401. Springer, Cham (2017)
6. Ishii, A., Kawahata, Y.: Opinion dynamics theory for analysis of consensus formation and division of opinion on the internet. In: Proceedings of The 22nd Asia Pacific Symposium on Intelligent and Evolutionary Systems (IES2018), pp. 71-76. arXiv:1812.11845 [physics.soc-ph]
7. Ishii, A.: Opinion dynamics theory considering trust and suspicion in human relations. In: Morais, D., Carreras, A., de Almeida, A., Vetschera, R. (eds.) Group Decision and Negotiation: Behavior, Models, and Support. GDN: Lecture Notes in Business Information Processing, 351st edn, pp. 193–204. Springer, Cham (2019)
8. Ishii, A., Kawahata, Y.: Opinion dynamics theory considering interpersonal relationship of trust and distrust and media effects. In: The 33rd Annual Conference of the Japanese Society for Artificial Intelligence, 2019, 33 (2019) Volume JSAI2019 2F3-OS-5a-05
9. French, J.R.P.: A formal theory of social power. Psychol. Rev. **63**, 181–194 (1956)
10. Harary, F.: A criterion for unanimity in French s theory of social power. In: Cartwright, D. (ed.) Studies in Social Power. Institute for Social Research, Ann Arbor (1959)

11. Abelson, R.P.: Mathematical models of the distribution of attitudes under controversy. In: Frederiksen, N., Gulliksen, H. (eds.) Contributions to Mathematical Psychology. Holt, Rinehart, and Winston, New York (1964)
12. De Groot, M.H.: Reaching a consensus. J. Amer. Statist. Assoc. **69**, 118–121 (1974)
13. Lehrer, K.: Social consensus and rational agnoiology. Synthese **31**, 141–160 (1975)
14. Chatterjee, S.: Reaching a consensus: some limit theorems. Proc. Int. Statist. Inst. pp. 159–164 (1975)
15. Tanimoto, K., Kita, H., Mitukuni, A.: Opinion choice model in public meeting by using evolutionary game theory infrasturcture planning review, vol. 18, no. 1, p. 8995 (2001)
16. Galam, S.: Rational group decision making: a random field Ising model at T = 0. Physica A **238**, 66 (1997)
17. Sznajd-Weron, K., Sznajd, J.: Opinion evolution in closed community. Int. J. Mod. Phys. C **11**(6), 1157–1165 (2000)
18. Sznajd-Weron, K., Tabiszewski, M., Timpanaro, A.M.: Phase transition in the Sznajd model with independence. Europhys. Lett. **96**, 48002 (2011)
19. Galam, S.: Application of statistical physics to politics. Physica A: Stat. Mech. Appl. **274**, 132–139 (1999)
20. Galam, S.: Real space renormalization group and totalitarian paradox of majority rule voting. Physica A: Stat. Mech. Appl. 285 **1–2**(15), 66–76 (2000)
21. Galam, S.: Are referendums a mechanism to turn our prejudices into rational choices? An unfortunate answer from sociophysics Chap. 19. In: Morel, L., Qvortrup, M. (eds.) The Rout-ledge Handbook to Referendums and Direct Democracy. Taylor & Francis, London (2017)
22. Galam, S.: The Trump phenomenon: An explanation from sociophysics. Int. J. Mod. Phys. B **31**, 1742015 (2017)
23. Jager, W., Amblard, F.: Uniformity, bipolarization and pluriformity captured as generic stylized behavior with an agent-based simulation model of attitude change. Comput. Math. Organ. Theory **10**, 295–303 (2004)
24. Jager, W., Amblard, F.: Multiple attitude dynamics in large populations. In: The Agent 2005 Conference on Generative Social Processes, Models and Mechanisms at The University of Chicago, 13-15 October 2005
25. Kurmyshev, E., Juárez, H.A., González-Silva, R.A.: Dynamics of bounded confidence opinion in heterogeneous social networks: concord against partila antagonism. Physica A **390**, 2945–2955 (2011)
26. Ishii, A., Arakaki, H., Matsuda, N., Umemura, S., Urushidani, T., Yamagata, N., Yoshida, N.: The 'hit' phenomenon: a mathematical model of human dynamics interactions as a stochastic process. New J. Phys. **14**, 063018 (2012)
27. Okano, N., Ishii, A.: Isolated, untrusted people in society and charismatic person using opinion dynamics. In: Proceedings of the IEEE/WIC/ACM International Conference on Web Intelligence, 14–17 October 2019, Thessaloniki, Greece (2019)
28. Ishii, A., Kawahata, Y.: Phase transition like behavior in consensus building using new opinion dynamics theory including both trust and distrust into human relationships to be published

Footstep Analysis for the Military Parade

Yohei Okugawa(✉), Masao Kubo, Hiroshi Sato, and Shun Sakai

Department of Computer Science, National Defense Academy of Japan, Yokosuka, Kanagawa, Japan
{em57037,masaok,hsato,em58031}@nda.ac.jp

Abstract. The highly synchronized parade can impress audiences how strong the troops seem to be; whereas, it is difficult to make the beautiful synchronized parade because of its complex collective behavior. However, there is no scientific research about what the important factor for the effective training is. One of the bottlenecks of the scientific approach is the difficulty for measurement of the synchronized level of the whole of the parade. In this paper, in order to measure the synchronized level of the parade quantitatively, we focused on the footsteps, which is one of audio information in the parade. When a parade is synchronized, we can estimate the tempo of the parade clearly because the footsteps of the parade are synchronized to the parade music. According to the proposed analysis method, we estimated the tempo from the three types of the footsteps: synchronized parade, not synchronized parade and disordered parade, and we compared the synchronized level of each parade. As a result, the more the footsteps are disordered, the lower the synchronized level are. In conclusion, the proposed method is applicable for the quantitative evaluation, and it is suggested that the audio information of the parade can be one of indicators for the quantitative evaluation of the parade.

Keywords: Collective behavior · Skill science · Military parade · Audio analysis

1 Introduction

1.1 The Attributes of the Military Parade

Parade, or marching, refers to the organized, uniformed, steady walking forward in rhythmic time. In a military organization, the parade is used to exhibit the apparent military strength. The military parade is a formation of soldiers whose movement is restricted by the movement of the close members, and each member in the parade needs to refer their movement each other. As we can see Fig. 1, in the beautiful military parade, their limbs are synchronized to the parade music.

The parade consists of the sum of local interaction that they adjust the walking movement with nearby members. However, the parade has different characteristics from flocking behaviors, such as pedestrian movement, where a member of them has no purpose about the whole shape of their swarm [1].

On the other hand, each member of the parade has the common purpose to synchronize each other. The member tries to fix small gaps between its neighbors as soon as

H. Sato et al. (Eds.): IES 2019, PALO 12, pp. 253–264, 2020.
https://doi.org/10.1007/978-3-030-37442-6_23

Fig. 1. The parade of the Russian Army (https://www.jiji.com/jc/d4?p=mrs165&d=d4_aaa)

possible. This strong motivation produces complex dynamics because they are coupled with each other tightly. This attribute makes them keep their synchronization difficult, and the training for the beautiful parade takes a long time [2].

1.2 The Current Issue of the Parade Training

In National Defense Academy of Japan (NDA), cadets conduct the parade training many times; however, their training needs a lot of participants, and the training is not based on scientific knowledge.

Figure 2 is an example of the parade evaluation sheet, which we use for the parade evaluation in NDA.According to the parade evaluation sheet, the parades are checked whether the categories are satisfied or not.

PARADE EVALUATION SHEET					
Category		Description		Satisfied	Not Satisfied
1	Dress Code	i	Name plates and badges are well fixed	✓	
		ii	They hold bags properly	✓	
		iii	Uniforms are dressed well	✓	
2	Turn	i	The arm swing are synchronized during the turn		✓
		ii	The lines are straight	✓	
		iii	They turn regularly, not in disorder		✓
3	Arm swing	i	The rows are straight	✓	
		ii	The cycle of arm swings are synchronized	✓	
		iii	The height of arm swing is good	✓	
		iv	The elbows are squared		✓
4	Salute	i	The salute order is loud	✓	
		ii	The form of salute is good	✓	
		iii	They turn their gaze on properly	✓	
		iv	They salute uniformly	✓	

Fig. 2. An example of the parade evaluation sheet in NDA

However, we do not have any quantitative evaluation method, and we have some problem with the current evaluation system. For example, the parade quality score of the small party tends to be higher than the big party's one, because of the heuristic evaluation, the evaluation seems not to be fair for each parade, and so on. The quality of the parades in NDA are already high because of the many trainings; therefore, the quantitative evaluation for the quality of the parade, which can figure out the slight difference among high skilled parades, is required. We do not know what element is important for the beautiful parade and for people to feel the strength from the parade unconsciously.

1.3 The Quantitative Evaluation for the Military Parade

On these backgrounds, we are working on quantitative evaluation for the military parade. In the previous study [3], we tried to measure the posture information of the parade with OpenPose [4], which is a posture estimation system with deep learning (Fig. 3). OpenPose can measure many kinds of posture data from one image simultaneously; therefore, it is much easier to get a large number of posture data than conventional motion analysis with motion capture [5, 6]. However, we could not get the whole of the posture data in each parade because, depending on the camera angle, there are some occlusions in the parade movie.

Fig. 3. The parade evaluation with OpenPose: We can get many posture data simultaneously, however, we cannot get the data of the back side because of the occlusion [3].

The audio information in the parade has many kinds of information about the parade, such as how strong the voice of order is, how well the footsteps are synchronized and so on. Moreover, there is no occlusion in the audio information of the parade. We expect the audio information contains important data which is not observed by the posture analysis. Therefore, as the first step of the audio analysis of the parade, in this paper, we use the footsteps, which is one of the audio information about the parade, for quantitative evaluation.

Audio information reveals hidden human status sometimes. For example, a voice information in diagnosis is used Pathologic condition analysis and Sensibility Technology (PST), which is for patient condition determination [7].

Figure 4 is the classification of the parade evaluation. In previous study, we focused on the synchronization of the arm swing with OpenPose. On the other hand, in this paper, we focus on the synchronization of the footsteps, which is circled with red line in Fig. 4. The Footsteps are most conspicuous in the parade, and it is suitable for the evaluation.

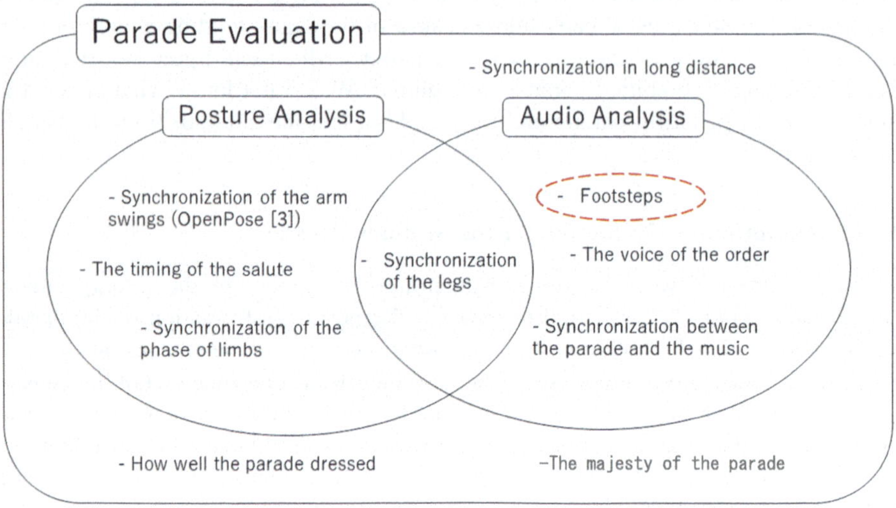

Fig. 4. The classification of the parade evaluation

This paper is organized as follows. After introducing the backgrounds of using the footsteps for the evaluation, we propose the footsteps analysis process to estimate the parade skill. After we explain the experiments with three types of the parade to verify using the footsteps for the parade skill, finally, we show our results, consideration and future tasks.

2 Proposed Method

In this section, we propose the algorithm of the quantitative evaluation method for the quality of the military parade using the footsteps. In the military parade, the members try to synchronize their movement to nearby members and the parade music. Therefore, when the parade is synchronized, the footsteps of each member are equalized.

2.1 The Background of the Tempo Analysis for the Military Parade

When the parade is synchronized, the footsteps are synchronized as well, and if the parade is not synchronized, the limbs of each member are not synchronized. Therefore, according to the footsteps, the sound of the parade is different between the synchronized one and not synchronized one. From these observations, when the parade is synchronized, the tempo of the footsteps will be clear, and when the parade is not synchronized, the tempo of the footsteps will be vague (Fig. 5).

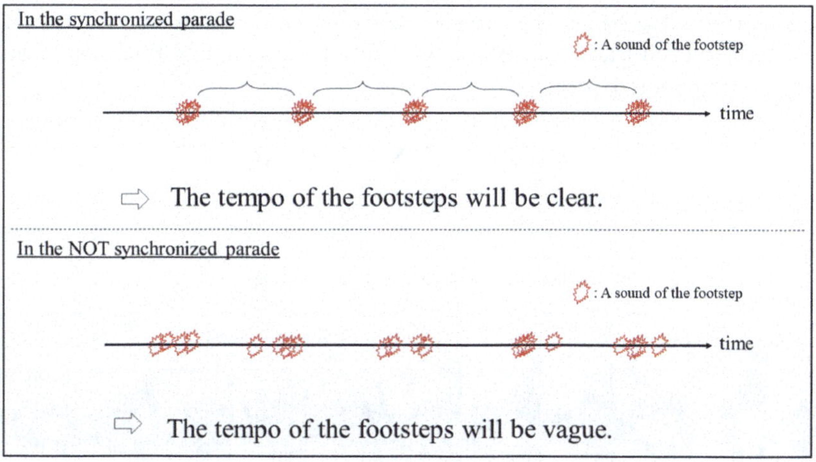

Fig. 5. Image of the difference of the footsteps

2.2 The Parade Skill from the Tempo Analysis

In this research, we focus on the footsteps, which is one of the audio information of the military parade. In the tempo analysis process, the transition of the volume is used to estimate the tempo. If the transition has periodic trends, we can analyze the tempo clearly by the frequency components from the volume transition.

If the parade is synchronized to the parade music, the footsteps are synchronized; therefore, we expect to confirm the same trend of the volume transition of the parade from the synchronized parade.

Figure 6 is the scheme of the tempo analysis process. We explain the tempo analysis process in order. First, we obtain the sound file from the parade. In this paper, the sound file is WAVE format, sampling frequency is 44.1 kHz, and the length of the sound data is 60 s. Next, in order to focus on only the footsteps, we mask the component of the background music from the audio file of the parade. Third, we analyze the tempo of the file, and at last, we calculate the parade skill by correlation coefficients between the results of the tempo analysis from the background music and the footsteps.

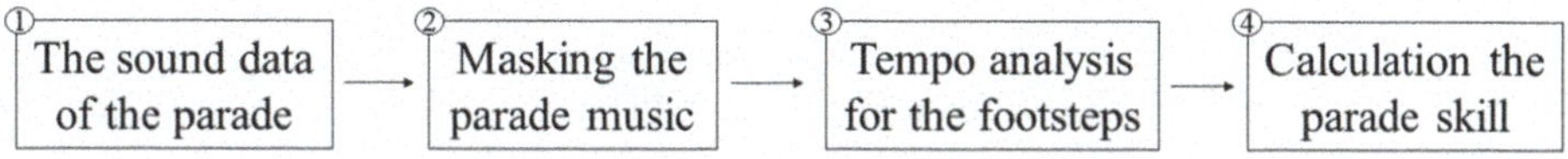

Fig. 6. Scheme of the tempo analysis process

2.3 Tempo Analysis for the Footsteps

We explain the tempo analysis process (Fig. 6-③) in order. Figure 7 is the overview of the tempo analysis process. In the beginning, we sample the parade sound data into same frame, which we set as 512. Then, we calculate the root mean squared value (RMS) of the amplitude in each frame (1). When the total of the sampling frame is N, and the amplitude in frame n is $V(n)$,

$$RMS_n = \sqrt{\frac{1}{N}\sum\nolimits_{n=0}^{N-1} V(n)^2} \tag{1}$$

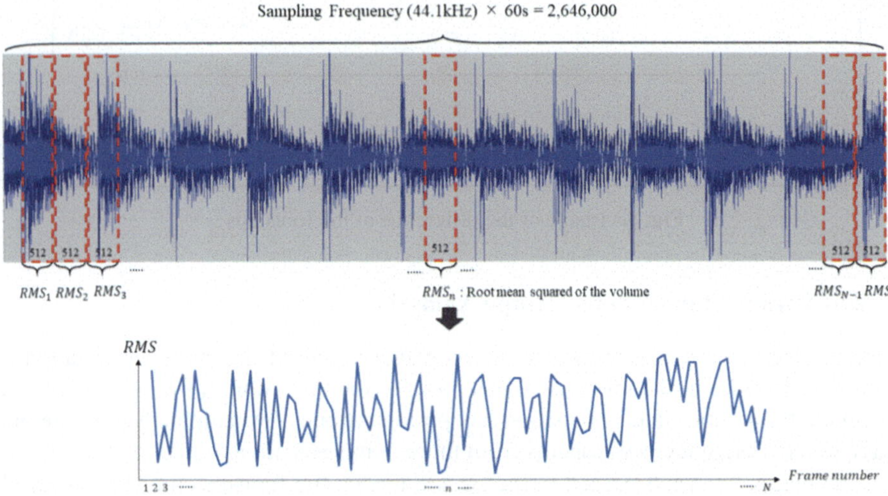

Fig. 7. Overview of the tempo analysis process

Next, we calculate the difference value from next frame, and the difference value is $RMS_i - RMS_{i-1}$. If the difference value is negative, it is set to 0 because we want to get the impulse point of the footsteps from the trend of the difference value. The increasing trend is periodic comparing to the decreasing trend.

Finally, we calculate the match rate R_{bpm}.

When $D(n)$ is the n th difference value, $f_{bpm} = Value\ of\ the\ tempo/60$, and $s = Sampling\ frequency\,(44100)/Sampling\ rate\,(512)$, the match rate R_{bpm} is as below (2).

$$R_{bpm} = \sqrt{a_{bpm}^2 + b_{bpm}^2} \tag{2}$$

$$a_{bpm} = \frac{1}{N}\sum\nolimits_{n=0}^{N-1} D(n)\cos\left(2\pi f_{bpm}\frac{n}{s}\right) \tag{3}$$

$$b_{bpm} = \frac{1}{N}\sum\nolimits_{n=0}^{N-1} D(n)\sin\left(2\pi f_{bpm}\frac{n}{s}\right) \tag{4}$$

Figure 8 shows an example of the proposed tempo analysis with a simple metronome sound and a white noise sound file. The triangle markers represent the result of the tempo analysis with the simple metronome sound file, and the round markers represent the result

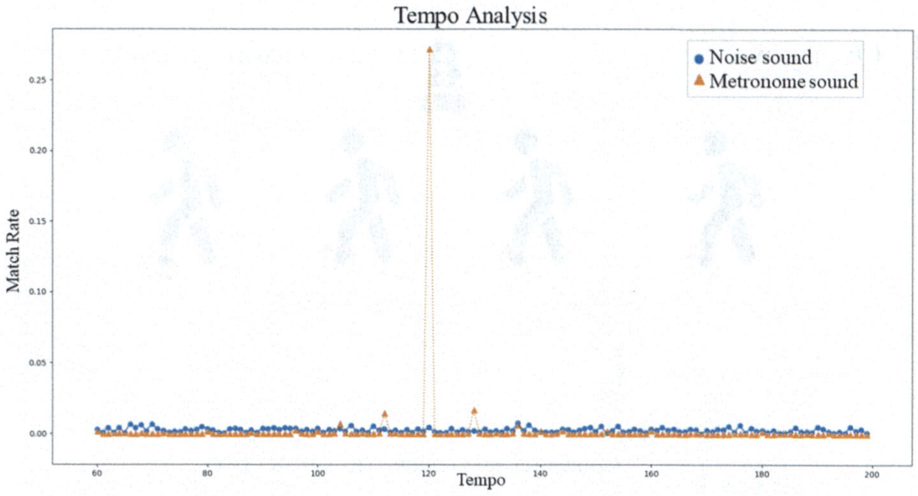

Fig. 8. An example of the tempo analysis from a metronome sound and a white noise

of the tempo analysis with the white noise sound. The information of the sound is as follows. The format is WAVE file, the sampling frequency is 44100, and the length is 60 s. The tempo of the metronome sound is 120 BPM.

In this figure, we can confirm the match rate of the 120 BPM is highest for the metronome sound, on the other hand, there is no peak of the match rate for the white noise; therefore, we could confirm the effectiveness of this tempo analysis process.

3 Experiment

In order to verify the proposed method is applicable, we had three types of experiments for four participants, who are familiar with the military parade. In these experiments, we focus on only the footsteps, which is the most conspicuous sound from the military parade. In order to analyze the footsteps, we asked four participants to conduct the military parade without stepping forward in a room and asked them to synchronize their parade to the metronome sound, where the tempo is 120 BPM (Fig. 9). We used the metronome as the parade music in the experiment. In order to mask the effects of the sound of the metronome, we set the note of the metronome as 800 Hz. We recorded the sound of the footsteps, and for sensitive recording, we used a condenser microphone (Rode NT1-A). We can control the input gain; therefore, it is not saturated if the number of parade member is large and the volume is loud. In our recording, we set sampling frequency as 44100, and the channel was monaural.

For comparing the different types of footsteps and environments, we set three types of experimental condition as follows,

- Synchronized parade: the four participants try to synchronize to the metronome sound for 60 s.

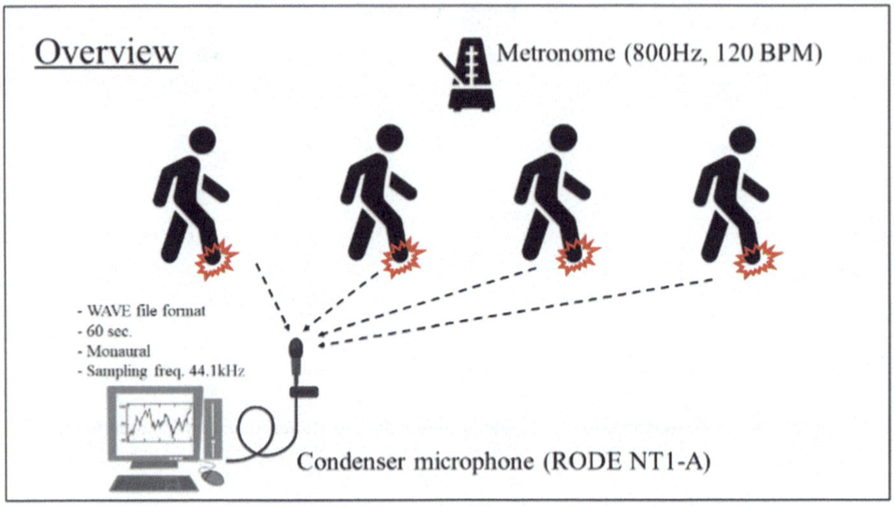

Fig. 9. The overview of the experiment

- Not synchronized parade: only one participant tries to disorder his footsteps, and the others try to synchronize to the metronome sound for 60 s.
- Disordered parade: each participant tries to disorder their footsteps for 60 s.

Figure 10 shows a picture of the experiment of "Disordered parade". As we can see, the limbs of each participant are not synchronized.

Fig. 10. A picture of the experiment of "Disordered parade"

Additionally, we had those experiments in two different condition; with noise and without noise.

When we evaluate the parade in real environment, the sound of the parade has many kinds of noise in open-air condition; therefore, we set the additional experiment as outdoor condition. In order to allow the noise, we opened the window of the room, which was used for the experiment.

Figure 11 is the difference of the two sound files; the upper one was obtained in the less noisy environment, and the lower one was obtained in the noisy environment. As you can see in this figure, the lower sound file has more noise than the upper one, and it seems to be vaguer for the tempo of the noisy sound file.

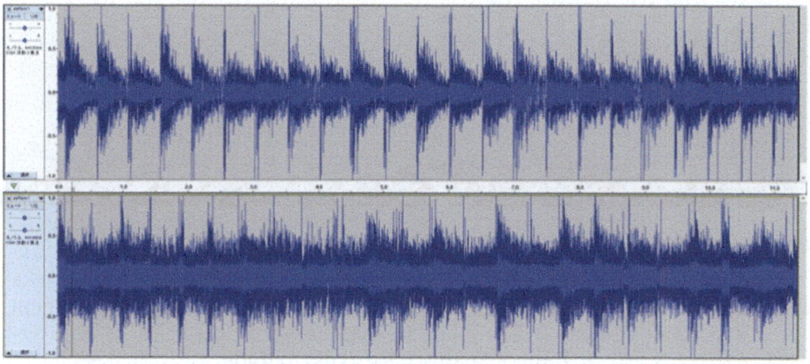

Fig. 11. The difference of the waveform between the two environments

4 Result

According to the tempo analysis method, we calculated the match rate, which is R_{bpm} in (2), of the three types of parades in less noisy environment. In order to compare, the results of the metronome sound and the parades are shown in Fig. 12. You can see the results of "Synchronized parade", "Not synchronized parade", and "Disordered parade", and the more the parade is disordered, the lower the match rate is. As I mentioned before, the metronome sound was masked before the tempo analysis; thus, there is no effect of the metronome sound.

Figure 13 is the results of three types of the parade in the noisy environment. As you can see, even if it is in the noisy environment, we obtained the results of the tempo analysis with same trend to the less noisy environment.

For the quantitative evaluation for the calculation of the parade skill, we adopt the correlation coefficient between the results of tempo analysis of the metronome sound and the parade sound.

In order to verify which correlation coefficient is better, we set two values; one is Pearson's correlation coefficient, and the other is cosine similarity. Table 1 and Fig. 14 shows the result. Whichever the environment is with noise or not, the more the parade is disordered, the worse each value is.

From the result, Pearson's correlation coefficient is more apparent to show the synchronized level. Therefore, it is suggested that the match rate of Pearson's correlation coefficient is applied to evaluate the synchronized level.

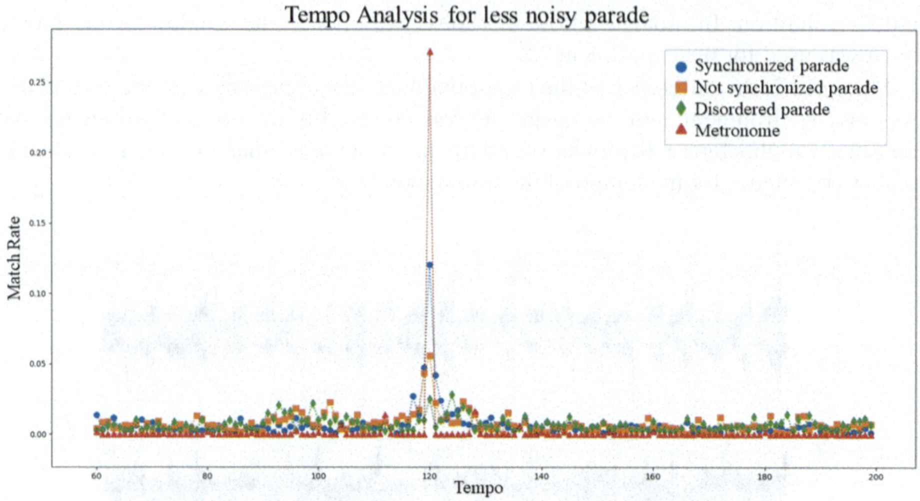

Fig. 12. The result of three types of the parade in the less noisy environment

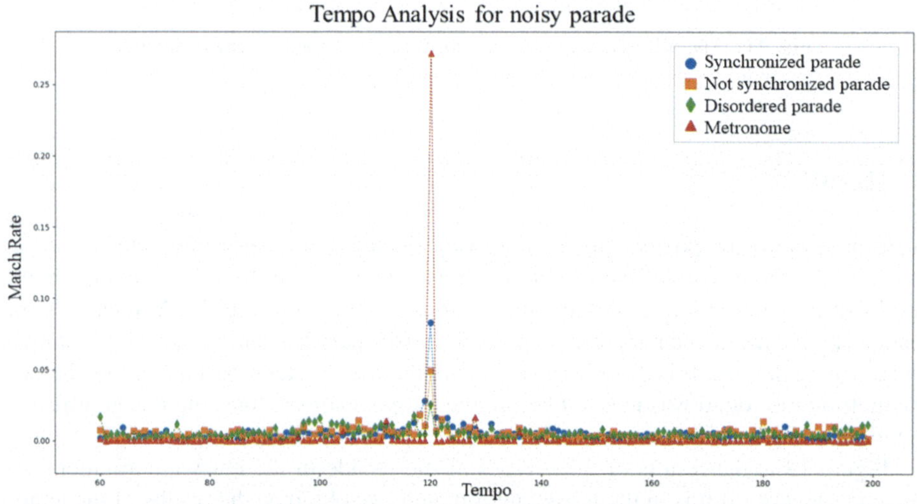

Fig. 13. The results of three types of the parade in the noisy environment

Table 1. The value of correlation coefficients

		Synchronized parade	Not synchronized parade	Disordered parade
Pearson's correlation coefficient	Less noisy	0.844	0.596	0.313
	Noisy	0.871	0.765	0.428
Cosine similarity	Less noisy	0.212	0.130	0.071
	Noisy	0.209	0.163	0.091

Fig. 14. Correlation coefficients of each parade in different environments

5 Conclusion

We proposed the quantitative evaluation for the parade skill by the footsteps, which is one of audio information. We use the tempo analysis process for calculating the estimate tempo of the footsteps based on that if the parade is synchronized, the tempo of the footsteps will be clear. Therefore, it is suggested that the correlation coefficient between the result of the metronome sound and the parade sound is applicable for estimation of the parade skill.

We had experiments with three types of the parade focusing on the footsteps in two environments; noisy environment and less noisy environment. We set the correlation coefficient between the estimate tempo of the parade sound and the metronome sound as the parade skill. As a result, the more the parade is disordered, the worse the correlation coefficients are. Although, in order to verify our evaluation method applied to the real parade, further experiments are needed. This method is only for the audio analysis; therefore, for the more precise evaluation, we need to combine posture analysis such as the evaluation method for the synchronization level of arm swings. As future tasks, in order to verify the scalability for this method, we need to confirm real parade.

References

1. Helbing, D., Kelsch, J., Molnar, P.: Modelling the evolution of human trail systems. Nature **388**, 47 (1997)
2. Okugawa, Y., Kubo, M., Shimohagi, T., Ishihara, T., Sato, H.: Analysis of a parade with the Kuramoto model For better performance without trainings. In: Proceedings of IEEE SMC, pp. 4005–4010 (2018)
3. Okugawa, Y., Kubo, M., Sato, H., Viet, B.D.: Evaluation for the synchronization of the parade with OpenPose. In: Proceedings of ICAROB (2019). https://doi.org/10.5954/icarob.2019.os17-2

4. Cao, Z., Simon, T., Wei, S.E., Sheikh, Y.: Realtime multi-person 2D pose estimation using part affinity fields. In: Proceedings of the 2017 IEEE Conference on Computer Vision and Pattern Recognition, pp. 7291–7299 (2017)
5. Inada, K., Ishikawa, T., Watanabe, H.: Coincidence evaluation of multiple people's dance using OpenPose. In: Proceedings of the 80th National Convention of IPSJ, no. 3, pp. 257–258 (2018)
6. Nakai, M., et al.: Prediction of basketball free throw scoring by OpenPose. In: Proceedings of the 32nd Annual Conference of the Japanese Society for Artificial Intelligence (2018). https://doi.org/10.11517/pjsai.JSAI2018.0_3Pin139
7. Hagiwara, N., Omiya, Y., Shinohara, S., Nakamura, M., Higuchi, M., Mitsuyoshi, S., Yasunaga, H., Tokuno, S.: Validity of mind monitoring system as a mental health indicator using voice. Adv. Sci. Technol. Eng. Syst. J. **2**, 338–344 (2017). https://doi.org/10.25046/aj020343

Sociophysics Theory for Conflict of Viewer Rating on Television Program
Proposal of New Index Instead of Viewer Rating

Akira Ishii[1(✉)], Daiki Nishiyama[1], and Yasuyuki Suzuki[2]

[1] Department of Applied Mathematics and Physics, Tottori University Koyama, Tottori 680-8552, Japan
ishii@tottori-u.ac.jp

[2] Video Research Ltd. 6-17, Sanban-cho, Chiyoda-ku, Tokyo 102-0075, Japan

Abstract. The competing viewers from multiple TV stations is a fierce competition. This competition is contested in the form of viewer rating of how much is viewed. In this study, the viewer's viewing behavior is formulated in the form of transport equation. Audiences gather in interesting scenes, and viewers leave from boring scenes. We propose a theory to quantify the fun of the program based on these viewing behaviors. The degree of interest from moment to moment according to this theory does not match the audience rating.

Keywords: Viewer rating · Degree of interest · Transport equation

1 Introduction

There are a lot of research for viewer rating [1–3]. However, the main interest is the relationship between program content and viewer rating. In particular, there are many studies on violence and sexual activity within the program.

However, the focus here is on the elucidation of how the variation in TV ratings is explained. Since the change in viewer rating can be explained by the inflow and outflow of viewers to the programs of each television station, it can be formulated as a transport equation for viewers.

What does the instantaneous viewer rating on TV programs mean? In general entertainment magazine/paper reports, the moment of highest moment rating is used to mean the most interesting moment of the program. However, the instantaneous maximum viewer rating at a certain time t also means that the viewer went away from the program because of the scene at that time t. However, if a scene that is not interesting at all continues, the viewer rating will continue to fall from before that, so the time t will not become the instantaneous highest viewer rating. So, the meaning of momentary highest viewer rating seems to be not simple.

In this paper, we build the equation of viewer behavior as a transport equation. We consider that the coefficients of this equation can be treated as an index of interest of each television program.

H. Sato et al. (Eds.): IES 2019, PALO 12, pp. 265–271, 2020.
https://doi.org/10.1007/978-3-030-37442-6_24

2 Theory

The changes of the number of viewers of a certain television program per minute is defined by the difference between the number of incoming viewers and the number of outgoing viewers as follows.

$$\Delta V(t) = N_{incoming}(t) - N_{outgoing}(t) \tag{1}$$

where $N_{incoming}(t)$ and $N_{outgoing(t)}$ are the number of incoming viewers and the number of outgoing viewers per minute. The number of incoming viewers are defined as the summation of the number of viewers who switch on the television for this TV program and the number of viewers who switch from other channel to this channel. The number of outgoing viewers are defined as the summation of the number of viewers who switch off the television at this TV program and the number of viewers who switch from this channel to other channel.

We can write down this as follows.

$$N^i_{incoming}(t) = N^i_{on}(t) + \sum_j N_{ji}(t) \tag{2}$$

$$N^i_{outgoing}(t) = N^i_{off}(t) + \sum_j N_{ij}(t) \tag{3}$$

where $N^i_{on}(t)$ is the number of viewers just switch on at the channel i at time t and $N^i_{off}(t)$ is the number of viewers switch off at the channel i at time t. N_{ij} is the number of viewers transfer from the channel i to the channel j at time t.

Here we define the degree of interest for the channel i at time t is Ami, t. $-Am(i,t)$ is the boring degree of the program of the channel i at time t. We consider that the number of viewers transfer from the channel i to j is

$$N_{ij}(t) = -Am(i,t)C_jV_i(t) \tag{4}$$

where $V_i(t)$ is the number of viewers of the channel i at time t and C_j is the probability to select the channel j at time t. This equation means that some viewers transfer from the channel i and move to the channel j because the channel i is boring at time t. In the following actual calculation, the probability C_j is calculated using the real viewer rating at that time. We assume that the positive $Am(i,t)$ means the program is very interesting and exciting at time t. The negative $Am(i,t)$ means the program is boring at time t.

Therefore, we can consider the change of the number of viewer of a certain TV channel is follows.

$$\Delta V_i(t) = \sum_j Am(i,t)C_jV_i(t) - \sum_j Am(j,t)C_iV_j(t) \tag{5}$$

The first term is the term of outgoing viewers from the channel i and the second term is the term of incoming viewers from other channels.

In the case of multiple stations in intense viewer rating, This formula Eq. (5) must be solved simultaneously for multiple TV stations. To obtain $Am(i,t)$ from Eq. (5) is very simple. We just solve the equation for each time for $Am(i,t)$ using the normal linear algebra. After the calculation, we obtain $Am(i,t)$ for number of TV stations. We call $Am(i,t)$ as the degree of interest for the channel i at time t. The degree of interest is the function of time.

3 Viewer Rating Data for Japanese Television

We obtain the viewer rating data of Japanese television from Video Research Ltd. In 1962, the same year that television came of age in Japan, Video Research Ltd. was established as a company specializing in TV audience ratings research.

More than 45 years, Video Research has helped marketers beat the competition and successfully tap this huge potential by providing them with the Edge of Information. Video Research is Japan's leading marketing and media research company, offering a comprehensive menu of services ranging from authoritative media research to syndicated and custom consumer marketing research. The media audience ratings research continues to set the standard for the industry and is relied on by virtually every major media company, advertising agency and advertiser in Japan.

Video Research supplies observation machine of the watching TV channel for 900 families in the greater Tokyo area. Since the controllers are supplied to each family members. Since the observation machine is handed over to each family member, the viewer rating can also be measured by age and gender.

In this research, we use the data measured by using the above machine. The real number of viewers are private on this paper.

Using the data from Video Research, we obtain the number of viewers for each television station per minute $V_i(t)$ and the number of transferred viewers between each channel $N_{ij}(t)$. Since the probability to select channel C_i can be obtained from $V_i(t)$ as

$$C_i(t) = \frac{V_i(t)}{N_{total}} \tag{6}$$

where N_{total} is the total number of observation machines deployed from Video Research. If $V_i(t) = N_{total}$, the viewer rating of the channel i is 100 %.

4 Results

Using the data of Video Research, we calculate the degree of interest for some television program and compare it with the number of viewers. The horizontal lines for each figures indicate the level zero of the degree of interest for the corresponding television program. Since the observation machine supplied from Video Research records the data per minute, the time resolution of the following figures is 1 min.

In Fig. 1, we show a typical example of the correlation between the degree of interest and the number of viewers. As we can see in the figure, at the period of positive degree of interest, the number of viewer is also increasing. At the period of negative degree of interest, the number of viewer is decreasing. It means that our degree of interest seems to be very natural as the degree of interest for the television program. In Fig. 1, we can also find that the peak of the number of viewer located between the two shaded periods do not correspond to the peak of the degree of interest. Thus, the maximum of the number of viewer is not equal to the maximum of the degree of interest.

In Fig. 2, we show two period corresponding to the negative degree of interest. In this two periods, the number of viewer is decreasing.

In Fig. 3, we show an example of calculation that the average of the number of viewer takes higher value after the high peak of the degree of interest. It also means that the degree of interest is related to the value of the number of viewer.

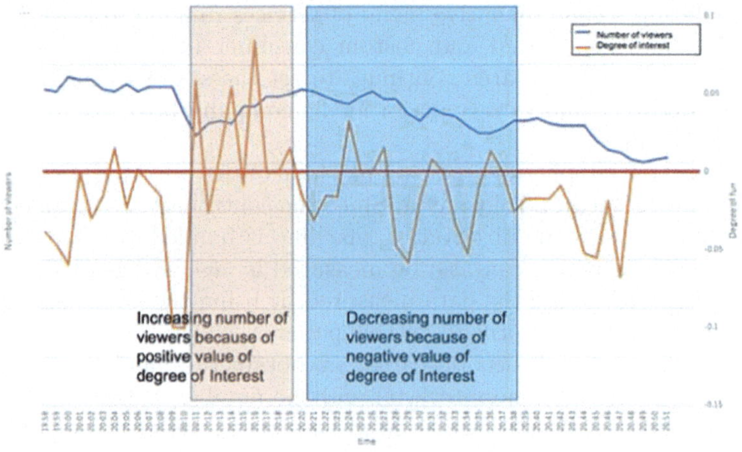

Fig. 1. The number of viewers and the numerical value of degree of interest for a day of 2019 spring on a television station A.

We obtain very interesting behavior for the peak of the number of viewer for the case of drama. This drama is very popular comedy drama of love romance at 2016. The actress of heroin and the actor of hero in this drama are very famous actress and actor in Japan. In Fig. 4, we show the number of viewer and the degree of interest for this drama of story 5. As indicated in the figure, there are several scene of hug of the heroin and hero. It is interesting that, for every scene of hugging in this drama, the degree of interest decreases at the time of the hugging, as we can see in the figure.

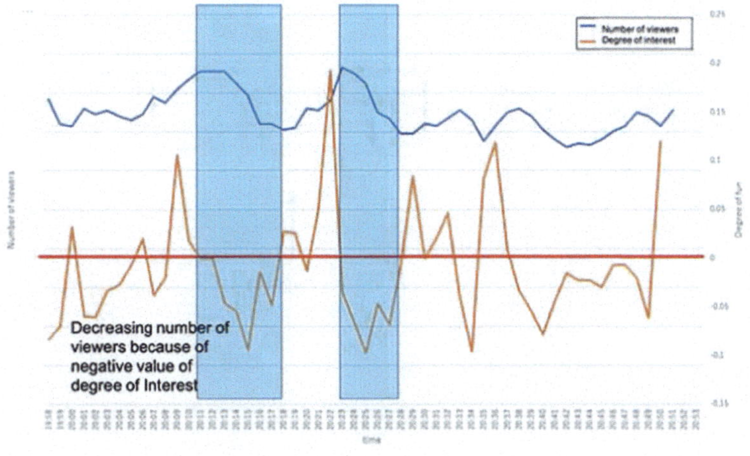

Fig. 2. The number of viewers and the numerical value of degree of interest for a day of 2019 spring on a television station B.

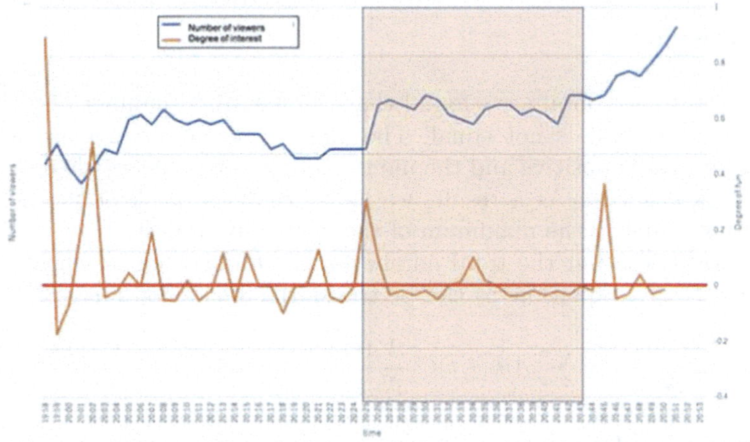

Fig. 3. The number of viewers and the numerical value of degree of interest for a day of 2019 spring on a television station C.

Usually, drama creators consider that the hugging between the popular actress and the popular actor is one of the best scene to increase the number of viewer. Indeed, we found weak peak of the number of viewer at the time of hugging. However, according to our analysis using the degree of interest, the scene of hugging does not seem interesting.

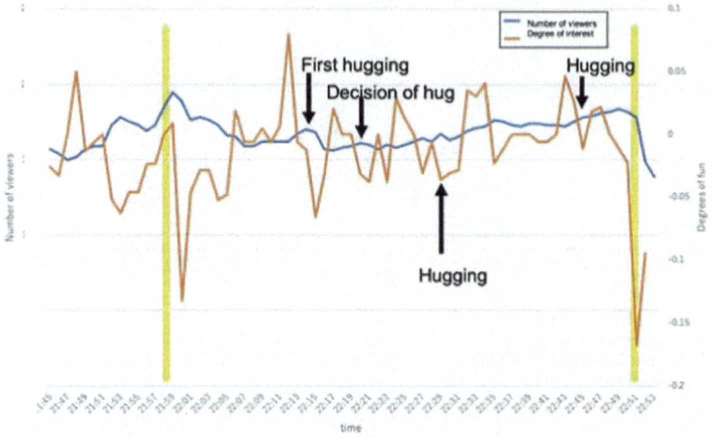

Fig. 4. The number of viewers and the numerical value of degree of interest for the popular love comedy drama. The arrow indicate the hugging scene.

5 Discussion

From the results we obtain, we found that the measured number of viewer and the degree of interest is not equal. The number of viewer increase when the degree of interest is positive and the number of viewer decrease when the degree of interest is negative. As a result, we found that the peak of the number of viewer is not equal to the maximum of the degree of interest.

Let us consider that the total number of the observation machine of Video Research is N_{total}. Equation (5) can be transformed as follows using N_{total},

$$\Delta \frac{V_i(t)}{N_{total}} = \sum_j Am(i,t) C_j \frac{V_i(t)}{N_{total}} - \sum_j Am(j,t) C_i \frac{V_j(t)}{N_{total}}. \tag{7}$$

Equation (7) is the equation for viewer rating. Thus, the amount $Am(i,t)$ is not the only amount in this case, but is regarded as an objective indicator of the fun of the television program. Therefore, we can use Eq. (7) as the calculation to obtain the objective indicator of the interest.

6 Conclusion

We formulate the viewer's viewing behavior in the form of transport equation. In this model, viewers gather in interesting scenes, and viewers leave from boring scenes. We propose a theory to quantify the fun of the program based on these viewing behaviors. The degree of interest from moment to moment according to this theory does not match the viewer rating. Calculated degree of interest can be used as an objective indicator for the fun of television programs.

Acknowledgements. We are grateful for Dr. Yasuko Kawahata for fruitful discussions.

References

1. Gensch, D., Shaman, P.: Models of competitive television ratings. J. Mark. Res. **17**, 307–315 (1980)
2. Rust, R.T., Alpert, M.I.: An audience flow model of television viewing choice. Mark. Sci. **3**, 83–178 (1984)
3. McCarty, J.A., Shrum, L.J.: The role of personal values and demographics in predicting television viewing behavior: implications for theory and application. J. Advertising **22**, 77–101 (2013)

Augmented Interaction Mechanism for Physically Consistent Collaboration Between Humans and Intelligent Agents in Network Management

Kazuto Sasai[1(✉)], Kazuya Kobayashi[2], and Tetsuo Kinoshita[3]

[1] Graduate School of Science and Engineering, Ibaraki University, 4-12-1, Nakanarusawa, Hitachi, Ibaraki, Japan
kazuto.sasai.z@vc.ibaraki.ac.jp

[2] Graduate School of Information Science, Tohoku University, 2-1-1, Katahira, Aoba-ku, Sendai, Japan

[3] Tohoku Univeristy, Sendai, Japan

Abstract. The growing complexity of networking in the ICT environment, the demands for non-professional administrators s increasing. Network management support systems consisting of intelligent agents are a possible solution for the situation. However, interaction support between humans and agents has not been fully improved. In this paper, we propose an augmented reality-based interface for the dialogue interaction with the agents. The proposed interaction mechanism improve the consistency of physical space and the agents, and therefore, the users can easily recognize the relationship with the agents. The result of evaluation implies that the proposed mechanism can support non-professional administrator and improve the quality of collaboration.

Keywords: Human-agent collaboration · Intelligent agent · Network management · Augmented reality · Dialogue agent

1 Introduction

Network systems are growing continuously according to the development of ICT technologies such as network and server virtualization, cloud computing, and IoT applications. On the other hand, the human resources to operate the network systems are small because of special knowledge and skills. Particularly, this problem becomes serious when emergent situations like huge disasters. As a solution, self-management functionality of network systems consists of autonomous agents which execute network and systems management operations behalf of the human administrators has been proposed [1]. However, the knowledge defines the action of the agents is limited, and the maintenance cost for the autonomic system unfortunately high. As a consequence, the autonomous agents are required

This research is supported by the Telecommunications Advancement Foundation.

H. Sato et al. (Eds.): IES 2019, PALO 12, pp. 272–283, 2020.
https://doi.org/10.1007/978-3-030-37442-6_25

to have the function which explores knowledge from the measurement data by themselves [2].

As a secondary solution, a collaboration between humans and agents (HAC: Human-Agent Collectives) [3] has been proposed. HAC shifts the unchangeable relationship between human and software tools to more flat collaboration, which involves the changeable role assignment regarding the situation an agent leads the users [4]. The idea of HAC is applied to the disaster situation and confirmed the effectiveness of its flexibility and robustness [5]. So far, we have introduced the concept of HAC into the autonomous agent-based network management and proposed it as the Knowledge-based Network Management Systems (KNMS) [6,7]. The KNMS property is to support human's network management tasks by providing professional knowledge stored in the NMS agents. According to the HAC concept, The agents in KNMS work with not only predefined action-rules but also human thinking in the deployment of the provided knowledge. However, the functional assembly of the HAC concept is not enough to support unprofessional administrators because of the complexity of information exchange interface. To enhance the flexibility of collaboration, we introduced the dialogue interface for NMS agents [8]. The dialogue interaction mechanism, which means natural language interface for agent message circulation in the HAC systems enables the KNMS to solve the gap between humans and agents.

Figure 1 shows a conceptual diagram of dialogue-based KNMS. Here, the idea of HAC indicates the mixed distribution of human administrative team and the agents managing the knowledge, information, and control of the target network and server systems. However, since the chat system does not support physical space relationship, the non-professional administrator can not easily

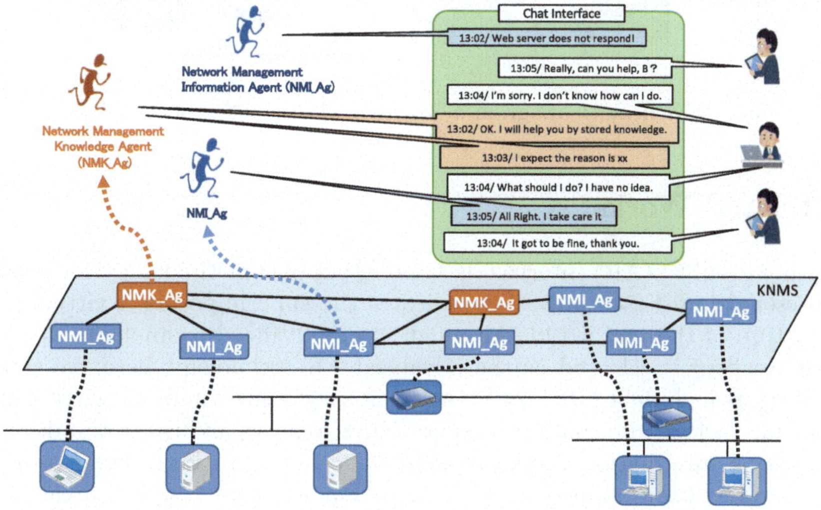

Fig. 1. Conceptual diagram of conversation based interaction scheme for knowledge-based network management system

recognize the physical location of the equipment handled by the agent. In this paper, we extend the dialogue interface by using the augmented reality interface, which we call "augmented interaction mechanism." Figure 2 shows a conceptual diagram of the augmented interaction mechanism for HAC. The augmentation of the human-agent interaction scheme to providing information superimposes the displaying messages on the object monitored by the server agents in an agent-based NMS. The conversation Agent mediates between the server agents and the augmented interaction interface by translating an ACL message to a natural language sentence.

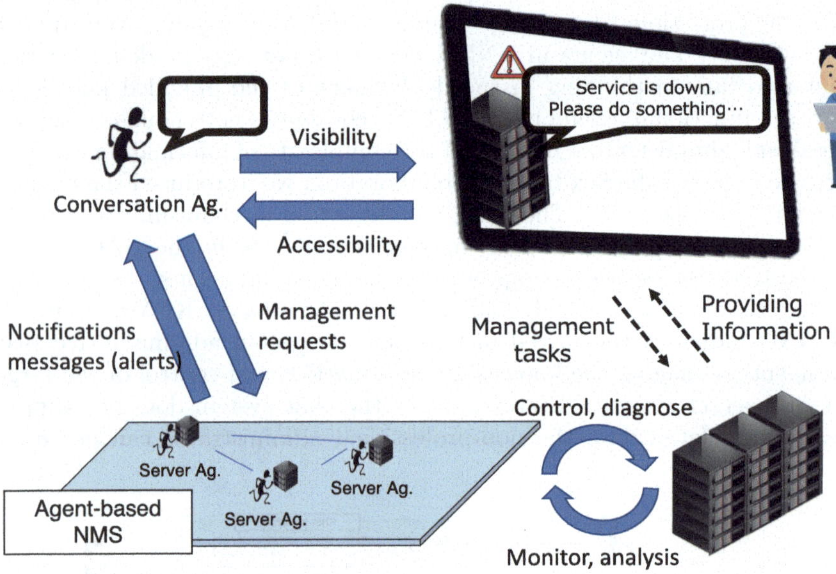

Fig. 2. Schematic diagram of augmented interaction mechanism

2 Related Works

Augmented reality (AR) interface is being a popular methodology for information presentation. Essentially, AR indicates the superimposing a virtual image on a picture of the real world. Although more advanced technology called MR (Mixed Reality) is released as the advanced AR technology, here, we call AR technology as including MR for simplicity. Since the superimposed virtual image can include various information presentation, it realizes an intuitive understanding of digital information displayed with the real image. From the reason, AR technology has been applied in the various management research area. In this section, we discuss the difference between existing studies and our proposal.

As a network and systems management support functions, AR is introduced to display networking information with real equipment figures [9]. The research group of [10] proposed to present security information. A more useful application is an information display for wireless communication system [11] because wireless communication is invisible from human eyes. The network information presentation is also applied in the field of home network management [12]. [13] proposed a method for the interactive message presentation with augmented reality. Although their approach seems to be very close to our proposal, the internal processing of the messages is quite different. Our method processes the natural language sentence and classifies them to select the agent to respond appropriately.

In this paper, we integrate the intelligent agents and augmented reality interface. Our approach proposes an advanced human-agent collaboration scheme such that the users can evoke the existence of the autonomous agents in the environment. This framework is conceptually supported by the idea of symbiotic computing [14], which enhances the presence of collaboration with the autonomous agents embedded in the environment. The asserted physical existence of the agent intelligence discussed in this paper is a possible solution for realizing flexibility, autonomy, and robustness of the network and systems management.

3 Proposal and Design

Figure 3 shows a design of augmentation mechanism and the detail of an agent-based NMS. An agent-based NMS consists of the three layers such as the server agent layer, the classification agent layer, and the functions layer, e.g., NMS function, chat function, and so on. The following sections explain the features of individual layers.

3.1 Server Agents

A server agent handles information about the target server and has an ID for mapping the AR marker to the data about the server. When receiving the message from the tablet agent, the server agent extracts the message with the corresponding ID tag and sends the message to classification agents. When the server agent receives a message from the classification agents, it generates ID tags into the message and relays it to the tablet agent.

3.2 Classification Agents

A classification agent has a function to extract the context from the received messages. The function includes natural language classification and keyword matching with regular expression. The natural language classification is used to extract the information for selecting the functions layer. For example, if a received message indicates a server failure, the message should transit to NMS agents, which can support the fault recovery process of the users with the defined knowledge. Else if the message is a kind of chat which means ordinal conversation like "Is it hot today?" the agent relay it to the chat agents.

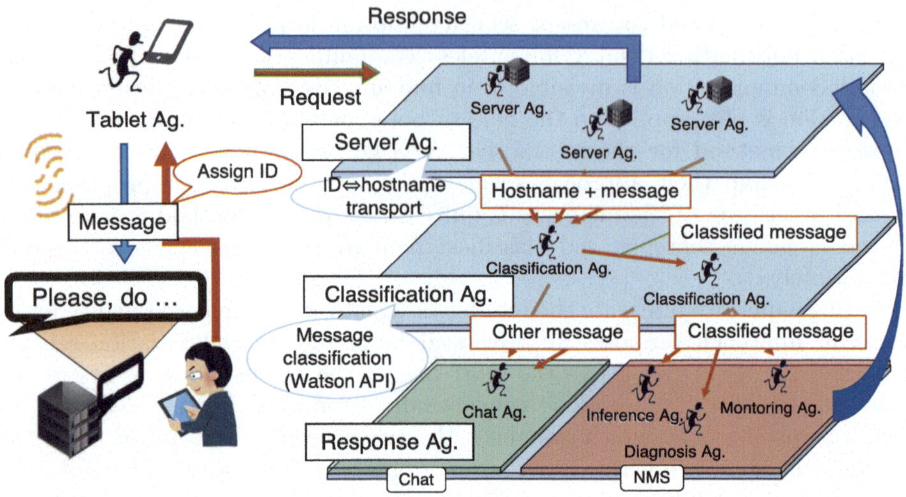

Fig. 3. Schematic diagram of the designed construction of proposal mechanism

3.3 Functions of Agent-Based NMS

The functionality of the agent-based NMS is constructed according to the concept of message routing inspired from the IP routing algorithm [8]. To handle a diverse of the message (conversation) types, we introduce the group of agents separated according to the purpose of conversations. In this paper, two types of conversations, failure recovery, and chat are considered. The support capability of the failure recovery task is based on the prototypical implementation of AIR-NMS [6].

3.4 Network Management Agent

Figure 4 shows the inference process to build the fault recovery methodology in a network management agent (NMAg). According to the inference process proposed in [6]. The process consists of the three steps, (1) retrieving measured *symptom*, (2) detecting essential *cause*, and (3) making a better counter *measure* which means recovery method. The knowledge of an agent has linked entities classified into the three steps.

3.4.1 Symptom Retrieval

The process of symptom retrieval is schematically shown in Fig. 6(a). When receiving an alert, the NMAg starts to support fault recovery task by inferring the possible cause of the failure. In most of the system failure situations, the visible symptoms are multiple possibilities of essential reason. For example, failure of network access can be caused by the trouble of the network switches, cables, NICs, and so on.

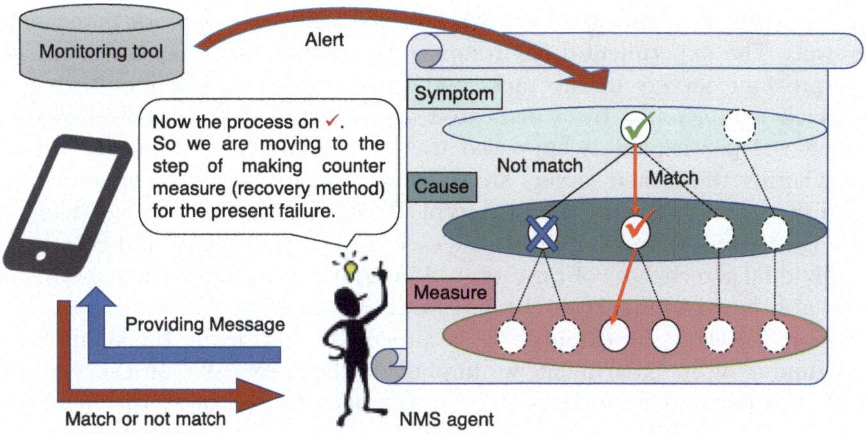

Fig. 4. A screenshot of augmented reality interface operated by tablet agent.

3.4.2 Cause Detection

Figure 5(b) shows the next step, the cause detection. From the predicted possible cause, the NMAg send corresponding check method as a message to the user interface. If the inference is decided as true by the users, the NMAg proceed the next step, and else if the cause is not true, the NMAg tests another possibility.

3.4.3 Measure Creation

Figure 5(c) shows the measure creation process according to the detected cause. When the previous step has been succeeded, the NMAg provides the appropriate countermeasure for the phenomenon causes the failure symptom and send the message. Finally, the user executes, and the recovery process is completed.

3.5 Tablet Agents

Figure 6 shows a screenshot of the prototypical implementation of the proposed interface. The racked servers are tagged by the AR marker associated with the server ID. Here, we introduce a balloon-type presentation for the message from the server agents. Where, the server ID:1 says "I'm not used now." because the server has been retired, the servers which have ID:2, ID:3, and ID:4 say "I'm fine." because there is no problem on the servers. Additionally, we put the text input area at the bottom of the display. The user can send a message via the text area by tapping the balloon of the target server.

4 Evaluation Experiment

4.1 Experimental Setup

We experiment with the participants of the human administrators who do not have much experience of network and server management. The participants consist of three graduate students who do not know the network management tasks well. The experimental environment is a server rack, as shown in Fig. 6. There are four servers in the rack, and these servers have a monitoring system which is controlled by a dedicated agent installed in another server. An experimental participant is supposed to be an administrator and is asked to check whether the racked servers are correctly operated or not. The administrator conducts the task with the prototypical interface installed in a tablet. The following scenario is an example of use case in the experiment, and corresponding individual screenshots of prototypical interface focusing on the experimental system are shown in Fig. 7(a), (b) and Fig. 8(a), (b).

Figure 7(a) shows an example of the process (1) retrieving the symptom. At the beginning of an experiment, we unplugged the LAN cable of a server in the rack. If the participant focuses the AR interface to the rack, the message of "Unavailable by ICMP ping" is displayed. Then, the user can input the message to ask how can he/she do at that time. Figure 7(b) displayed the screen when the server agent of ID:2 replied to the request of the user. From the inference process explained in the previous section, the possibility of the cable problem raises, and the server agent suggests to check the LAN cable connection. If the user answers as that is true, the process goes to the next step of measure presentation. Figure 8(a) indicates the process of presenting the corresponding countermeasure. Here, the reconnection of the cable is suggested to the user. Finally, as shown in Fig. 8(b), the current status of the server ID:2 is provided from the monitoring tool, the displayed message has been changed to "Recovered from failure. The status is healthy." The message indicates the failure is recovered.

4.2 User Experience

We conduct the questionnaire after the problem-solving experiment. Table 1 shows a list of questionnaires against the average scores of them. Since the score has five degrees, the result indicates a relatively good tendency in most of the entities. Q1, Q3, and Q4 show the high values of 4.0, 3.7, and 4.0, respectively. From the results, it is confirmed that the prototypical system achieves some effectiveness for support to the failure recovery operation of the users. In contrast, Q2 is comparatively low score. The reason is inferred as that the technical understanding of the failure requires some experience of the network management. Nonetheless, the participants succeeded 100%, and it implies the proposal achieved the support functionality for non-professional users. Although it requires a control experiment comparing the proposal with the previous interaction method with chat-based interface to prove the advantage, we remains for the future work. As a helpful information, in the failure recovery experiments in previous study, some participants could not complete the operation given from the agents, but there is no participant fail to recover in the experiment of this paper.

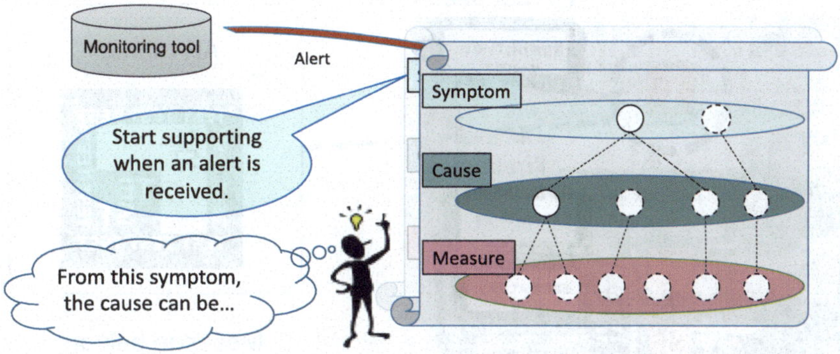

(a) Retrieving measured symptom

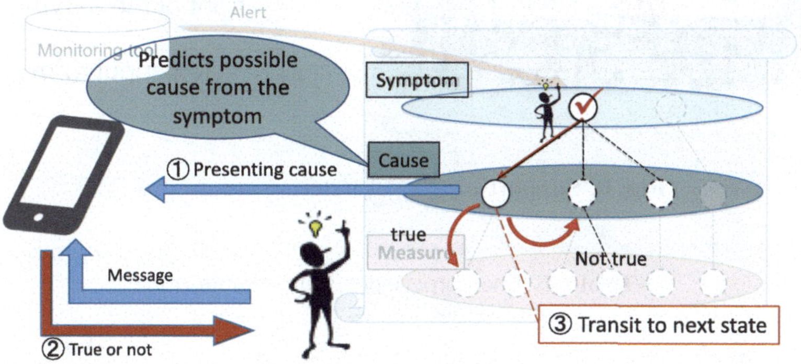

(b) Detecting essential cause

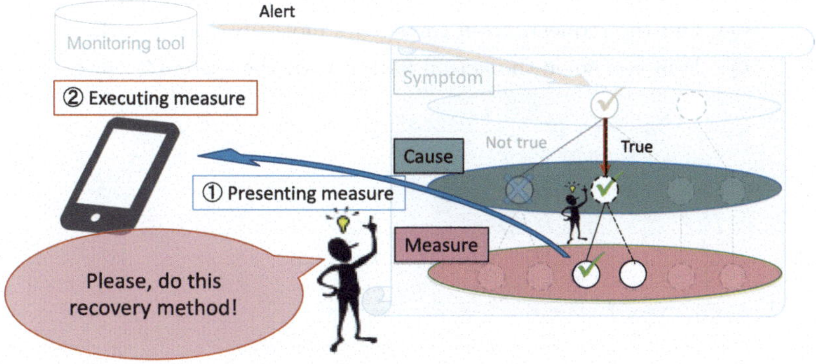

(c) Making a better countermeasure

Fig. 5. Schematic diagram of the knowledge structure in an network management agent

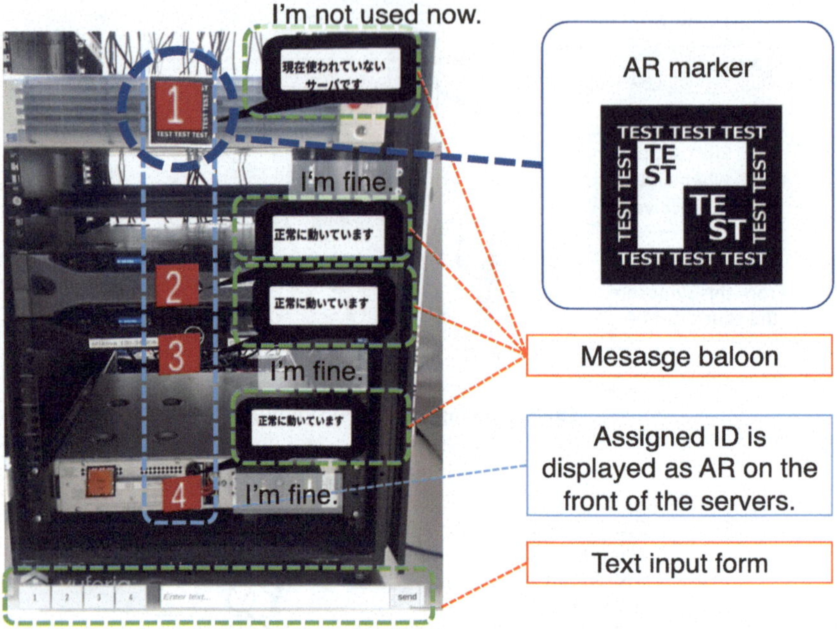

Fig. 6. Support process of failure recovery task

Table 1. User experience (Questionnaire)

Question	Avg.
Q1. Did you understand which server is failed?	4.0
Q2. Was the error notification easy to recognize?	2.3
Q3. Did you understand the massages from the agents?	3.7
Q4. Did the requests from the agents motivate you?	4.0
Q5. Will you solve the actual failure with the system?	3.0

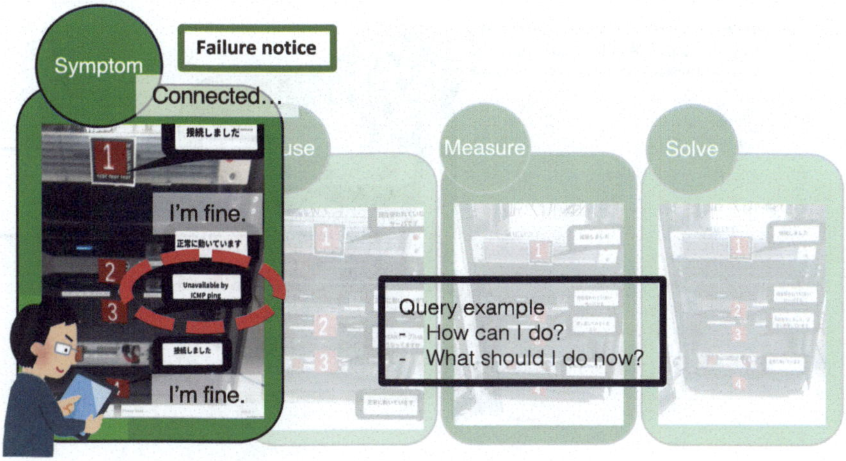

(a) Failure notice with symptom

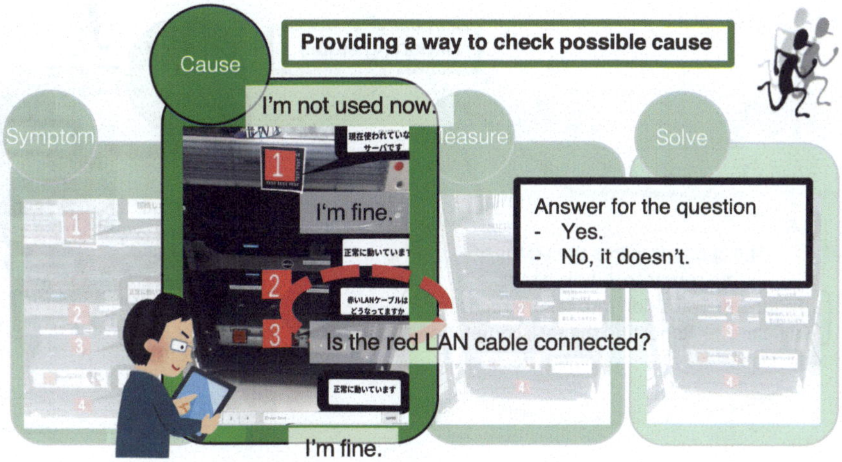

(b) Providing a way to check possible cause

Fig. 7. An example of the evaluation experiment

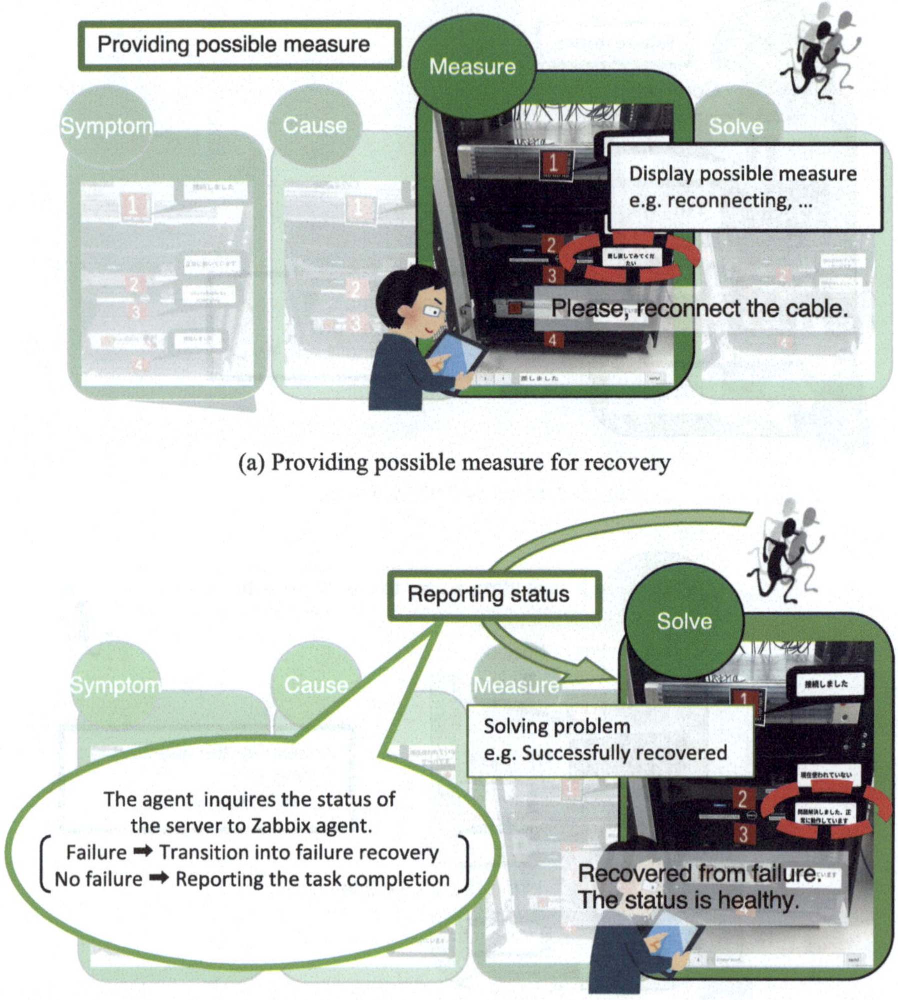

(a) Providing possible measure for recovery

(b) Reporting status after execute measure

Fig. 8. An example of the evaluation experiment

5 Conclusion

In this paper, we propose an augmented reality-based human-agent interaction interface by using balloons and text input field to provide the intuitive collaboration using natural language processing between the administrator and the autonomous intelligent agents for network management. We experimented the failure recovery task using the rack servers as an evaluation of the effectiveness of the proposal. As a result, the interface can adequately support the collaboration of non-professional users because they easily recognize which server do they interact. In the previous study, the users have to exchange massage via the chat interface. Since the interface is an entirely virtual environment, and then it is difficult to detect, which is the server for the administrator who does not have enough information. Consequently, the augmented interaction interface enhance the recognition of the users for the existence of intelligent agents and that can improve the quality of collaborative work between humans and autonomous agents.

References

1. Samaan, N., Karmouch, A.: Towards autonomic network management: an analysis of current and future research directions. IEEE Commun. Surv. Tutor. **11**(3), 22–36 (2009)
2. Kalegele, K., Sasai, K., Takahashi, H., et al.: Four decades of data mining in network and systems management. IEEE Trans. Knowl. Data Eng. **27**(10), 2700–2716 (2015)
3. Jennings, N.R., Moreau, L., Nicholson, D., et al.: Human-agent collectives. Commun. ACM **57**(12), 80–88 (2014)
4. Picard, W.: Resilient robust human-agent collectives: a network perspective. PRO-VE 2015: Risk and Resilience of Collaborative Networks **463**, 79–87 (2015)
5. Ramchurn, S.D., Wu, F., Jiang, W., et al.: Auton. Agent. Multi-Agent Syst. **30**, 82 (2016)
6. Sasai, K., Sveholm, J., Kitagata, G., et al.: A practical design and implementation of active information resource based network management system. Int. J. Energy Inf. Commun. **2**(4), 67–86 (2011)
7. Tanimura, Y., Sasai, K., Kitagata, G., et al.: Knowledge-based network management system for movable and deployable ICT resource unit. J. Comput. Commun. **5**(7), 135–151 (2017)
8. Sasai, K., Hoshino, T., Fukutani, R., et al.: Collaboration mechanism between human administrators and agent-oriented network management systems. In: 2019 IEEE International Conference on Big Data and Smart Computing (BigComp), Kyoto, Japan, pp. 1–5 (2019)
9. Haramaki, T., Nishino, H.: A network topology visualization system based on mobile AR technology. In: 2015 IEEE 29th International Conference on Advanced Information Networking and Applications, Gwangiu, pp. 442–447 (2015)
10. Suganya, S., Raajan, N.R., Priya, M.V., et al.: Augmented reality in security systems. Res. J. Inf. Technol. **5**(2), 256–261 (2013)
11. Sato, K., Sakamoto, N., Shimada, H.: Visualization and management platform with augmented reality for wireless sensor networks. Wirel. Sens. Netw. **7**, 1–11 (2015)
12. Lee, J., Kim, J.I., Kim, J., et al.: A unified remote console based on augmented reality in a home network management. In: 2007 Digest of Technical Papers International Conference on Consumer Electronics, Las Vegas, NV, pp. 1–2 (2007)
13. Hamid, A.K.A., Wada, K., Kawahara, Y., et al.: A network troubleshooting system using augmented reality technology. IPSJ SIG Technical Report 2010-UBI-26, pp. 1–6 (2010)
14. Suganuma, T., Sugawara, K., Shiratori, N.: Symbiotic computing: concept, architecture and its applications. In: Indulska, J., Ma, J., Yang, L.T., Ungerer, T., Cao, J. (eds.) Ubiquitous Intelligence and Computing, UIC 2007. Lecture Notes in Computer Science, vol. 4611. Springer, Heidelberg (2007)

Author Index

H. Sato et al. (Eds.): IES 2019, PALO 12, pp. 285–286, 2020.
https://doi.org/10.1007/978-3-030-37442-6

The manufacturer's authorised representative in the EU is Springer Nature Customer Service Centre GmbH, Europaplatz 3, 69115 Heidelberg, Germany. If you have any concerns regarding our products, please contact ProductSafety@springernature.com

Printed and bound by CPI Group (UK) Ltd, Croydon, CR0 4YY
05/10/2026
02228145-0010